**EMIS DATA REVIEWS SERIES 21**
Series Editor: Prof. B.L. Weiss



Properties, Processing and Applications of

# Indium Phosphide

**ELECTRONIC MATERIALS INFORMATION SERVICE**
**Other books in the EMIS Data Reviews Series from INSPEC:**

| | |
|---|---|
| No. 6 | **Properties of Indium Phosphide** (1991) |
| No. 7 | **Properties of Aluminium Gallium Arsenide** (1993) |
| No. 8 | **Properties of Lattice-Matched and Strained Indium Gallium Arsenide** (1993) |
| No. 10 | **Properties of Narrow Gap Cadmium-based Compounds** (1994) |
| No. 13 | **Properties of Silicon Carbide** (1995) |
| No. 14 | **Properties of Metal Silicides** (1995) |
| No. 15 | **Properties of III-V Quantum Wells and Superlattices** (1996) |
| No. 17 | **Properties of Wide Bandgap II-VI Semiconductors** (1997) |
| No. 18 | **Properties of Porous Silicon** (1997) |
| No. 19 | **Properties of Amorphous Silicon and its Alloys** (1998) |
| No. 20 | **Properties of Crystalline Silicon** (1999) |
| No. 21 | **Properties, Processing and Applications of Indium Phosphide** (1999) |
| No. 22 | **Properties, Processing and Applications of Glass and Rare-Earth-Doped Glasses for Optical Fibres** (1998) |
| No. 23 | **Properties, Processing and Applications of Gallium Nitride and Related Semi-Conductors** (1999) |
| No. 24 | **Properties of Silicon Germanium and SiGe:Carbon** (2000) |
| No. 25 | **Physical Properties of Liquid Crystals, Volume 1: Nematics** (2000) |
| No. 26 | **Properties, Growth and Applications of Diamond** (2000) |
| No. 28 | **Properties of Lithium Niobate** (2002) |
| No. 29 | **Properties of Amorphous Carbon** (2002) |

Details of books in the EMIS Data Reviews Series are available at www.theiet.org/publications/booksandconferences

Properties, Processing and Applications of

# Indium Phosphide

Edited by T.P. Pearsall
Corning S.A., France

The Institution of Engineering and Technology

Published by The Institution of Engineering and Technology, London, United Kingdom



First published 2000 (0 85296 949 X)
Paperback edition 2006



The Institution of Engineering and Technology
Michael Faraday House
Six Hills Way, Stevenage
Herts, SG1 2AY, United Kingdom

www.theiet.org



**British Library Cataloguing in Publication Data**
A catalogue record for this product is available from the British Library

**ISBN (10 digit) 0 86341 640 3**
**ISBN (13 digit) 978-0-86341-640-8**

Printed in the UK by Hobbs the Printers Ltd, Hampshire
Reprinted in the UK by Lightning Source UK Ltd, Milton Keynes

# Contents

# Introduction

It is a pleasure to present this latest EMIS book on InP-related materials and devices. Since the previous EMIS book on this subject, published in 1991, there has been significant growth in the use of InP, but the context defined by device applications has changed. The challenge was to prepare a new book that includes material that is essential to the use of InP in these new applications, yet which is complementary where possible to the old book.

Initially, it was believed that InP would find its niche in microwave applications as a transistor or as a Gunn-effect oscillator. It is increasingly used as a substrate for epitaxial growth to support GaInAsP lasers for optical fibre telecommunications and to fabricate GaInAs/AlInAs pseudomorphic high electron mobility transistors which have proved to be a more desirable choice for microwave generation and amplification. There are applications where InP is used for its own optoelectronic properties. A good example is solar cells, where the near optimum spectral response of InP plus its superior resistance to radiation damage combine to make InP the best available choice, based on performance, for power generation in space vehicles.

Unfortunately, the choice of InP is sometimes restricted by its cost. InP substrates remain significantly more costly that those made from GaAs, and even more so when compared to silicon. In addition, the technology of InP substrate manufacture is much more difficult than that for either GaAs or silicon. This situation is not simply the result of lower market demand, but is linked fundamentally to the high vapour pressure of phosphorus that creates an obstacle to the synthesis of single crystal boules of larger diameters. While 8-inch silicon substrates and 6-inch GaAs substrates are the rule in commercial fabrication, InP substrates are still 2-inch, and are likely to remain there until new concepts of single-crystal growth are developed.

There is potential relief for the high cost of InP substrates from wafer-bonding to silicon. The smart-cut method, for example, would permit the same InP wafer to generate many InP substrates on Si. However, for the moment these will need to be produced on 2-inch Si wafers which may become collectors' items as the mainstream of the silicon microelectronics industry moves toward increasingly larger wafer size. Additional information on wafer-bonding technology can be found in *Properties of Crystalline Silicon*, edited by Robert Hull (INSPEC, 1999).

Applications are presented first in this volume followed by the relevant properties and fabrication technologies. Applications requirements drive the developments in fabrication technologies, particularly for doping, etching, passivation, and metallisation. Limits to what can be achieved are defined by materials properties, both for InP, and the related alloy GaInAsP. The list of applications is not intended to be exhaustive, but rather to illustrate new directions where InP technology is evolving: integrated optoelectronic circuits for wavelength division multiplexing, microwave circuits, and photovoltaics.

We owe a debt of thanks to the authors who have prepared these Datareviews. Their work collected here will help advance InP technology to the next level of achievement. I would like to acknowledge in particular the considerable contribution of John Sears, whose continual support, encouragement, and persistence have guided this volume from an idea to the book you are holding in your hands. His leadership of the project to develop Datareviews Series that cover the field of electronic and optoelectronic materials is resulting in a powerful research and development tool for us all.

Thomas P. Pearsall
Corning, SA, Avon, France
October, 1999

# Contributing Authors


| | |
|---|---|
| S.Adachi<br>2.1-2.6 | Gunma University, Dept. of Electronic Engineering<br>Kiryu-shi, Gunma 376, Japan |
| A.R.Adams<br>4.2-4.5 | University of Surrey, Dept. of Physics<br>Guildford, Surrey GU2 5XH, UK |
| R.K.Ahrenkiel<br>3.1-3.7 | Solar Energy Research Institute<br>1617 Cole Blvd., Golden, CO 80401,USA |
| S.G.Bailey<br>1.7 | NASA Lewis Research Center<br>MS 302-1, Cleveland, OH 44135, USA |
| P.D.Bertone<br>7.5 | CSELT, Optoelectronic Devices<br>via G.Reiss Romoli 274, 10148 Turin, Italy |
| M.R.Brozel<br>5.7 | UMIST, Centre for Electronic Materials<br>PO Box 88, Manchester M60 1QD, UK |
| Y.Chen<br>1.5 | TRW, Electronics and Technology Division<br>MS D-1302, One Space Park, Redondo Beach, Ca 90278, USA |
| H.Eisele<br>1.1 | University of Michigan<br>Solid State Electronics Laboratory, 1301 Beal Avenue<br>Ann Arbor, MI 48109-2122, USA |
| G.I.Haddad<br>1.1 | University of Michigan<br>Solid State Electronics Laboratory, 1301 Beal Avenue<br>Ann Arbor, MI 48109-2122, USA |
| I.Harrison<br>7.4 | University of Nottingham, Dept. of Electrical and Electronic Engineering<br>Nottingham, NG7 2RD, UK |
| M.Horita<br>1.6 | KDD R & D Laboratories, Inc., Optoelectronics Group<br>2-1-15 Ohara, Kamifukuoka-shi, Saitama 356-8502, Japan |
| Q.H.Hua<br>5.6 | Tianjin Electronic Materials Institute<br>PO Box 5,Tianjin 300192, China |
| S.H.Jones<br>1.2 | University of Virginia, Applied Electrophysical Laboratory<br>Charlottesville, VA 22903, USA |
| J.D.Lambkin<br>4.6 | International Microelectronics Research Centre<br>Lee Maltings, Prospect Row, Cork, Ireland |
| D.Lancefield<br>3.3, 4.2-4.5 | University of Surrey, Dept. of Physics<br>Guildford, Surrey GU2 5XH, UK |
| Y.Matsushima<br>1.6 | KDD R & D Laboratories, Inc., Optoelectronics Group<br>2-1-15 Ohara, Kamifukuoka-shi, Saitama 356-8502, Japan |
| O.R.Monteiro<br>7.6 | Lawrence Berkeley National Laboratory, MS53-103<br>1 Cyclotron Road, Berkeley, CA 94720, USA |

| | |
|---|---|
| A.R.Peaker<br>6.1-6.8, 6.10 | UMIST, Centre for Electronic Materials<br>PO Box 88, Manchester M60 1QD, UK |
| A.F.Phillips<br>4.6 | International Microelectronics Research Centre<br>Lee Maltings, Prospect Row, Cork, Ireland |
| B.J.Sealy<br>7.2 | University of Surrey<br>School of Information and Electronic Engineering<br>Guildford, Surrey GU2 5XH, UK |
| C.Sirtori<br>5.4 | Thomson CSF, Laboratoire Central de Recherches<br>BP10, 91401Orsay, France |
| L.Spiekman<br>1.3 | JDS-Uniphase Netherlands BV<br>Prof. Holstaan 4, 5656 AA Eindhoven, The Netherlands |
| D.C.Streit<br>1.5 | TRW, Electronics and Technology Division<br>MS D-1302, One Space Park, Redondo Beach, Ca 90278, USA |
| S.Tanaka<br>1.6 | KDD R & D Laboratories, Inc., Optoelectronics Group<br>2-1-15 Ohara, Kamifukuoka-shi, Saitama 356-8502, Japan |
| P.J.Topham<br>1.4 | Sony Semiconductor Europe<br>The Crescent, Jays Close, Basingstoke, RG22 4DE, UK |
| G. van den Hoven<br>1.3 | JDS-Uniphase Netherlands BV<br>Prof. Holstaan 4, 5656 AA Eindhoven, The Netherlands |
| M.F.Zybura<br>1.2 | University of Virginia, Applied Electrophysical Laboratory<br>Charlottesville, VA 22903, USA |


In producing the present volume Datareviews by the following authors of *Properties of indium phosphide* (INSPEC, 1991) were reproduced, merged or adapted:

**F.R.Bacher, I.Bloomer, J.P.Fillard, A.R.Forouhi, A.M.Hennel, Q.H.Hua, G.Irmer, G.W.Iseler, A.Katz, B.Lambert, J.M.Lussert, R.Murray, E.P.O'Reilly, A.Sibille, R.A.Stradling and H.J.von Bardeleben.**

# Abbreviations

| | |
|---|---|
| ADC | analogue to digital converter |
| AE | activation energy |
| AFM | atomic force microscopy |
| AM | amplitude modulated |
| AM0 | air mass zero |
| APMOVPE | atmospheric pressure metal-organic vapour phase epitaxy |
| AS | admittance spectroscopy |
| ASE | amplified spontaneous emission |
| | |
| BE | band-edge |
| BER | bit error rate |
| BH | buried heterostructure |
| BJT | bipolar junction transistor |
| | |
| CATV | cable television |
| CBE | chemical beam epitaxy |
| CLEFT | cleavage of a lateral epitaxial film for transfer to another substrate |
| CML | current mode logic |
| CPU | central processing unit |
| CW | continuous wave |
| | |
| D-A | donor-acceptor |
| DALA | disorder-activated longitudinal acoustic |
| DALO | disorder-activated longitudinal optical |
| DATA | disorder-activated transverse acoustic |
| DATO | disorder-activated transverse optical |
| DBR | distributed Bragg reflector |
| DC | direct current |
| DEMUX | demultiplexer |
| DFB | distributed feedback |
| DGF | dynamic gradient freeze |
| DH | double heterostructure |
| DHBT | double heterojunction bipolar transistor |
| DI | deionised water |
| DLOS | deep level optical spectroscopy |
| DLTS | deep level transient spectroscopy |
| DMZ | dimethylzinc |
| DRO | dielectric resonator oscillator |
| DSL | dilute Sirtl and light |
| | |
| EBIC | electron beam induced current |
| ECL | emitter coupled logic |
| EDX | energy dispersive X-ray analysis |
| EPD | etch pit density |
| EPR | electron paramagnetic resonance |
| ER | etch rate |
| ESR | electron spin resonance |
| | |
| FET | field effect transistor |
| FIR | far infrared |
| FK | Franz-Keldysh |

| | |
|---|---|
| FSR | free spectral range |
| FWHM | full width at half maximum |
| | |
| GCSOA | gain-clamped semiconductor optical amplifier |
| GSMBE | gas source molecular beam epitaxy |
| HBT | heterojunction bipolar transistor |
| HEMT | high electron mobility transistor |
| HH | heavy-hole |
| HJ | heterojunction |
| HPL | hot photoluminescence |
| | |
| IC | integrated circuit |
| IF | intermediate frequency |
| IMPATT | impact avalanche transit-time |
| IPE | internal photoemission |
| IR | infrared |
| ISBT | intersubband transition |
| ISE | in-situ etching |
| ITO | indium tin oxide |
| | |
| JFET | junction field effect transistor |
| | |
| LA | longitudinal acoustic |
| LEC | liquid encapsulation Czochralski |
| LED | light emitting diode |
| LH | light-hole |
| LNA | low noise amplifier |
| LO | longitudinal optical |
| LP-MOVPE | low pressure metal-organic vapour phase epitaxy |
| LPE | liquid phase epitaxy |
| LST | light scattering tomography |
| LVM | local vibrational mode |
| | |
| MBE | molecular beam epitaxy |
| MCTS | minority carrier transient spectroscopy |
| MESFET | metal semiconductor field effect transistor |
| MIM | monolithic interconnected module |
| MIS | metal insulator semiconductor |
| MISFET | metal insulator semiconductor field effect transistor |
| MMIC | monolithic microwave integrated circuit |
| MOCVD | metal-organic chemical vapour deposition |
| MOMBE | metal-organic molecular beam epitaxy |
| MOVPE | metal-organic vapour phase epitaxy |
| MP | microprecipitate |
| MPL | magneto-photoluminescence |
| MQW | multiple quantum well |
| MSM-PD | metal-semiconductor-metal photodetector |
| | |
| NEPM | non-local empirical pseudopotential method |
| NF | noise figure |
| NRZ | non return to zero |
| | |
| OADM | optical add and drop multiplexer |
| ODENDOR | optically detected nuclear double resonance |

| | |
|---|---|
| ODLTS | optical deep level transient spectroscopy |
| ODMR | optically detected magnetic resonance |
| OEIC | optoelectronic integrated circuit |
| OMVPE | organometallic vapour phase epitaxy |
| | |
| PAE | power added efficiency |
| PC | photocurrent |
| PDC | phonon dispersion curve |
| PE | photoetching |
| PEEL | preferentially etched epitaxial liftoff |
| PICTS | photoinduced current transient spectroscopy |
| PL | photoluminescence |
| PLE | photoluminescence excitation |
| PRBS | pseudo-random binary sequence |
| PSG | phosphosilicate glass |
| PV | potential versus a saturated calomel reference electrode |
| | |
| QC | quantum cascade |
| QE | quantum efficiency |
| QW | quantum well |
| QWIP | quantum well infrared detector |
| | |
| RF | radio frequency |
| RIE | reactive ion etching |
| RT | room temperature |
| RTA | rapid thermal annealing |
| | |
| SCH-MQW | separate confinement heterostructure multiple quantum well |
| SD | single drift |
| SEM | scanning electron microscopy |
| SHBT | single heterojunction bipolar transistor |
| SHG | second harmonic generation |
| S-I | substitutional-interstitial |
| SI | semi-insulating |
| SI-BH | semi-insulating - buried heterostructure |
| SIMS | secondary ion mass spectrometry |
| SL | superlattice |
| SMSR | sidemode suppression ratio |
| SO | spin-split-off |
| SOA | semiconductor optical amplifier |
| SPC | surface photocurrent |
| SPM | self-phase modulation |
| SPV | surface photovoltage |
| SRH | Shockley-Read-Hall |
| | |
| TA | transverse acoustic |
| TBP | tertiarybutylphosphine |
| TCL | transmission cathodoluminescence |
| TE | transverse electric |
| TED | transferred electron device |
| TEM | transmission electron microscopy |
| TEO | transferred electron oscillator |
| TM | transverse magnetic |
| TMI | trimethylindium |

| | |
|---|---|
| TO | transverse optical |
| TPV | thermophotovoltaic |
| TRPL | time resolved photoluminescence |
| TSC | thermally stimulated current |
| UPS | ultraviolet photoelectron spectroscopy |
| USV | ultrasound velocity |
| UV | ultraviolet |
| VCO | voltage controlled oscillator |
| VECCS-OADM | vertically and contra-directionally coupled semiconductor - optical add and drop multiplexer |
| VPE | vapour phase epitaxy |
| WDM | wavelength-division multiplexed |
| XMA | X-ray microanalysis |
| XPS | X-ray photoelectron spectroscopy |
| XRD | X-ray diffraction |

CHAPTER 1

# IMPORTANCE OF InP PROPERTIES IN DEVICES

1.1 IMPATT diodes using InP
1.2 InP transferred electron devices
1.3 InP-based alloys as optical amplifiers and lasers
1.4 InP bipolar electronic and optoelectronic circuits
1.5 InP-based HEMT low-noise and power amplifiers
1.6 InP in integrated optics
1.7 InP as a solar cell material

# 1.1 IMPATT diodes using InP

H. Eisele and G.I. Haddad

November 1998

## A INTRODUCTION

Impact avalanche transit-time (IMPATT) diodes are the most powerful semiconductor devices known to date for power generation in the millimetre-wave region [1-3]. Two mechanisms, i.e. avalanche formation and carrier drift, jointly cause a phase delay between the RF voltage and RF current of around 180°, which is equivalent to a dynamic negative resistance between the diode terminals. This negative resistance is utilised to compensate loss, i.e. a positive resistance, in a resonant circuit which then in turn generates RF power. As shown in FIGURE 1, an IMPATT diode is reversed-biased with a high-field region near the pn-junction and an adjacent low-field region. In the high-field region, electrons (and holes) gain enough energy that their interaction with electrons in the valence band creates new hole-electron pairs. These pairs subsequently join the same process and an avalanche forms. The so-generated bunch of carriers drifts through the low-field region and induces a current in the outer circuit. The simplified voltage and current waveforms of FIGURE 2 show that the ionisation avalanche reaches its peak when the RF voltage decreases back to zero. This peak is delayed by 90° from the peak in the RF voltage, and carrier drift then adds further delay to the peak in the induced current.

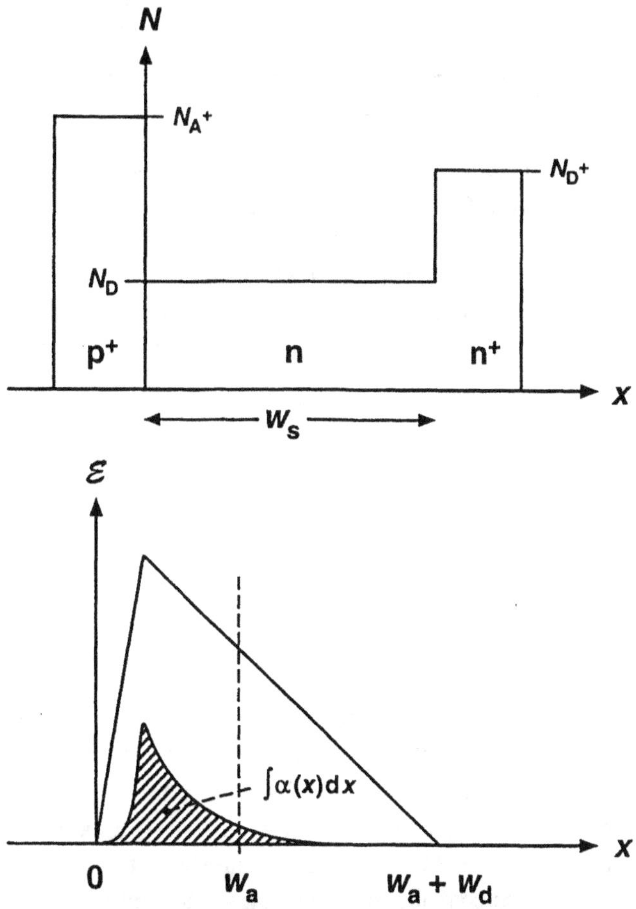


FIGURE 1 Schematic structure and electric field profile of a single-drift flat-profile IMPATT diode.

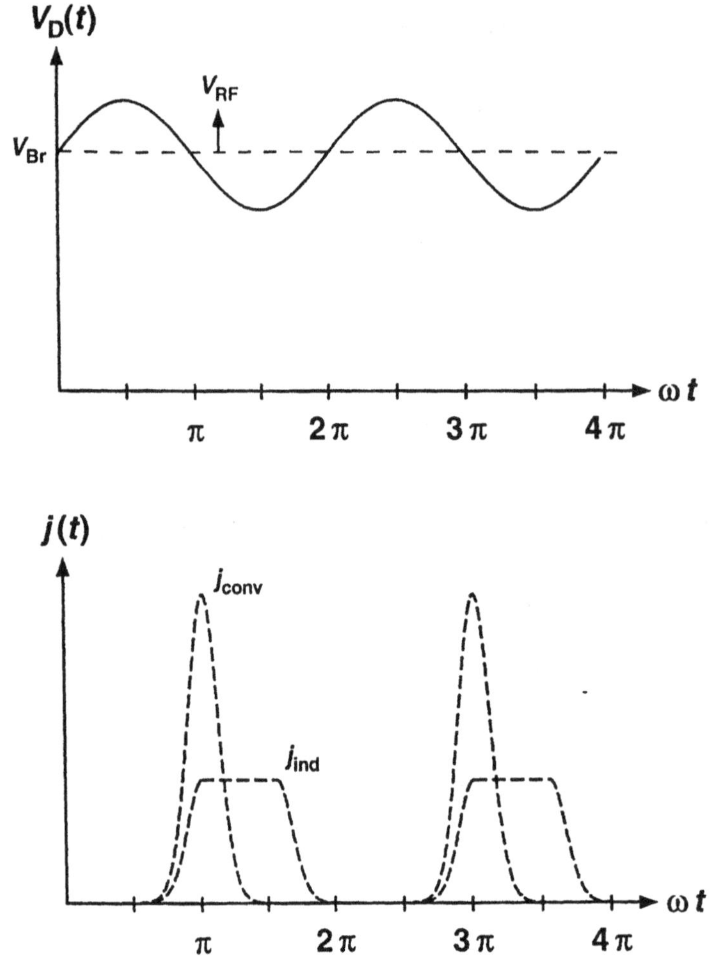


FIGURE 2 Voltage and current waveforms in a single-drift IMPATT diode ($V_{Br}$: breakdown voltage).

IMPATT diodes are typically highly symmetrical mesa-type structures on integral or diamond heat sinks. In most cases, therefore, the current distribution can be assumed constant across the diode, and a one-dimensional quasi-static drift-diffusion model accurately describes the carrier dynamics in the active region if energy relaxation effects are to be neglected. In this model, the current continuity equations are [1]

$$\frac{\partial j_n}{\partial x} = q\frac{\partial n}{\partial t} + q(R_n - G_n) \tag{1}$$

$$\frac{\partial j_p}{\partial x} = -q\frac{\partial p}{\partial t} - q(R_p - G_p) \tag{2}$$

where q is the elementary charge, n and p the concentrations of electrons and holes, and $G_{n,p}$ and $R_{n,p}$ the generation and recombination rates of electrons and holes, respectively. The field-dependent impact ionisation rates $\alpha_n$ and $\alpha_p$ of electrons and holes give the most significant contribution to the generation rate, and recombination can be neglected in the active region. However, at the very high field E (>1 MV $cm^{-1}$) in some millimetre-wave IMPATT diodes, interband tunnelling from the valence band to the conduction band at a rate $g_T$ also contributes to the carrier generation, and

$$qG_n = qG_p = \alpha_n(E)|j_n| + \alpha_p(E)|j_p| + qg_T(E) \tag{3}$$

Electron and hole currents, $j_n$ and $j_p$, are

$$j_n = qnv_n + qD_n \frac{\partial n}{\partial x} \tag{4}$$

$$j_p = qnv_p - qD_p \frac{\partial p}{\partial x} \tag{5}$$

with the electron and hole drift velocities, $v_n$ and $v_p$, and the electron and hole diffusion coefficients $D_n$ and $D_p$, respectively. The total current j through the diode consists of the induced (particle) current and the displacement current

$$j = j_n + j_p + \varepsilon_r \varepsilon_0 \frac{\partial E}{\partial t} \tag{6}$$

Poisson's equation gives the field distribution across the diode for acceptor and donor concentrations $N_A$ and $N_D$ and a dielectric constant $\varepsilon_r\varepsilon_0$ of the material:

$$\frac{\partial E}{\partial x} = \frac{q}{\varepsilon_r \varepsilon_0}\left(p - n + N_D - N_A\right) \tag{7}$$

and, integrated across the total length $L = w_a + w_d$,

$$V_D - V_{bi} = -\int_0^{w_a + w_d} E dx \tag{8}$$

where $V_D$ is the voltage between the diode terminals ($V_{bi}$: built-in voltage [1]).

Breakdown for equal ionisation rates $\alpha = \alpha_n = \alpha_p$ occurs at

$$\int_0^{w_a + w_d} \alpha(x) dx = 1 \tag{9}$$

with the majority of the carrier generation in the avalanche region $0 < x < w_a$ as illustrated with the shaded area in FIGURE 1.

Computer programs solve EQNS (1) - (9), and render accurate predictions of the large-signal performance of IMPATT diodes [4-9]. For very short active regions, quantum-mechanical effects need to be included, and typically energy-momentum models are a good compromise between accuracy in the predictions and required CPU time [5]. However, a simple large-signal analysis is sufficient in order to compare the fundamental potential and power generation capabilities of different material systems on a more general basis. This analysis is based on an approximation for the waveforms of FIGURE 2 [2,10] and gives a frequency-dependent figure of merit $M_P$ [1,3]

$$M_P = \left(\frac{E_C v_s}{f}\right)^2 \tag{10}$$

that only depends on the breakdown field $E_C$ and the saturated carrier drift velocity $v_s$ of the material. The oscillator circuit imposes a lower limit on the reactive part $X_C$ of the diode impedance of typically around -10 Ω [10] and, as a consequence, on the RF output power

$$P_{RF} \propto \frac{M_P}{|X_C|} \tag{11}$$

In this large-signal analysis, the DC-to-RF conversion efficiency η reaches an optimum at a drift region length $w_d$

$$w_d = \frac{3}{8}\frac{v_s}{f} \tag{12}$$

and also most strongly depends on the widths of the avalanche region $w_a$ with respect to the drift region $w_d$ [10]

$$\eta \propto \frac{1 + w_a / w_d}{1 + 2w_a / w_d} \tag{13}$$

TABLE 1 lists relevant properties and compares the figure of merit $M_P$ of three material systems from which IMPATT diodes were successfully fabricated, i.e. Si, GaAs and InP [3]. $M_P$ clearly indicates superiority of the InP material system over the GaAs material system, but not over Si. Nonetheless, several material parameters are more favourable in InP than in Si.

TABLE 1 Properties of semiconductors used in IMPATT diodes [3].

| | Si | GaAs | InP |
|---|---|---|---|
| Bandgap (eV) (@ 300 K) | 1.12 | 1.42 | 1.34 |
| Electron mobility (@ 300 K, $cm^2/V\ s$) | 1400 | 8500 | 4600 |
| Hole mobility (@ 300 K, $cm^2/V\ s$) | 450 | 400 | 140 |
| Breakdown field $E_C$ ($10^6$ V/cm) (@ $N_D \sim 10^{17}\ cm^{-3}$) | 0.61 | 0.65 | 0.75 |
| Thermal conductivity (W/cm K) | 1.25 | 0.46 | 0.68 |
| Sat. electron drift vel. $v_s$ ($10^7$ cm/s) (@ $E > 5 \times 10^5$ V/cm) | 1 | 0.6 | 0.75 |
| Dielectric constant | 11.8 | 12.8 | 12.6 |
| Electron $P_{RF}$ figure of merit relative to Si $(E_C v_s)^2$ | 1 | 0.4 | 0.9 |

## B RELEVANT MATERIAL PARAMETERS

Short avalanche regions (see EQN (13)) for high DC-to-RF conversion efficiencies and contributions from both electron and hole drift regions for maximum RF power levels [2] are achieved with diode structures as shown in FIGURE 3. Favourable material parameters for these structures are high ionisation rates at high electric fields for both electrons and holes as well as high electron and hole drift velocities. FIGURE 4 compares the field-dependent measured and extrapolated electron and hole ionisation rates of the semiconductor materials Si [11], GaAs [8,12,13] and InP [1,8,14]. InP exhibits the highest ionisation rates for both electrons and holes above electric fields E of approximately 800 kV cm$^{-1}$, and therefore the doping profile of the avalanche region can be designed to achieve maximum confinement. However, interband tunnelling becomes more significant for E > 1 MV cm$^{-1}$ [8,15] and needs to be considered. Interband tunnelling generates carriers much earlier in the RF cycle than impact ionisation; thus this premature carrier generation severely degrades the DC-to-RF conversion efficiency of an otherwise optimum diode structure [1,10,15]. Tunnelling rates are too low in Si [16] to be of concern, but relevant and of similar magnitude in InP and GaAs [8,9]. As can be seen from FIGURE 4, electron and hole ionisation rates in GaAs tend to saturate above 1 MV cm$^{-1}$

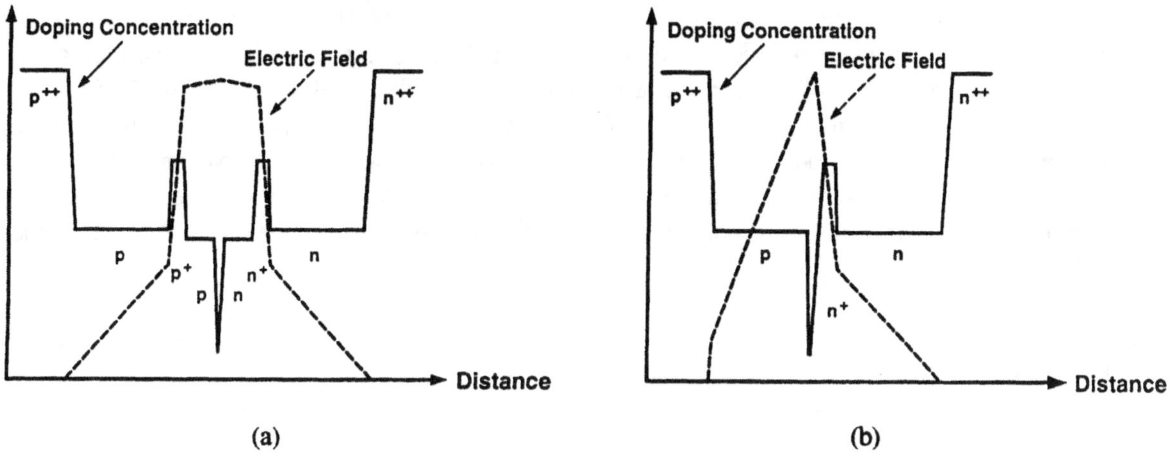


FIGURE 3 Schematic doping profiles for high-power, high-efficiency, double-drift IMPATT diodes: (a) double-drift Read low-high-low; (b) double-drift hybrid Read flat-high-low.

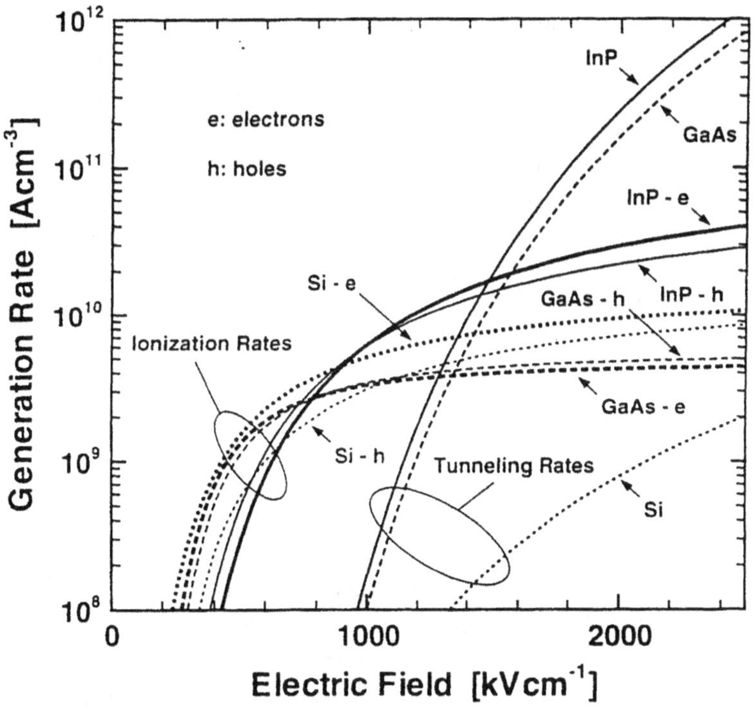


FIGURE 4 Field-dependent generation rates from impact ionisation of electrons and holes, $\alpha_n J$ and $\alpha_p J$, as well as from tunnelling, $qg_T$, in the semiconductor materials InP [8,9,14,20], GaAs [9,12,13] and Si [11] at 300 K and a current density J of 25 kA cm$^{-2}$.

[9,13,17] and a sizeable fraction of carriers is generated by interband tunnelling. Since tunnelling needs to be avoided in high-efficiency diodes, well confined avalanche regions are difficult to achieve in GaAs IMPATT diodes at high millimetre-wave frequencies.

This moderate confinement limits the achievable DC-to-RF conversion efficiencies and, because of the lower thermal conductivity than in Si and InP, also the attainable RF power levels in GaAs IMPATT diodes. This disadvantage is also exacerbated by low saturated electron and hole drift velocities in GaAs above 100 kV $cm^{-1}$ as illustrated in FIGURE 5 [9,13,18,19]. According to EQN (12), lower drift velocities $v_s$ require shorter drift regions $w_d$, which mandate smaller diode areas for the same $X_C$ levels and, ultimately, reduce $P_{RF}$. The pronounced roll-off in both RF power levels and DC-to-RF conversion efficiencies above 60 GHz [3] points to these physical limits in the power generation capabilities of GaAs IMPATT diodes. Neither Si nor InP suffers from similar limitations, and well-confined avalanche regions can be utilised even above 100 GHz although much lower hole ionisation rates in Si need to be taken into account in a diode design. In InP, ionisation rates of electrons and, in particular, of holes are much higher than in Si at relevant electric fields. As a consequence, the avalanche region in InP IMPATT diodes is well-confined even in a simple flat-profile structure and allows efficient operation as confirmed by extensive diode simulations and experimental results. However, low ionisation rates at lower electric fields cause breakdown voltages to be higher in InP than in Si and GaAs at the same doping levels, which requires thermal limitations to be taken more seriously in InP than in Si.

Similar to GaAs, InP exhibits a region of negative differential electron mobility. However, the small number of experimental studies on InP IMPATT diode performance continues to impede a conclusion as to whether this effect can be utilised effectively at medium millimetre-wave frequencies.

## C PREDICTED DIODE PERFORMANCE

In FIGURE 6, the general trend in predicted RF power levels and DC-to-RF conversion efficiencies of optimised InP double-drift IMPATT diodes is plotted against operating frequency. These predictions assume continuous-wave (CW) operation and are based on extensive diode simulations [8,9,21] that took the discussed material parameters of InP and realistic real parts of the load impedance $Z_L$ of around 1 Ω as well as appropriate series resistances from the ohmic contacts into account. These predictions agree well with those from other large-signal diode simulations [4,6,17] and also confirm the power generation capabilities as predicted by $M_P$ from the simple large-signal analysis. For a comparison, the best results published on Si and GaAs double-drift diodes and taken from references in [3] are included in FIGURE 6. RF power levels of, for example, near 1 W are predicted around 100 GHz and are similar to those from Si IMPATT diodes. However, much higher DC-to-RF efficiencies of, for example, near 20% around 100 GHz are expected as a result of more favourable material parameters.

## D EXPERIMENTAL RESULTS

Early work on InP IMPATT diodes around 1980 focused on operation at microwave and lower millimetre-wave frequencies. TABLE 2 summarises the best results in CW mode [22,23] and TABLE 3 gives the results under long pulses of 0.5 - 1.0 μs duration [22-28]. Fabrication technologies were similar to those employed in the fabrication of InP Gunn devices and included (a) substrate thinning to minimise skin effect problems and (b) electroplated integral heat sinks for heat removal. Since the material for these studies was grown before the advent of advanced growth techniques such as molecular beam epitaxy (MBE), metalorganic chemical vapour deposition (MOCVD), metalorganic

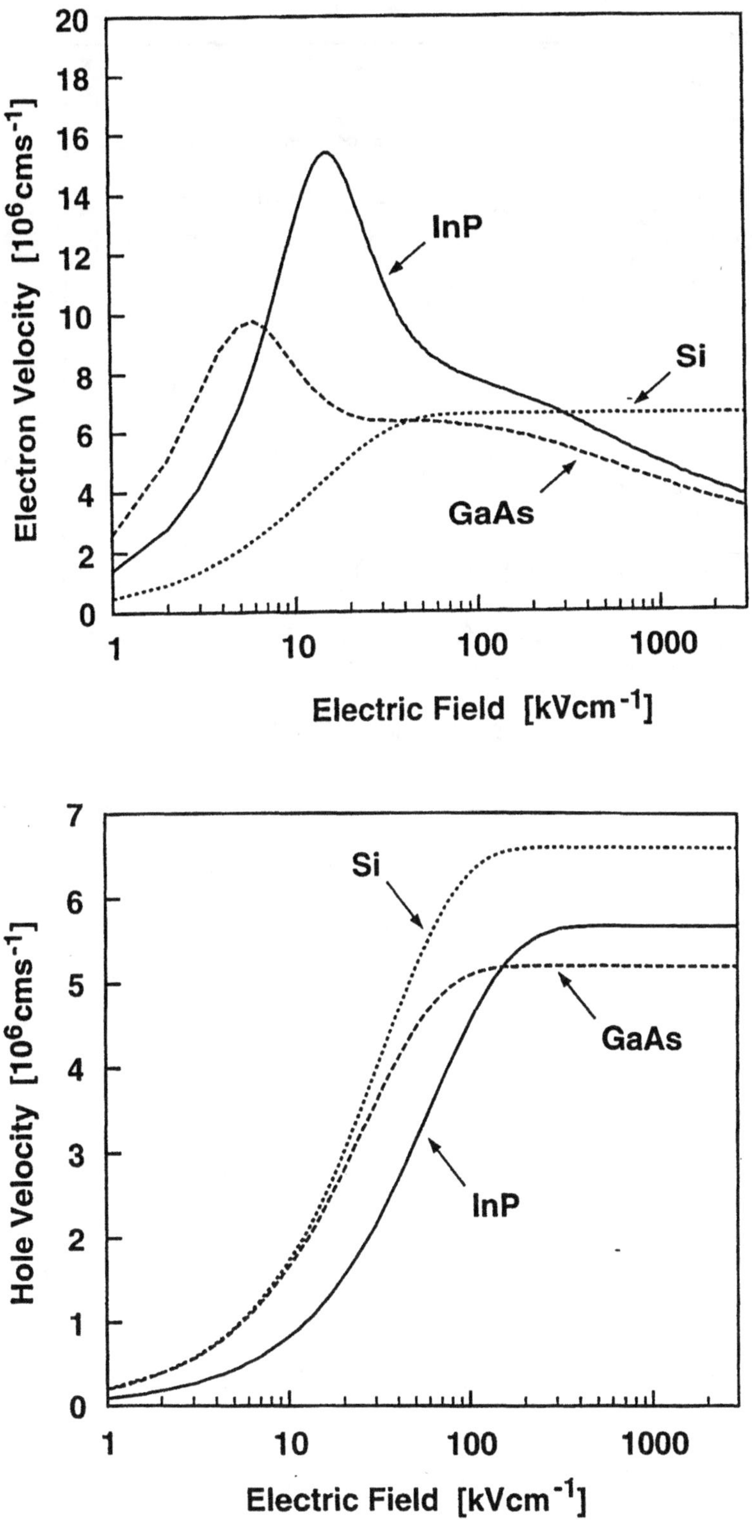


FIGURE 5 Field-dependent electron and hole drift velocities in the semiconductor materials InP [9], GaAs [9] and Si at 500 K.

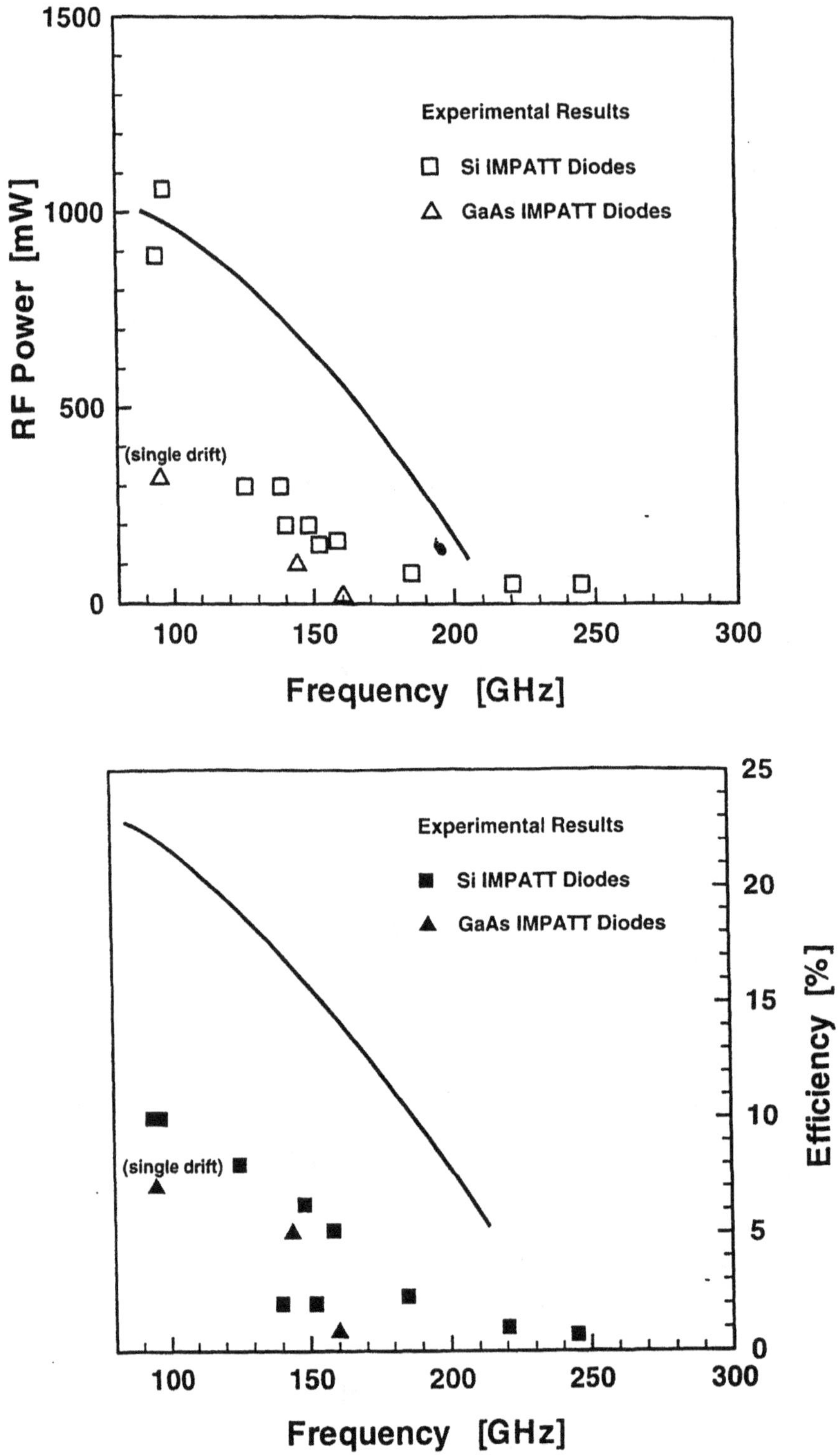


FIGURE 6 Comparison of performance predicted for InP double-drift IMPATT diodes with measured performance of Si and GaAs double-drift IMPATT diodes.

TABLE 2 Performance of InP IMPATT diodes in continuous-wave (CW) operation below 60 GHz (SD: single drift, F: flat profile, R: Read profile).

| Frequency | RF power | Efficiency | Structure | Laboratory | Ref |
|---|---|---|---|---|---|
| 9.78 GHz | 1.6 W | 11.1% | SDF | TRW | [22] |
| 9.78 GHz | 1.7 W | 10.8% | SDF | TRW | [22] |
| 40 GHz | 2.0 W | 8% | SD | TRW | [23] |

TABLE 3 Performance of InP IMPATT diodes in a pulsed mode below 60 GHz (SD: single drift, F: flat profile, R: Read profile).

| Frequency | RF power | Efficiency | Structure | Duration | Duty cycle | Laboratory | Ref |
|---|---|---|---|---|---|---|---|
| 10.8 GHz | 5 W | 14% | SDF | 500 ns | 10% | TRW | [22] |
| 10.8 GHz | 6.1 W | 13.7% | SDF | 500 ns | 10% | TRW | [22] |
| 11.8 GHz | 20.5 W | 16.4% | SDR | 1 μs | 1% | TRW | [24] |
| 11.8 GHz | 22 W | 15% | SDR | 1 μs | 1% | TRW | [24] |
| 27 GHz | 3.6 W | 5.8% | SDF | 500 ns | 1% | Varian | [25] |
| 27.1 GHz | 9.2 W | 9.3% | SDR | 500 ns | 1% | Varian | [25] |
| 29.5 GHz | 7.7 W | 10.5% | SDR | 500 ns | 1% | Varian | [25] |
| 32 GHz | 9.3 W | 10.6% | SD | 500 ns | 1% | TRW | [23] |
| 35 GHz | 10 W | 10% | SDR | 500 ns | 1% | - | [28] |
| 33 GHz | 4.5 W | 8.3% | SDR | 500 ns | 1% | Varian | [25] |
| 45 GHz | 2.5 W | 5.1% | SDR | 500 ns | 1% | Varian | [25] |
| 53 GHz | 1.7 W | 3% | SDR | - | 1% | TRW | [26,27] |
| 54 GHz | 1.5 W | 4% | SDR | - | 1% | TRW | [27] |
| 60 GHz | 3 W | 6% | SDR | 500 ns | 1% | - | [28] |

molecular-beam epitaxy (MOMBE) and chemical-beam epitaxy (CBE), high-quality p-doped InP layers and lattice-matched InGaAs layers were not yet available and diode structures were limited to medium-power single-drift profiles. As a result, the performance of InP IMPATT diodes remained somewhat below that of their Si and GaAs counterparts. This performance was also limited by poor ohmic contacts on p-type material.

More recent research explored the potential and capabilities of InP IMPATT diodes at V-band (50 - 75 GHz) and W-band (75 - 110 GHz) frequencies [21]. Single-drift and double-drift profiles were obtained by CBE and MOCVD. The fabrication process was adopted from a well established technology for InP Gunn devices on integral heat sinks [29], which is based on selective etchants and lattice-matched InGaAs as stop-etch layer [2]. Single-drift diodes from a wafer with a thin, highly p doped InP layer and a 0.45 μm long active region with $N_D = 1.5 \times 10^{17}$ cm$^{-3}$ showed very promising results. Typical RF power levels around 80 mW were measured in standard full-height top-hat WR-15 and WR-10 waveguide cavities [2] at oscillation frequencies from about 60 GHz to 80 GHz. Corresponding DC-to-RF conversion efficiencies typically ranged from 3% to 4%. The highest oscillation frequency observed was 87 GHz [21]. FIGURE 7 shows RF output power and oscillation frequency of the best diode as a function of the DC input power. As a sign of thermal limitation, no saturation in the RF power was observed up to the maximum applied input power of 2.2 W. This diode yielded RF power levels (and corresponding DC-to-RF conversion efficiencies) of more than 110 mW (more than 5%) at 64.7 GHz, 82 mW (3.7%) at 79 GHz and 55 mW (2.4%) at 84.8 GHz, and the RF output power was thermally limited at each frequency [21].

Although the specific series resistance from the ohmic contacts was estimated to be approximately $4 \times 10^{-6}$ Ω cm$^{-2}$, RF output power and oscillations around 63.5 GHz were detected above a low threshold current density $J_{th}$ of 2 kA cm$^{-2}$ [21]. This value is much lower than corresponding values in the range of 4 - 6 kA cm$^{-2}$ measured with similar GaAs single-drift flat-profile IMPATT diodes on

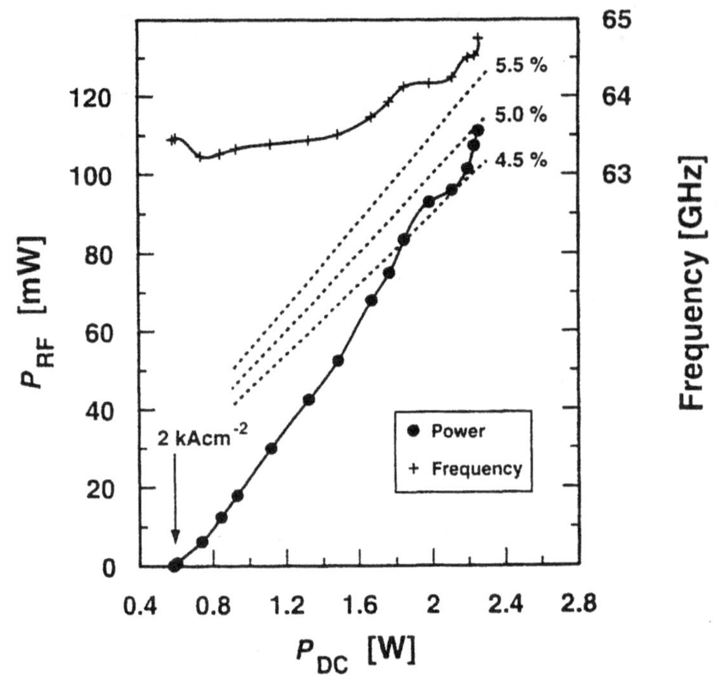


FIGURE 7 Measured performance of an InP single-drift flat-profile IMPATT diode on an integral heat sink around 60 GHz [21].

integral heat sinks around 60 GHz [30,31] (it also should be noted that, for an equitable comparison at the same frequency, $J_{th} \propto f^3$ around the optimum frequency needs to be taken into account). These GaAs diodes typically exhibit much lower specific series resistances of around $2 \times 10^{-6}\ \Omega\ cm^{-2}$. Lower threshold current densities indicate higher impedance levels at the same current densities and, therefore, larger diode areas can be chosen resulting in higher RF power levels.

Simulations accurately predicted the above measured diode performance around 65 GHz and provided a valid extrapolation to the yet unattained operation of an InP IMPATT diode on a diamond heat sink. Although the same high series resistance of $4 \times 10^{-6}\ cm^{-2}$ was taken into account and the operating junction temperature was limited to a safe value of 500 K, RF power levels of more than 1.5 W with corresponding DC-to-RF conversion efficiencies above 11% were predicted around 65 GHz. This performance of a simple single-drift flat-profile structure is competitive with the performance of a Si double-drift hybrid Read IMPATT diode at the same frequency and a comparable operating junction temperature [32]. This predicted performance again highlights the favourable InP material parameters as well as the excellent confinement of the avalanche region.

## E CONCLUSION

As a result of favourable material parameters, InP IMPATT diodes are well suited for high-power applications at millimetre-wave frequencies. However, considerable improvements in the area of ohmic contacts to p-doped InP, of tight doping control and of heavily p-doped regions without outdiffusion are necessary to exploit the potential of the InP material system to its full extent. High DC input power densities also mandate the use of diamond heat sinks for efficient heat removal in these applications. The appropriate technology can be adopted from InP Gunn devices on diamond heat sinks [2,3], where state-of-the-art performance was demonstrated.

**REFERENCES**


[1] S.M. Sze [ *Physics of Semiconductor Devices 2nd Edition* (John Wiley & Sons, New York, 1981) ]

[2] H. Eisele, G.I. Haddad [ in *Modern Semiconductor Device Physics* Ed. S.M. Sze (John Wiley & Sons, New York, 1997) ch.6 p.343-407 ]

[3] H. Eisele, G.I. Haddad [ *IEEE Trans. Microw. Theory Tech. (USA)* vol.MTT-46 (1998) p.739-46 ]

[4] D. Lippens, E. Constant, M.R. Friscourt, P.A. Rolland, G. Salmer [ *IEEE Electron Device Lett. (USA)* vol.EDL-3 (1982) p.213-5 ]

[5] R.K. Mains, G.I. Haddad [ *Infrared Millim. Waves (UK)* vol.10 (1983) p.111-231 ]

[6] C. Dalle, D. Degrugillier, J.L. Nieruchalski, D. Lippens, P.A. Rolland [ *Proc. IEEE/Cornell Conference on Advanced Concepts in High Speed Semiconductor Devices and Circuits* Ithaca, New York, 29-31 July 1985 (IEEE, 1986) p.351-9 ]

[7] C.-C. Chen, R.K. Mains, G.I. Haddad, H. Eisele [ *Proc. IEEE/Cornell Conference on Advanced Concepts in High Speed Semiconductor Devices and Circuits* Ithaca, New York, 2-4 August 1993 (IEEE, 1993) p.194-202 ]

[8] C.-C. Chen, R.K. Mains, G.I. Haddad, H. Eisele [ *Proc. 5th Int. Symp. on Space Terahertz Technology* Ann Arbor, Michigan, 10-13 May 1994 (IEEE, New York, 1995) p.648-58 ]

[9] C.-C. Chen [ PhD Thesis, University of Michigan, 1994 ]

[10] B. Culshaw, R.A. Giblin, P.A. Blakey [ *Avalanche Diode Oscillators* (Taylor & Francis Ltd., London, 1978) ]

[11] R. Van Overstraeten, H. de Man [ *Solid-State Electron. (UK)* vol.13 (1970) p.583-608 ]

[12] G.E. Bulman, V.M. Robbins, G.E. Stillman [ *IEEE Trans. Electron Devices (USA)* vol.ED-32 (1985) p.2454-66 ]

[13] H. Eisele [ *Int. J. Infrared Millim. Waves (USA)* vol.12 (1991) p.345-54 ]

[14] L. Umebu, A.N.M.M. Choudhury, P.N. Robson [ *Appl. Phys. Lett. (USA)* vol.36 (1980) p.302-3 ]

[15] M.E. Elta, G.I. Haddad [ *IEEE Trans. Electron Devices (USA)* vol.ED-25 (1978) p.694-702 ]

[16] J.-F. Luy, F. Kuehne [ *IEEE Trans. Electron Devices (USA)* vol.ED-36 no.3 (1989) p.589-95 ]

[17] P.A. Rolland, M.R. Friscourt, D. Lippens, C. Dalle, J.L. Nieruchalski [ *Proc. Int. Workshop on Millimeter Waves* Rome, Italy, 2-4 April 1986, p.125-77 ]

[18] T.H. Windhorn, L.W. Cook, T.J. Roth, G.E. Stillman [ in *Physics of Submicron Structures - Proceedings of a Workshop* Eds H.L. Grubin, K. Hess, G.J. Iafrate, D.K. Ferry (Plenum Press, New York, USA, 1984) p.253-9 ]

[19] K. Brennan, K. Hess [ *Phys. Rev. B (USA)* vol.29 no.10 (1984) p.5581-90 ]

[20] K. Taguchi, T. Torikai, Y. Sugimoto, K. Makita, H. Ishihara [ *J. Appl. Phys. (USA)* vol.59 no.2 (1986) p.476-81 ]

[21] H. Eisele, C.-C. Chen, G.O. Munns, G.I. Haddad [ *IEEE MTT-S Int. Microw. Symp. Dig. (USA)* (1996) p.529-32 ]

[22] J.J. Berenz, F.B. Fank, T.L. Hierl [ *Electron. Lett. (UK)* vol.14 (1978) p.683-4 ]

[23] G.R. Thoren [ *RCA Rev. (USA)* vol.45 (1984) p.557-78 ]

[24] J.J. Berenz, P.A. Blakey [ *Proc. 8th Biennial Cornell Conference on High-Speed Semiconductor Devices and Circuits* Ithaca, New York, August 1981 (IEEE, New York, USA, 1981) p.395-404 ]

[25] F.B. Fank, M. Vichr, I.V. Zubeck, D.R. Tringali, J.C. Crowley [ *Proc. 8th Biennial IEEE/Cornell Conference on High-Speed Semiconductor Devices and Circuits* Ithaca, New York, August 1981 (IEEE, 1981) p.385-94 ]

[26] E.J. Haugland [ *Microw. RF (USA)* vol.23 no.8 (1984) p.100-7 ]

[27] J.A. Stubstad, M. Vichr, F.B. Fank [ *Proc. IEEE/Cornell Conference on High Speed Semiconductor Devices and Circuits* Ithaca, New York, 15-17 August 1983 (IEEE, 1983) p.38-46 ]

[28] Y.E. Ma [ *Proc. SPIE (USA)* vol.544 (1985) p.95-102 ]

[29] R. Kamoua, H. Eisele, G.I. Haddad [ *Solid-State Electron. (UK)* vol.36 (1993) p.1547-55 ]

[30] X. Zhang [ PhD Thesis, Technical University of Munich, 1984 ]

[31] W. Bogner [ PhD Thesis, Technical University of Munich, 1993 ]

[32] J.-F. Luy, F. Schäffler, M. Schlett [ *Proc. 22nd European Microwave Conference* Espoo, Finland, 24-27 August 1992, p.485-90 ]

# 1.2 InP transferred electron devices

S.H. Jones and M.F. Zybura

January 1999

## A INTRODUCTION

The millimetre and submillimetre-wave regions of the electromagnetic spectrum continue to be areas of increasing interest. Millimetre waves correspond to the frequencies between 30 GHz and 300 GHz, while the submillimetre-wave range is defined to be the region between 300 GHz and 3 THz. Millimetre-wave systems find use in such diverse areas as emission spectra analysis in radio astronomy, atmospheric and oceanic remote sensing, auto collision avoidance radar systems, high performance site-to-site communication systems, plasma diagnostics and surveillance radar. Presently, the most serious obstacle to the development of these systems is the lack of low noise solid-state sources. The prominent sources used today in these systems are GaAs and InP transferred electron devices (TEDs), the InP TED being used almost exclusively at frequencies above 100 GHz. In this Datareview we describe the state-of-the-art performance that is presently available in InP TEDs.

## B BACKGROUND

TEDs and transferred electron oscillators (TEOs) were first experimentally demonstrated by Gunn [1] and therefore are often referred to as Gunn diodes. The operation of TEOs is possible given the devices' unique negative resistance properties associated with the transfer of electrons to low velocity conduction bands at high electric fields (voltage bias). Essentially, when a voltage is applied to the TED, the conduction electrons gain energy and transfer to the upper valleys that have lower electron mobility; this causes the average electron velocity to decrease with increasing field creating a negative differential resistance. An excellent review of TED physics and technology can be found in [2].

There are essentially two types of InP TED. The first is the Gunn diode comprised of an n-type active layer with average doping near $1 \times 10^{16}$ cm$^{-3}$. This layer is typically 1 to 3 μm long and sandwiched between the heavily doped cathode and anode contacting layers [3-6]. Usually, the doping profile of the active layer is graded with a slope of 1.5 - 2.0 from cathode to anode. This non-uniform doping distribution aids the formation of accumulation layers and dipole domains that create current instabilities and negative resistance. These $N^+NN^+$ InP Gunn diodes are generally operated at 4 - 6 V DC bias and 0.5 - 1.0 A of DC current. The second style InP TED is the current limiting cathode device which has a similar structure, but there is not a heavily doped cathode layer; the cathode contact is formed directly on the active layer [7]. This device has a very shallow metal-semiconductor contact junction that limits current, similar to a low barrier Schottky diode. The current limiting devices have very uniform doping profiles, and have negative resistance associated with the stable modulated depletion layer [8]. The $NN^+$ TEDs are generally operated at 9 - 10 V DC bias and 0.15 - 0.3 A of DC current. The df/dV for these devices can be positive or negative as a function of heatsink temperature by adjusting the cathode characteristics [7].

Both devices are fabricated as single discrete diodes with the anode and cathode contacts on opposite sides of the device. Special consideration must be given to heat sinking and packaging the devices since 2 - 5 W of DC power (or heat) is typically dissipated within the diode. The cathode contact diameters typically range from 50 to 100 μm and directly control the DC current during operation. Various device designs and fabrication techniques are described in [2-7]. Most InP TEOs in operation today are of the J.E. Carlstrom cavity style [9] which has proven to be extremely reliable.

## C COMPARISON OF GaAs AND InP TEDS

The most complete comparison of GaAs and InP TED performance for 30 GHz to 300 GHz has been given by Eisele and Haddad [5,10]. These results are given in FIGURE 1. As shown, the InP diodes generally produce about twice as much power and have twice the efficiency as the GaAs devices. Also, there have been very few reported results for modest operation of GaAs devices at 100 GHz or higher, while a number of researchers have demonstrated good performance with InP diodes from 140 to 300 GHz.

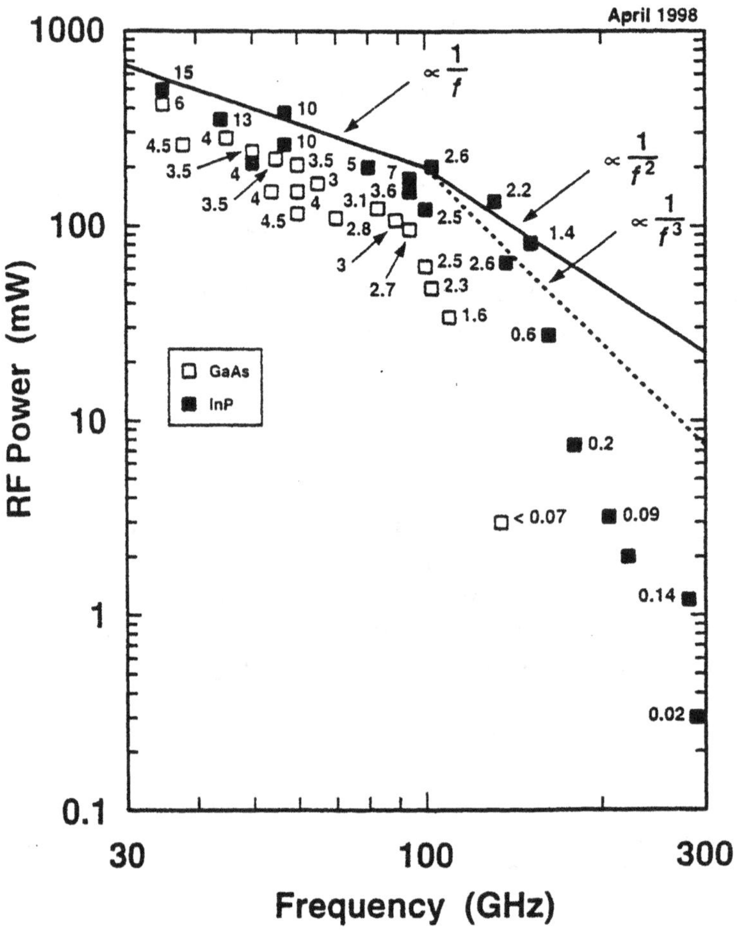


FIGURE 1 State-of-the-art RF power levels for TEDs under CW operation are shown. The numbers next to the symbols indicate DC-RF efficiency in % [5].

The improved performance of InP diodes over GaAs diodes stems from three basic differences in material properties [11]. First, the energy relaxation times for GaAs are approximately 0.4 - 1.5 ps while for InP they are 0.2 - 0.75 ps [5]. These faster relaxation times directly translate to faster electron transfer to the upper valleys, and ultimately improved oscillator efficiency at high frequency. Secondly, the thermal conductivity of GaAs is 0.46 W/cm K, and for InP it is 0.68 W/cm K. For a given DC power dissipation and similar device geometry, the InP device will run cooler than the GaAs device. Since the electron mobility and differential negative resistance are both reduced with the temperature, the InP devices have an inherent advantage as a result of the improved thermal conductance. Thirdly, the ratio of the peak low-field electron velocity to the minimum high-field electron velocity is between 1.5 and 2.0 times larger for InP versus GaAs. For a given RF voltage, this directly translates to larger RF current magnitudes and higher RF power.

## D RESEARCH RESULTS FOR 100 - 300 GHz InP TEOS

Some of the best TEO results above 100 GHz are given in TABLE 1. The performance is divided between the $N^+NN^+$ devices and the current limiting $NN^+$ style devices. As shown, 200 mW of power can be generated at about 100 GHz, while as much as 1 mW of RF power has been reported near 300 GHz.

TABLE 1 Summary of reported RF performance for InP TEOs [5,13].

| Device style | TEO frequency | RF output power | Ref |
|---|---|---|---|
| $N^+NN^+$ | 100 GHz | 200 mW | [4,5] |
| $N^+NN^+$ | 132 GHz | 130 mW | [4,5] |
| $N^+NN^+$ | 152 GHz | 80 mW | [4,5] |
| $N^+NN^+$ | 188 GHz | 6 - 8 mW | [3] |
| $N^+NN^+$ | 269 GHz | 0.7 mW | [5,10] |
| $N^+NN^+$ | 279 GHz | 1 mW | [5,10] |
| $N^+NN^+$ | 280 GHz | 1.2 mW | [5,10] |
| $NN^+$ | 122 GHz | 85 mW | [12-14] |
| $NN^+$ | 127 GHz | 65 mW | [12-14] |
| $NN^+$ | 137 GHz | 42 mW | [12-14] |
| $NN^+$ | 140 GHz | 60 mW | [7,12-14] |
| $NN^+$ | 146 GHz | 49 mW | [12-14] |
| $NN^+$ | 151 GHz | 29 mW | [12-14] |
| $NN^+$ | 167 GHz | 6 mW | [12-14] |
| $NN^+$ | 170 GHz | 8 mW | [12-14] |

## E COMMERCIAL InP TEO RESULTS FOR 30-140 GHz

The leading manufacturer of GaAs and InP TEDs is Litton Solid State, San Jose, CA, USA. Given in TABLE 2 are commercially available $NN^+$ style current limiting TEDs presently offered for sale and being used in numerous radar and receiver applications. These diodes are based on the research and development outlined in [7,8]. As shown, high reliability and commercially available diodes produce

TABLE 2 Specifications for commercially available InP TEDs manufactured by Litton Solid State, San Jose, CA, USA [7].

| Part number | Operating frequency min./max. (test freq.) (GHz) | Min. CW output power (mW) | Typ. bias voltage (V) | Typ. bias current (mA) | Efficiency (%) |
|---|---|---|---|---|---|
| Fundamental mode | | | | | |
| LSA-9110-S1 | 30/40 (35+/-2) | 250 | 12.0 | 200 | 11.0 |
| LSA-9110-S2 | 30/40 (35+/-2) | 300 | 12.0 | 220 | 12.0 |
| LSA-9110-S3 | 30/40 (35+/-2) | 350[6] | 12.0 | 240 | 13.0 |
| LSQ-9119-S1 | 40/50 (44+/-2) | 150 | 11.0 | 170 | 9.0 |
| LSQ-9119-S2 | 40/50 (44+/-2) | 200 | 11.0 | 200 | 10.0 |
| LSQ-9119-S3 | 40/50 (44+/-2) | 250[6] | 11.0 | 250 | 11.0 |
| LSE-9120-S1 | 50/60 (56+/-2) | 150 | 8.0 | 260 | 8.0 |
| LSE-9120-S2 | 50/60 (56+/-2) | 200[6] | 8.5 | 280 | 9.0 |
| Second harmonic mode | | | | | |
| LSW-9177-S1 | 75/85 (77+/-2) | 60 | 10.0 | 180 | 4.5 |
| LSW-9177-S2 | 75/85 (77+/-2) | 70 | 10.0 | 220 | 5.0 |
| LSB-9122-S14 | 85/110 (94+/-2) | 45 | 10.0 | 160 | 3.5 |
| LSB-9122-S13 | 85/110 (94+/-2) | 60 | 10.0 | 170 | 4.0 |
| LSB-9122-S15 | 85/110 (94+/-2) | 70 | 10.0 | 180 | 4.0 |
| LSB-9122-S16 | 85/110 (94+/-2) | 80 | 10.0 | 180 | 5.0 |

350 mW at 40 GHz, 200 mW at 60 GHz and as much as 80 mW at 110 GHz. Litton Solid State also offers 140 GHz devices that produce 20 - 40 mW at 140 GHz (LST-9123s4).

The most notable manufacturer of 70 - 140 GHz TEOs is J.E. Carlstrom Company, Chicago, IL, USA [9]. Mechanically tuned and high reliability oscillators incorporating InP TEDs from Litton Solid State are available over the 70 - 140 GHz regime with bandwidths as large as 30 GHz. Over the 110 - 147 GHz regime these oscillators generally provide 20 - 40 mW of output power.

## REFERENCES


[1] J.B. Gunn [ *Solid State Commun. (USA)* vol.1 (1963) p.88-91 ]

[2] B.G. Bosch [ *Gunn Effect Electronics* (John Wiley & Sons, New York, USA, 1975) ]

[3] A. Rydberg [ *IEEE Electron Device Lett. (USA)* vol.11 no.10 (1990) p.439-41 ]

[4] H. Eisele, G. Haddad [ *IEEE Microw. Guid. Wave Lett. (USA)* vol.5 (1995) p.385-7 ]

[5] H. Eisele, G. Haddad [ *IEEE Trans. Microw. Theory Tech. (USA)* vol.46 no.6 (1998) p.739-46 ]

[6] H. Eisele [ *Proc. 9th Int. Symp. on Space Terahertz Technology* Pasadena, CA, USA, March 1998 ]

[7] J.D. Crowley et al [ *Electron. Lett. (UK)* vol.30 no.6 (1994) p.499-500; see also www.littonsolidstate.com/device/gd3594.mtm ]

[8] M.F. Zybura, S.H. Jones, B.W. Lim, J.D. Crowley, J.E. Carlstrom [ *Solid-State Electron. (UK)* vol.39 no.4 (1996) p.547-53 ]

[9] J.E. Carlstrom, R.L. Plambeck, D.D. Thornton [ *IEEE MTT (USA)* vol.33 (1985) p.610-9 ]; [ J.E. Carlstrom Co., Chicago, IL, USA ]

[10] H. Eisele [ in *Encyclopedia of Electrical and Electronics Engineering* Ed. J.G. Webster (Wiley, New York, USA, to be published Feb. 1999) ]

[11] B.K. Riley [ *J. Appl. Phys. (USA)* vol.48 no.2 (1977) ]

[12] S.H. Jones, M.F. Zybura, J.E. Carlstrom, T. O'Brien [ *Proc. 8th Int. Symp. on Space Terahertz Technology* Harvard Univ., Boston, MA, 1997 (IEEE, 1997) ]

[13] S.H. Jones, M.F. Zybura, J.E. Carlstrom, T. O'Brien [ *IEEE Trans. Electron Devices (USA)* vol.46 no.1 (January 1999) p.17-23 ]

[14] M.F. Zybura [ PhD Dissertation, University of Virginia, Charlottesville, VA, January 1996 ]

# 1.3 InP-based alloys as optical amplifiers and lasers

G. van den Hoven and L. Spiekman

June 1999

## A OPTICAL GAIN IN COMPOUND SEMICONDUCTOR MATERIALS

Amplification in semiconductor materials such as InP is most well-known for its application in semiconductor lasers. Pumping a semiconductor material with an electrical current can create a population inversion. Passing a light signal through the semiconductor causes amplification through stimulated emission processes [1]. In a semiconductor laser, the amplifying material is combined with optical feedback elements (for example mirrors) to create a resonating cavity in which light propagates back and forth (Section D). In a semiconductor optical amplifier (SOA) the signal light passes through the amplifier only once and special care is taken to eliminate optical feedback (Section C).

FIGURE 1 shows the energy band structure of a direct bandgap semiconductor such as InP [2]. The band structure represents the collection of energy states which electrons in the semiconductor can occupy. For a certain range of electron energies, no states exist. The size of the forbidden energy gap or bandgap between the valence and conduction band is determined by material properties such as composition and temperature. In thermal equilibrium, the states in the valence band are (nearly) all filled; only a small number of electrons reside in the conduction band due to thermal excitation, leaving behind an equal number of positively charged holes in the valence band. When passing an electrical current through the semiconductor, the electrons in the conduction band travel towards the high potential (positive) side, and the holes in the valence band travel towards the low potential (negative) side. The conduction properties of the semiconductor crystal can be strongly influenced by adding impurities or doping: p-type doping enhances the hole concentration in the valence band; n-type doping enhances the electron concentration in the conduction band.

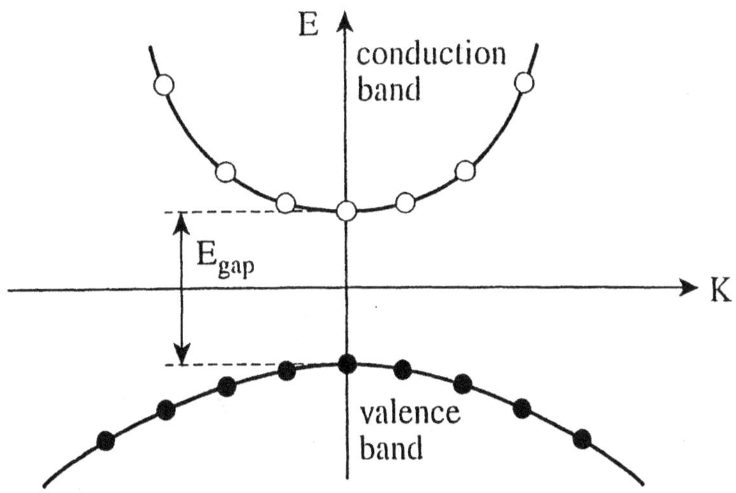


FIGURE 1 Energy band structure of a direct bandgap semiconductor. In thermal equilibrium, the valence band is (almost) filled, and the conduction band is (almost) empty. A forbidden gap (the so-called bandgap) exists between valence and conduction band, in which no electron states can exist.

Consider the InP structure shown in FIGURE 2, which consists of a sandwich of p-type material (Zn-doped InP) on top of an undoped (intrinsic) layer on top of n-type material (S-doped). The material composition of the central or active layer is chosen differently from the surrounding layers in order to create a lower bandgap in the active layer. This can be achieved by using a quaternary alloy such as InGaAsP. Applying a forward voltage over this heterostructure diode will cause electrons from the n-type material and holes from the p-type material to travel towards the active layer. Here, the electrical carriers accumulate as they are trapped in the low bandgap potential well. If sufficient current is applied, large concentrations of electrons and holes build up in the active layer, leading to population inversion. Photons passing through the active layer can stimulate radiative recombination of the carriers, resulting in amplification of the light. In addition, carriers can also decay spontaneously, leading to amplified spontaneous emission (ASE), or they can decay through non-radiative processes. The balance between the applied current and the decay of carriers determines the carrier concentration.

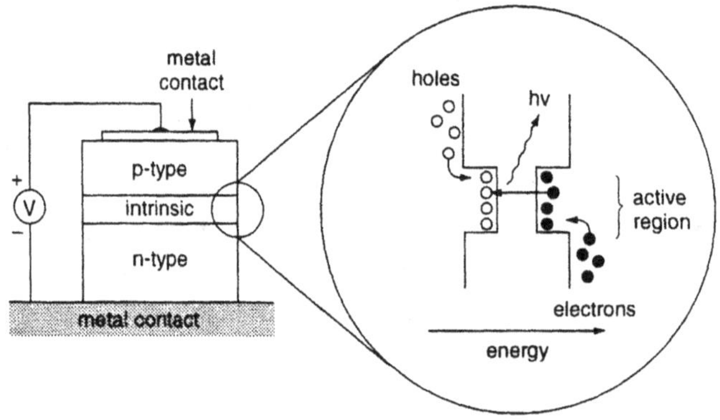


FIGURE 2 Basic structure of a semiconductor active layer. A thin active region of small bandgap material is sandwiched between two layers of larger bandgap material, that are respectively n- and p-doped. Upon application of a forward voltage over the thus formed diode, electrons and holes from these doped regions travel to the active region, where they are trapped in the low bandgap potential well.

## B HETEROJUNCTION DEVICE STRUCTURE

An InP-based optical amplifier has to meet several requirements in order to function efficiently.

(1) Electrical carriers (electrons and holes) must be transported to the active layer.

(2) The carriers must be confined within the active layer in order to achieve a population inversion. This also implies that the carrier lifetime must be sufficiently long.

(3) The light beam to be amplified has to sufficiently overlap with the inverted active layer in order to achieve optical gain.

(4) The light beam must be guided through the structure so that no light is lost along the length of the active layer.

(5) Unwanted optical reflections must be avoided.

(6) Signal light must be efficiently coupled into and out of the amplifier chip.

A typical SOA device structure is shown in FIGURE 3 [3,4]. In this example, the device is based on an (n-type) InP substrate. The active layer consists of a stack of InGaAsP quantum wells separated by thin barrier layers (quantum wells will be discussed below). The active layer, and in particular the quantum wells, have a bandgap lower than that of InP in order to trap or confine electrical carriers. On top of the active layer, a p-type InP layer has been grown. The device is completed with metal contacts. Note that the active layer is restricted in the lateral dimension as well. On either side, it is surrounded by isolating Fe-doped InP, ensuring that all the electrical current will flow through the active layer. This results in high current densities which are necessary to achieve high carrier concentrations in the active layer. Lastly, in order to suppress optical feedback the active stripe has been placed at an angle with respect to the front and back facets [5,6]. In addition, the facets are anti-reflection coated and so-called window regions have been included in which the signal light diverges before passing through the facets.

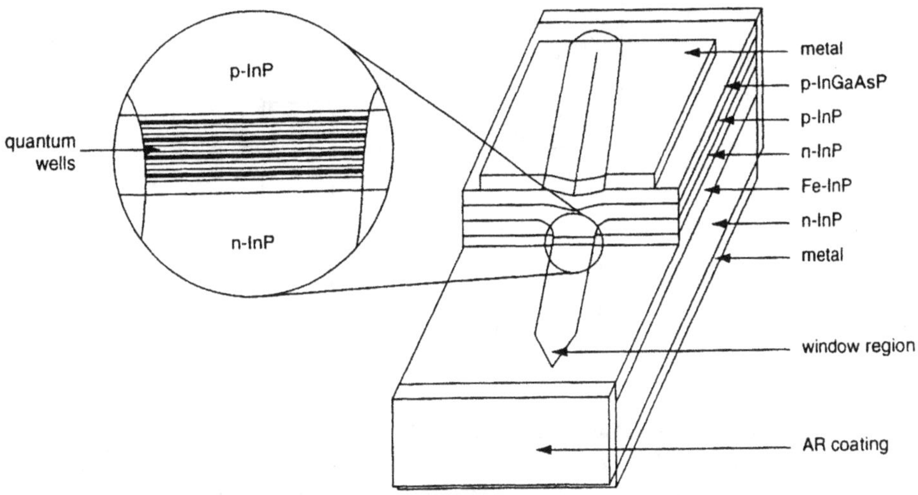


FIGURE 3 Typical SOA device structure. In addition to the vertical layer structure as already shown in FIGURE 2, the current as well as the light are also confined in the horizontal direction due to the isolating Fe-InP barrier layers on both sides. The active stripe is placed on the chip at an angle to avoid reflections from the facets. For the same reason, window regions and anti-reflection coatings are used.

In addition to confining electrical carriers, the active layer also confines light passing through the structure. The low bandgap with respect to the surrounding materials implies a larger refractive index for the active layer. This leads to waveguiding within the active region. For efficient amplification, the waveguided light beam in the amplifier must have sufficient overlap with the active layer. Also, it is of paramount importance that the waveguided light is coupled efficiently in and out of the amplifier chip, usually to single mode optical fibre. This implies that the amplifier should operate as a single mode waveguide, with circular beam waist that can be matched to that of the single mode fibre. In general, careful design of the material composition and layer thicknesses is necessary to obtain both the desired electrical properties and optical properties.

Using epitaxial growth of semiconductor layers, different types of active layer can be grown. The simplest type of active layer consists of a bulk semiconductor layer with a bandgap smaller than the surrounding layers [7-9]. The second type of active layer is the quantum well [10,11]. FIGURE 4 shows the Fermi-Dirac distribution for electrons and holes in (a) a bulk active layer, and (b) a quantum well. The corresponding density-of-states functions are also shown. The parabolic band structure of the bulk layer together with the Fermi-Dirac distribution governing the occupation of states by electrons lead to the given carrier concentration versus energy. In turn this determines the light emission spectrum and parameters such as the noise figure. In quantum wells the electrical carriers are restricted in one dimension, as the thickness of the layer approaches the De Broglie wavelength of the electrons. This results in a square shaped density of states function. Combined with the Fermi-Dirac distribution, this leads to a significantly different carrier concentration versus energy compared to bulk material.

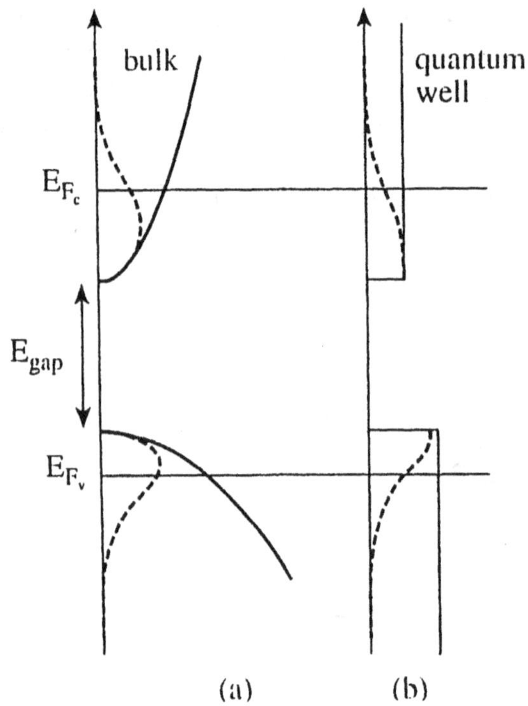


FIGURE 4 Fermi-Dirac distributions (dashed lines) for electrons and holes in (a) bulk semiconductor material and (b) semiconductor quantum wells. The Fermi energies are indicated. The density-of-states functions are shown as solid lines. Note that the graphs have been drawn to scale and the total number of electrical carriers integrated over the energy is constant.

## C InP-BASED OPTICAL AMPLIFIERS

Three key performance parameters describe the InP based semiconductor optical amplifier: optical gain, saturation output power and noise figure. Clearly it is not possible to amplify an input signal of arbitrary power; if the input power is too high the amplifier gain will saturate. Similarly, if the input signal is too low, it cannot be discriminated from the (amplified) spontaneous emission.

### C1 Optical Gain and Saturation Behaviour

Light propagating through an optical amplifier structure like the one shown in FIGURE 3 can stimulate the recombination of electrons and holes, thereby stimulating the emission of light. In order for this to occur the wavelength of the light must coincide with the energy difference between the electrons and holes in the active layer of the semiconductor. As it propagates, the light power is enhanced. The higher the concentration of electrons and holes (carriers), the larger the enhancement factor or gain. The gain coefficient g describes the process of optical amplification:

$$g = \Gamma A_g (N - N_{tr}) \qquad (1)$$

Here N is the carrier concentration; $N_{tr}$ is the carrier concentration at transparency, i.e. the carrier concentration for which population inversion is achieved. $A_g$ is the differential gain coefficient (often expressed as dg/dN), and is determined by the material properties of the semiconductor active layer. $\Gamma$

describes the fraction of the light mode that interacts with the active layer; only this fraction is subject to gain. Upon traversing an amplifier of length L, the integrated amplification or amplifier gain G is equal to:

$$G = \exp((g - \alpha)L) \qquad (2)$$

where $\alpha$ stands for the internal or excess loss in the amplifier.

Important for applications of the optical amplifier is the behaviour of the gain as a function of the power of the amplified signal light. FIGURE 5 shows a measurement of this. It is clear that in the limit of low signal powers, the optical gain is constant. In this regime the gain is referred to as the small-signal gain. At higher signal powers the gain deviates from the small-signal value. The signal has become so powerful that it eats into the carrier concentration, thereby compressing the gain coefficient shown in EQN (1). This so-called saturation regime is best characterised by the 3 dB saturation power $P_{-3\ dB}$: the signal power at which the gain is 3 dB less than the small signal gain. This point is indicated in FIGURE 5.

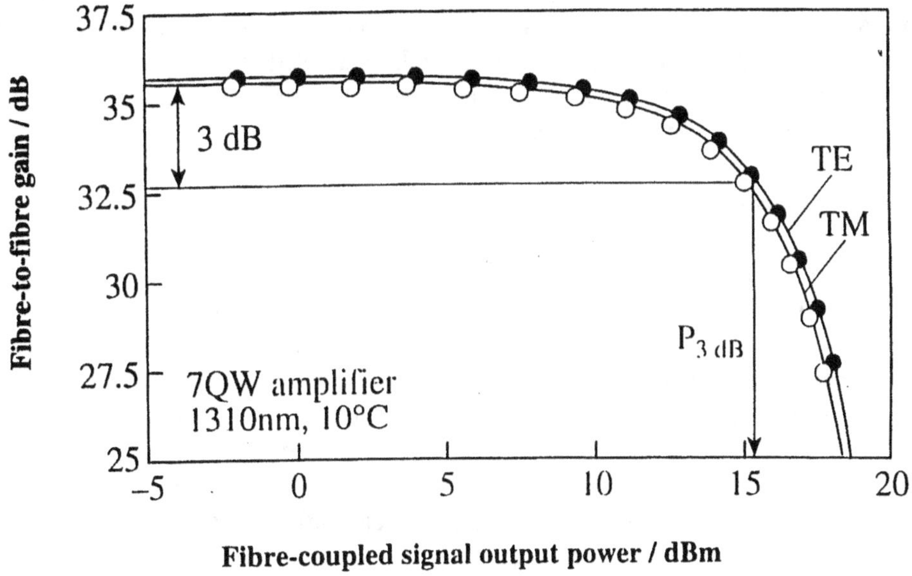


FIGURE 5 Gain compression or saturation in a seven quantum well semiconductor optical amplifier operated at 1310 nm. When a small signal is amplified, this device has a gain of 36 dB. When the input signal becomes larger, the limited number of carriers available causes saturation of the gain. At an output power of +15 dBm, the gain is saturated by 3 dB. The data are shown for both TE and TM input signals.

## C2 Noise

An active layer pumped with electrical carriers will also produce spontaneous emission noise through radiative recombination of carriers. In an amplifying structure the fraction of spontaneous emission coupled into the waveguiding mode will be amplified along the length of the amplifier resulting in amplified spontaneous emission (ASE). The beating of spontaneously emitted photons with signal photons leads to amplifier noise. The noise figure $F_{sig\text{-}sp}$, defined as the signal-to-noise ratio at the input over the signal-to-noise ratio at the output (assuming a shot noise limited input signal), for such signal-spontaneous beat noise can be expressed as follows [4]:

$$F_{sig\text{-}sp} = (S/N_{input})/(S/N_{output}) = (2\ \rho_{ase,par})/(gh\nu) = (2\ n_{sp}K_x)/(\eta_{in}) \qquad (3)$$

with S/N the signal-to-noise ratio, $\rho_{ase,par}$ the ASE power spectral density having the same polarisation as the amplified signal, g the optical gain, $n_{sp}$ the population inversion parameter, $K_x$ the excess noise factor, and $\eta_{in}$ the input coupling efficiency. Essentially, the noise figure is determined by the ratio between ASE power and gain, which is in turn determined by the level of population inversion. Clearly, the input coupling efficiency should be high in order not to degrade the signal intensity.

### C3 Polarisation Dependence

For many amplifier applications it is of importance that the gain be independent of the polarisation of the input signal light. Semiconductor active layers are in general sensitive to the polarisation for two reasons. First, the overlap of transverse electric (TE) light with the active layer is different from transverse magnetic (TM) light [13]. Second, crystal strain, which is commonly applied in semiconductor active layers, breaks the symmetry of the crystal leading to different optical gain for TE and TM light [11]. To obtain polarisation independent operation of the amplifier several techniques can be used. In bulk layers, with no crystal strain applied, the layer can be designed to have a square cross-section leading to equal overlap of TE and TM light [7,8]. Alternatively, the application of strain in a non-square bulk layer can be used to compensate for the differences in overlap [9,14]. In quantum wells, strain can be used to fabricate layers which produce only TE or TM gain. Combining such quantum wells in one active layer can also lead to polarisation insensitive operation [10]. Since the square cross-section solution requires sub-micron waveguide widths, which are difficult to fabricate, usually one of the latter two solutions is chosen.

### C4 Dynamic Behaviour

Until now we have discussed the performance of semiconductor optical amplifiers under the assumption of continuous wave (CW) operation. In practice, optical signals carrying information are modulated in time; usually intensity modulation is applied to the signal. This means that the gain itself will adjust to the changing signal intensity, but only as long as the carrier concentration can react to the time varying signal. The time constant governing the carrier concentration, or gain recovery time, determines whether this is the case. For the SOA, the carrier lifetime is of the order of 0.1 ns; hence the gain recovery time is short with respect to typical data rates ($10^9$ $s^{-1}$). Therefore, different signal intensity levels (say the 'ones' and 'zeros' in a digital signal) receive different optical gains, leading to signal distortion. Note this is only significant when operating close to saturation conditions, i.e. at signal power levels close to $P_{-3\,dB}$.

The dynamic behaviour of the SOA leads to several application advantages and restrictions. Most importantly, for amplification at data frequencies lower than the gain recovery time, the SOA should not be operated in the saturation regime as this leads to signal distortion. This puts a limitation on the maximum amplifier output power. In addition, the gain modulation can lead to self-phase modulation (SPM), which in systems can cause frequency modulation or chirping. Notwithstanding these restrictions, SOAs have many useful applications as amplifiers, as will be discussed later. For more demanding applications, gain-clamping of SOA structures by introduction of optical feedback presents a solution to the issue of gain modulation (see Section C5). There are also applications which actually require amplifiers to react fast to changing signal powers or to switch. In addition, some applications such as wavelength conversion and signal regeneration utilise the amplifier non-linearity.

### C5 Gain-Clamping

One of the main disadvantages of semiconductor optical amplifiers is the effect of gain compression or saturation. This effect may be overcome by the use of gain-clamped semiconductor optical amplifiers (GCSOAs), in which the optical gain is constant over a large range of signal power. Gain clamping can be achieved by introducing wavelength selective feedback, e.g. from a distributed Bragg reflector

(DBR), to the amplifying medium. The device lases at the corresponding wavelength as soon as the round trip gain (material gain plus mirror losses) equals unity (see Section D). At this point, the carrier concentration is fixed and hence the optical gain is clamped at a certain value.

FIGURE 6 shows a schematic of a GCSOA. The DBR mirrors are chosen at a wavelength outside the signal band, so as not to interfere with the signal. The gain versus output power diagram is shown in FIGURE 7 [15-18]. The behaviour in this diagram is distinctly different from that shown in FIGURE 5. The gain of the GCSOA is flat up to a certain output power level, where the gain drops steeply. At this power level the signal depletes the active layer of its carriers to such an extent that laser oscillation can no longer be sustained.

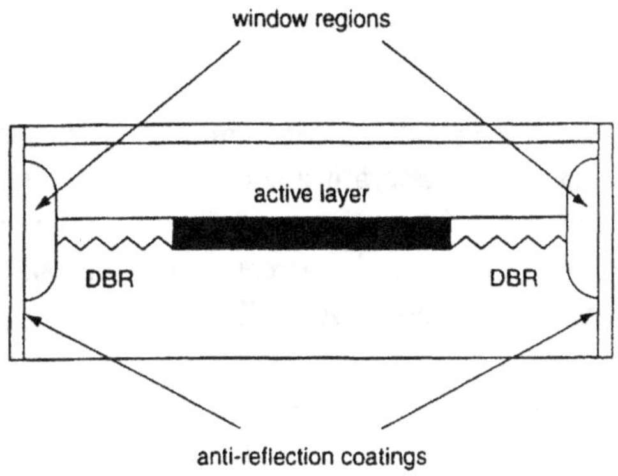


FIGURE 6 Schematic diagram of a gain-clamped semiconductor optical amplifier. Distributed Bragg reflectors are located on both sides of the active stripe, causing the device to behave as a laser at a well-defined wavelength. The lasing action clamps the carrier concentration at a fixed level, and thus also clamps the optical gain.

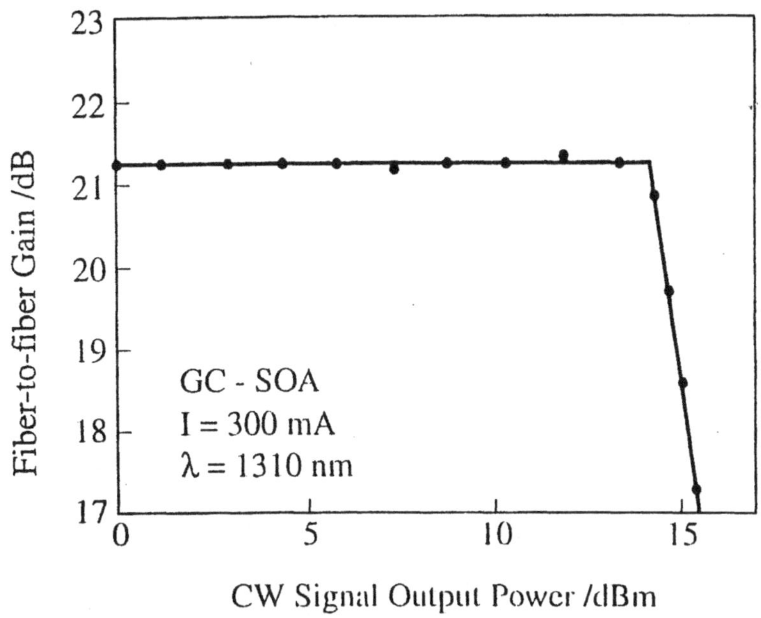


FIGURE 7 Gain versus output power of a gain-clamped semiconductor optical amplifier. As long as lasing action is sustained, the gain is clamped at a fixed value of 21 dB. The available carriers are shared by the lasing action and the amplification. At too high amplified signal power, the lasing action stops (goes below threshold) and the gain of the amplifier drops steeply.

## D InP-BASED LASERS

The signals used to transport data in optical networks are usually generated by semiconductor lasers. A laser is essentially an optical amplifier that is brought into resonance using optical feedback. The very optical reflections that one tries so hard to avoid in an SOA, are required to obtain laser operation. The simplest way to fabricate a semiconductor laser is to take an SOA that has its stripe at right angles with respect to the facets, and leave the facets as cleaved, i.e. without anti-reflection coating. This type of laser is called a Fabry-Perot laser. When electrical current is injected into this structure, it will start to lase as soon as the net gain of the amplifier has become high enough to overcome the reflection losses at the facets:

$$G = \exp((g - \alpha)L) = 1/R \qquad (4)$$

with R the reflection coefficient, and the other symbols as in Section C1. This is called the lasing condition, and expresses that, for an operating laser, the round trip gain and round trip losses balance each other. The current at which lasing action starts is called the threshold current. When the injection current is increased above the threshold current, all additional carriers are used immediately in stimulated emission, keeping the gain constant, and the lasing condition valid at all times.

In order to obtain the spectrum of the laser, we must also consider the round trip phase of the light. This must be an integer multiple of $2\pi$ to obtain constructive interference. Therefore, in a Fabry-Perot laser, lasing will occur at wavelengths $\lambda$ for which

$$2nL\,(2\pi/\lambda) = 2\pi\, m \qquad (5)$$

in which nL is the optical length of the laser, and m is an integer. This results in a spectrum as shown in FIGURE 8, consisting of peaks at the allowed lasing wavelengths (the so-called 'longitudinal modes' of the laser), in an envelope determined by the wavelength-dependent gain curve of the optical amplifier.

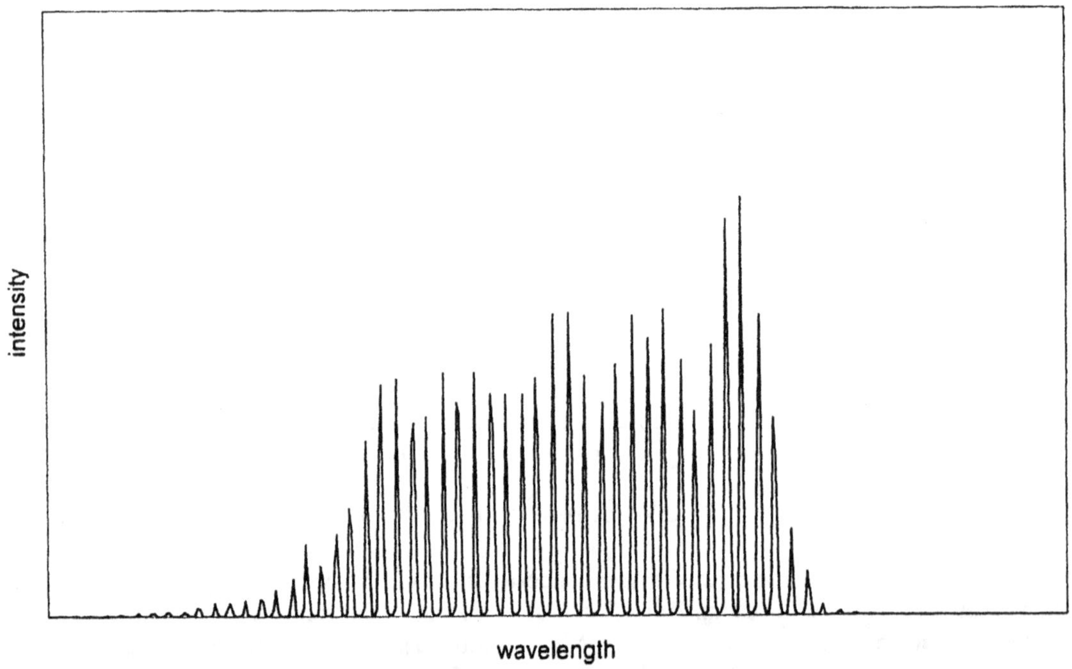


FIGURE 8 Emission spectrum of a Fabry-Perot laser. The characteristic wide spectrum is caused by the many longitudinal modes that fit in the laser cavity. The spectrum shown belongs to a laser intended for pumping an erbium doped fibre: hence the wavelength of 1480 nm.

The glass fibres used in optical telecommunication are dispersive, meaning that different wavelengths travel through the fibre at different speeds. This severely limits the applicability of multimode lasers such as the Fabry-Perot laser in this field. One would prefer a 'single-mode' laser, that emits light only at one single wavelength. This can be realised by making the optical feedback wavelength-selective. A much-used method is to form a Bragg grating along the full length of the SOA waveguide. This grating reflects light in a narrow wavelength band around the Bragg wavelength. Lasing action will start due to round-trip reflections between the Bragg grating and one of the chip facets. Since the feedback supplied by the Bragg grating is distributed over the entire length of the amplifier, this type of laser is known as a distributed feedback (DFB) laser. The narrow wavelength band reflected by the grating selects a single longitudinal Fabry-Perot mode. The single mode spectrum, as shown in FIGURE 9, makes this the laser of choice for telecom applications.

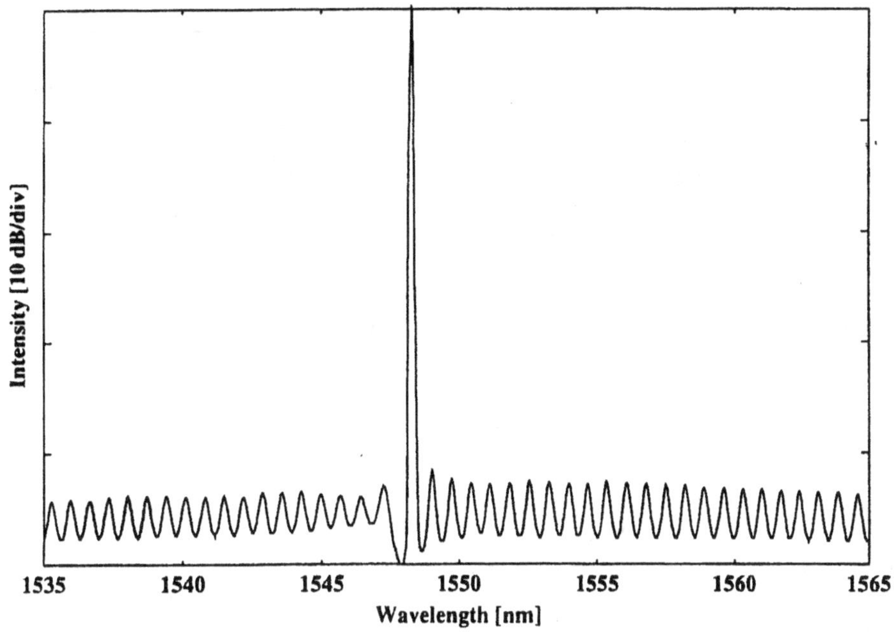


FIGURE 9 Emission spectrum of a distributed feedback (DFB) laser. A single longitudinal mode is selected by the Bragg grating. The Fabry-Perot modes of the cavity are also visible, suppressed by more than 40 dB. The single mode spectrum of the DFB laser makes it extremely suitable for telecom applications.

## E APPLICATIONS OF SEMICONDUCTOR OPTICAL AMPLIFIERS

### E1 Digital Transmission Systems

FIGURE 10 shows a diagram of a digital transmission system employing SOAs. Three amplifier types perform three basic functions required in such a system: power boosting [19], in-line amplification, and pre-amplification [20]. Power boosting is used to increase the laser transmitter power. In-line amplifiers are installed at regular intervals to compensate for optical losses encountered in the fibre. Pre-amplification is necessary to enhance the receiver sensitivity. The different functionalities lead to different requirements on the SOA performance.

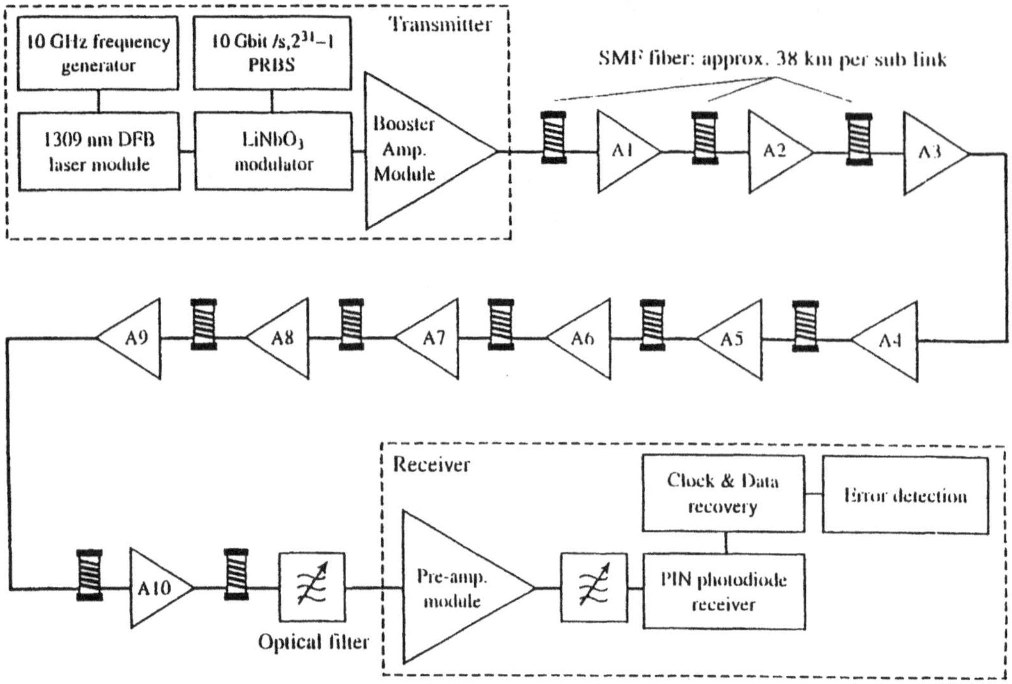


FIGURE 10 Digital transmission system employing semiconductor optical amplifiers. The signal traverses ten sublinks of 38 km each. The optical loss of each sublink (~20 dB at 1310 nm) is compensated by an amplifier. SOAs are also present in the transmitter, to boost the power injected into the transmission line, and in the receiver, acting as a pre-amplifier.

The feasibility of SOAs in digital transmission systems is demonstrated in FIGURE 11, which shows the bit-error-rate for the system shown in FIGURE 10. The system consists of a 10 Gbit/s RZ 1310 nm transmitter with a 5% extinction ratio connected to an SOA booster giving 0 - 2 dBm average output power (7 - 9 dBm peak power). The signal is passed through 420 km of standard single mode fibre with SOAs at 38 km intervals. At the receiver, the signal is amplified by a high-gain optical SOA pre-amplifier before detection. Details may be found in [20]. Similar experiments have been carried out elsewhere [21,22], including experiments in the field [23].

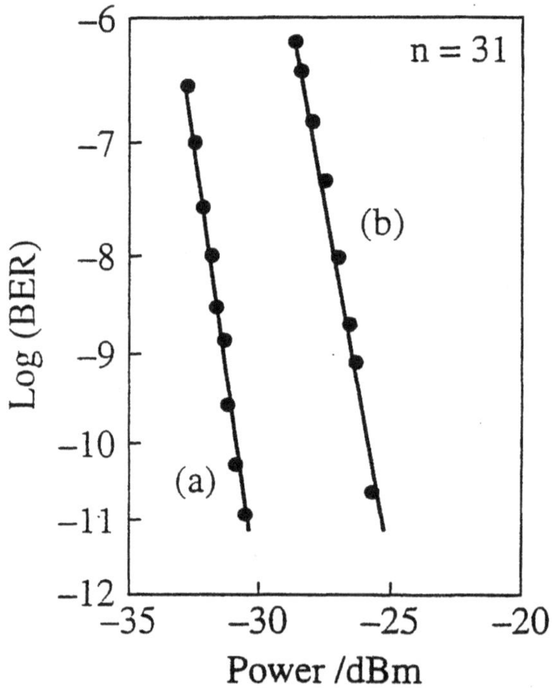


FIGURE 11 Bit-error-rate results obtained from the system shown in FIGURE 10, plotted against the received power, (a) directly after the transmitter, and (b) after 400 km, at the end of the transmission line. Error free transmission was obtained, with a power penalty of 5 dB caused by accumulated noise.

Wavelength division multiplexing (WDM) of signals in order to increase transmission capacity is very beneficial when applied in optically amplified systems. Instead of using a regenerator for each separate signal, all signal wavelengths are amplified in the same optical amplifier. This implies that cross talk between the wavelength channels must be avoided. In conventional SOAs, cross gain modulation severely limits the performance in a WDM system. The power dependence of the optical gain leads to gain modulation in phase with the signal. As different signals pass through the amplifier simultaneously, each signal is subject to the gain modulation caused by the other signals. Hence, cross gain modulation is detrimental in a WDM system. In gain-clamped amplifiers (Section C5), on the other hand, the gain is constant independent of signal power. Experiments have shown error-free amplification of 16 × 10 Gbit/s channels in a GCSOA device [24].

## E2 Analogue Transmission Systems

Analogue transmission of optical signals is primarily used for the transport and distribution of cable television (CATV). Each channel is amplitude modulated (AM) at a given carrier frequency; the carriers are frequency multiplexed to enable high signal transmission capacity. Typically, analogue systems transport frequencies between 50 and 800 MHz. The AM nature of the signal places high demands on the linearity of components in the system. Optical amplifiers operating in an analogue transmission system must meet the stringent requirements on linearity. Conventional SOAs cannot meet these needs when operated at high powers because of gain compression (FIGURE 5). On the other hand, highly linear gain versus signal power is possible for gain-clamped amplifiers discussed in Section C5 and shown in FIGURE 7. The remaining non-linearity in gain-clamped amplifiers is mainly due to spatial hole burning effects discussed in [18]. Such non-linearities can be tackled by adjusting the ratio between the mirror reflectivities to achieve a laser light intensity distribution in the amplifier that matches the signal intensity.

## E3 Other Applications

One of the advantages of the SOA is its ability to switch at speeds comparable to typical data rates. This ability gives the SOA added functionality in systems which require switching between different optical paths or simple data switching. As optical communication systems evolve from simple point-to-point systems to intricate networks such functionality is becoming an important feature. Basic switching and signal modulation using SOAs has been demonstrated [25-29]. An important application of switches may be found in the all-optical cross connect. Here N input channels can be configured to any of N output channels, requiring $N^2$ switching elements. Switching capabilities together with the possibility of large scale integration make SOAs ideally suited for such applications.

Other applications of SOAs may be found in the area of wavelength conversion and signal regeneration. These applications are based on non-linearities in the SOA. In a signal SOA, cross-gain modulation [30] or four-wave mixing [31] may be used to accomplish wavelength conversion. Cross-phase modulation may also be used by monolithic integration of the SOA in an interferometric photonic integrated circuit [32]. FIGURE 12 shows a wavelength converter based on a Mach-Zehnder

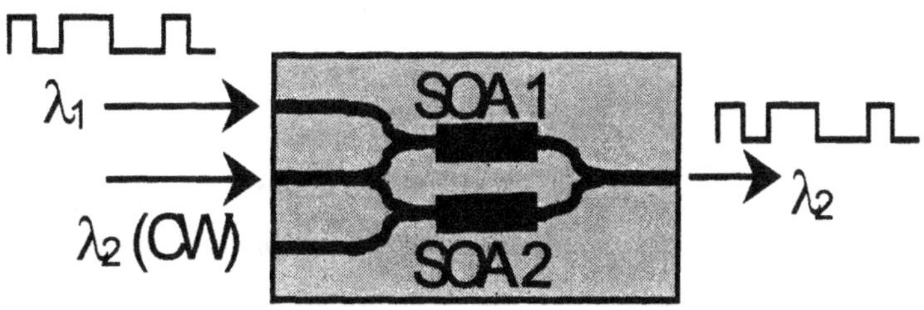


FIGURE 12 A Mach-Zehnder interferometer wavelength converter. The input signal at $\lambda_1$ is injected in the top SOA, while the bottom SOA only sees CW light. This results in a modulation of the carrier density in the top SOA, which leads to phase modulation of $\lambda_2$ in the upper arm, that is converted to intensity modulation at the output of the interferometer.

interferometer. In this device, the input signal at $\lambda_1$ is injected in the top SOA. This results in a modulation of the carrier density, which imposes a phase modulation on $\lambda_2$ in the upper arm. This phase modulation is converted to intensity modulation by the interferometer. Monolithic integration of such a structure with a DFB laser supplying $\lambda_2$ has also been reported [33].

## REFERENCES

[1] A. Yariv [ *Optical Electronics* (Holt-Saunders, New York, 1985) ]

[2] C. Kittel [ *Introduction to Solid State Physics* (Wiley, New York, 1986) ]

[3] L.F. Tiemeijer, P.J.A. Thijs, T. van Dongen, J.J.M. Binsma, E.J. Jansen, S. Walczyk [ *Tech. Dig. (USA)* vol.18 (1995) p.PD1-5 ]

[4] L.F. Tiemeijer, P.J.A. Thijs, T. van Dongen, J.J.M. Binsma, E.J. Jansen [ *J. Lightwave Technol. (USA)* vol.14 (1996) p.1524-33 ]

[5] L.F. Tiemeijer [ *Proc. 21st Eur. Conf. on Opt. Commun.* Brussels, 1996, Tu.B.2.1, p.259-66 ]

[6] A.E. Kelly, I.F. Lealman, L.J. Rivers, S.D. Perrin, M. Silver [ *Electron. Lett. (UK)* vol.32 (1996) p.1835-6 ]

[7] S. Kitamura, K. Komatsu, M. Kitamura [ *IEEE Photonics Technol. Lett. (USA)* vol.7 (1995) p.147-8 ]

[8] Ch. Holtmann, P.-A. Besse, T. Brenner, H. Melchior [ *IEEE Photonics Technol. Lett. (USA)* vol.8 (1996) p.343-5 ]

[9] P. Doussiere [ *Tech. Dig. (USA)* vol.11 (1996) p.220-3 ]

[10] L.F. Tiemeijer et al [ *Appl. Phys. Lett. (USA)* vol.62 (1993) p.828 ]

[11] P.J.A. Thijs, L.F. Tiemeijer, J.J.M. Binsma, T. van Dongen [ *IEEE J. Quantum Electron. (USA)* vol.30 (1994) p.477-99 ]

[12] S. Shimada, J. Ishio (Eds.) [ *Optical Amplifiers and Their Applications* (Wiley, Chichester, 1994) ch.3 ]

[13] T.D. Visser, B. Demeulenaere, J. Haas, D. Lenstra, R. Baets, H. Blok [ *J. Lightwave Technol. (USA)* vol.14 (1996) p.885-7 ]

[14] J.-Y. Emery et al [ *Proc. 22nd Eur. Conf. on Opt. Commun.* Oslo, 1996, WeD.2.3, p.3.165-8 ]

[15] Ch. Holtmann, P.-A. Besse, H. Melchior, D.L. Williams [ *Proc. 21st Eur. Conf. on Opt. Commun.* Brussels, 1995, Th.B.3.7, p.1031-4 ]

[16] L.F. Tiemeijer, G.N. van den Hoven, P.J.A. Thijs, T. van Dongen, J.J.M. Binsma, E.J. Jansen [ *IEEE Photonics Technol. Lett. (USA)* vol.8 (1996) p.1453-5 ]

[17] P. Doussiere et al [ *Proc. 22nd Eur. Conf. on Opt. Commun.* Oslo, 1996, WeD.2.4, p.3.169-72 ]

[18] J.L. Pleumeekers et al [ *Proc. Optical Amplifiers and Their Applications* Victoria B.C., 1997, WB5, p.224-7 ]

[19] L.F. Tiemeijer et al [ *IEEE Photonics Technol. Lett. (USA)* vol.7 (1995) p.1519-21 ]

[20] P.I. Kuindersma et al [ *Proc. 22nd Eur. Conf. on Opt. Commun.* Oslo, 1996, TuD.2.1, p.2.165-8 ]

[21] A. Shipulin, G. Onishchukov, P. Reidel, D. Michaelis, U. Peschel, F. Lederer [ *Electron. Lett. (UK)* vol.33 (1997) p.507-9 ]

[22] R.C.J. Smets et al [ *Proc. Opt. Fiber Commun.* Dallas, 1997, paper ThH2, p.269-70 ]

[23] J.J.E. Reid et al [ *Proc. 23rd Eur. Conf. on Opt. Commun.* Edinburgh, 1997, paper Mo3A.5, p.83-6 ]

[24] M. Bachmann et al [ *Electron. Lett. (UK)* vol.32 (1996) p.2076-8 ]

[25] M.D. Feuer et al [ *IEEE Photonics Technol. Lett. (USA)* vol.8 (1996) p.1175-7 ]

[26] C. Tai et al [ *J. Lightwave Technol. (USA)* vol.14 (1996) p.525-33 ]

[27] F. Masetti et al [ *J. Lightwave Technol. (USA)* vol.14 (1996) p.979-98 ]

[28] J.-Y. Emery et al [ *Proc. Optical Amplifiers and Their Applications* Victoria B.C., 1997, TuC2, p.112-5 ]

[29] S. Kitamura, H. Hatakeyama, T. Kato, M. Yamaguchi, K. Komatsu [ *Proc. Optical Amplifiers and Their Applications* Victoria B.C., 1997, TuC3, p.116-9 ]

[30] B. Glance, J.M. Wiesenfeld, U. Koren, A.H. Gnauck, H.M. Presby, A. Jourdan [ *Electron. Lett. (UK)* vol.28 (1992) p.1714-5 ]

[31] R. Ludwig, G. Raybon [ *Electron. Lett. (UK)* vol.30 (1994) p.338 ]

[32] T. Durhuus, C. Joergensen, B. Mikkelsen, R.J.S. Pedersen, K.E. Stubkjaer [ *IEEE Photonics Technol. Lett. (USA)* vol.6 (1994) p.53-5 ]

[33] L.H. Spiekman, U. Koren, M.D. Chien, B.I. Miller, J.M. Wiesenfeld, J.S. Perino [ *IEEE Photonics Technol. Lett. (USA)* vol.9 (1997) p.1349-51 ]

# 1.4 InP bipolar electronic and optoelectronic circuits

P.J. Topham

August 1998

## A INTRODUCTION

The development of the InP heterojunction bipolar transistor (HBT) has made rapid progress both for optoelectronic integration and also in high performance electronic ICs. These devices now give the highest operating frequencies of any bipolar technology. The InP HBT shares many of the advantages of the relatively well-developed GaAs/GaAlAs HBT [1], namely a high current gain transition frequency ($f_T$), high transconductance, good output current drive and a semi-insulating substrate. In this review we highlight the specific advantages and, where appropriate, the limitations of the InP HBT and how these can be translated into benefits for different circuit applications.

The InP based HBTs have additional advantages over GaAs HBTs because of the wide range of compounds that are lattice-matched to InP enabling the selection of materials having the best properties for each layer, such as GaInAs for the base layer leading to lower voltage and hence lower power operation. The choice of materials structures is discussed in Section B1 along with the other advantages of low surface recombination velocity and good thermal conductivity.

The original motivation for studying InP based HBTs was for integration with optoelectronic components such as lasers and photodiodes for fibre optic communications at 1.3 and 1.55 μm wavelengths. These components were fabricated in the GaInAsP materials system, lattice matched to InP. Progress on optoelectronic integration of HBTs is summarised in Section D. More recently the transport properties of the alloys on InP have produced record transition frequencies - these results are detailed in Section B2. The small bandgap of GaInAs has been utilised in high speed, low power consumption digital circuits as detailed in Section C1.

The inherent materials advantages of InP and lattice-matched compounds have to be weighed against the fabrication difficulties in this materials system. Experimenters have worked on one of two lattice matched systems on InP. The more common system and the one chiefly used for optoelectronic devices is GaInAsP. More recently the tri-metal system of GaInAlAs has been investigated. Growth of GaInAsP was originally by liquid phase epitaxy (LPE) but this is being supplanted by metal organic chemical vapour deposition (MOCVD) and molecular beam epitaxy (MBE). A problem in both materials systems, but especially in GaInAsP, is compositional grading between different alloys.

Processing of InP based HBTs follows that for the optoelectronic devices, using compositionally selective wet etches - which work reliably when the correct selection of alloy is made for each layer. The most critical etch is when removing the emitter layer to expose the thin base layer. Fortunately the etches used for the InP or AlInAs emitter have a high selectivity for the GaInAs base. A major disadvantage when processing InP HBT circuits is the inability to form semi-insulating material by ion implantation damage [2] which means that devices have to be isolated by mesa etching which gives an uneven topography for subsequent metallisation. Nevertheless, impressive levels of integration have been achieved in InP HBT circuits, particularly in the logic circuits described in Section C1 and the data converters described in Section C3.

## B PERFORMANCE

### B1 DC Characteristics

The most important feature in determining the DC characteristics of all HBTs is the base/emitter heterojunction. The narrow bandgap base layer determines the transport properties of the transistor, which for all the InP HBTs discussed here is GaInAs. This allows the injection of a given current density at a much lower base/emitter voltage than in GaAs based HBTs [3], which in practical terms translates to a lower operating voltage for many circuits. The wider bandgap emitter layer can reduce reverse injection into the emitter to a negligible level, and it is this that allows the base layer to be doped much more heavily than the emitter, giving a low base resistance (important for high frequency gain: see Section B2) and reducing the output conductance. The choice of wide bandgap emitter material is between InP or AlInAs. An abrupt heterojunction gives satisfactory injection into the base but with a slightly higher base/emitter voltage [4] than a compositionally graded junction. AlInAs is often chosen because it is easier to grade the junction using this material than InP. It is usual with an abrupt junction to include a GaInAs undoped spacer layer between the base and emitter layers because if the base dopant diffuses into the emitter layer during growth, then the p/n junction is formed in the wide bandgap emitter layer and the heterojunction action is lost.

The base/collector junction can either be a homojunction, using a GaInAs collector, or another heterojunction giving a double heterojunction bipolar transistor (DHBT). The advantage of a DHBT is that the structure is almost symmetrical which reduces the saturation voltage, down to values as low as 50 mV [5]. At the same time, the collector layer can be chosen for a high breakdown field which is why InP is the preferred choice. AlInAs is not used for DHBTs as the saturated electron velocity is only about $4 \times 10^6$ cm/s [6] compared with about $10^7$ cm/s for InP [7]. The chief difficulty with the DHBT is that even a small conduction band discontinuity at the base/collector junction impedes the flow of electrons and increases the potential across the base/collector junction. This degrades both the DC and microwave performance of the transistor and increases the saturation voltage - so losing much of the advantage in using the double heterojunction in the first place. Several approaches have been used to eliminate this discontinuity, such as a compositional superlattice grade [5] or a delta doped layer of $n^+$ InP [8] to provide a high electric field at the base/collector junction.

The minimum operating voltage of the transistor, often referred to as the knee voltage, is also strongly dependent on low emitter and collector resistances. These in turn are largely controlled by the ohmic contact resistance which can be made small in InP HBTs due to the very low contact resistivity to heavily doped n-type GaInAs. The emitter contact is particularly critical as any parasitic resistance limits the transconductance.

A great advantage of InP HBTs is the low surface recombination velocity of GaInAs which is more than two orders of magnitude smaller than that of GaAs [9]. The importance of this is that it allows transistors to be scaled to narrower emitter widths while maintaining adequate current gain [10]. Emitter widths as small as 0.3 μm have given current gains exceeding 50 [11]. Scaling transistors to small geometries not only improves the high frequency performance but also reduces the operating current which in turn allows lower power operation and enables increased levels of integration.

### B2 Microwave

The high electron mobility and hence diffusivity of GaInAs [12] gives a short transit time across the base layer while the high electron velocity at moderate fields in InP (FIGURE 1) minimises the overall transit time in DHBTs, giving them the highest transition frequency ($f_T$) of any bipolar transistor, with values up to 230 GHz [8]. The high $f_T$ combined with the low base resistance mentioned above give a corresponding maximum frequency of oscillation ($f_{max}$) of 147 GHz. Increasing the collector thickness reduces capacitance, with a slight increase in transit time, to give an $f_{max}$ of 300 GHz [8] with a corresponding $f_T$ of 170 GHz. Parasitic capacitance can be further reduced by the transferred substrate process [13] giving an $f_{max}$ of 520 GHz with a corresponding $f_T$ of 134 GHz.

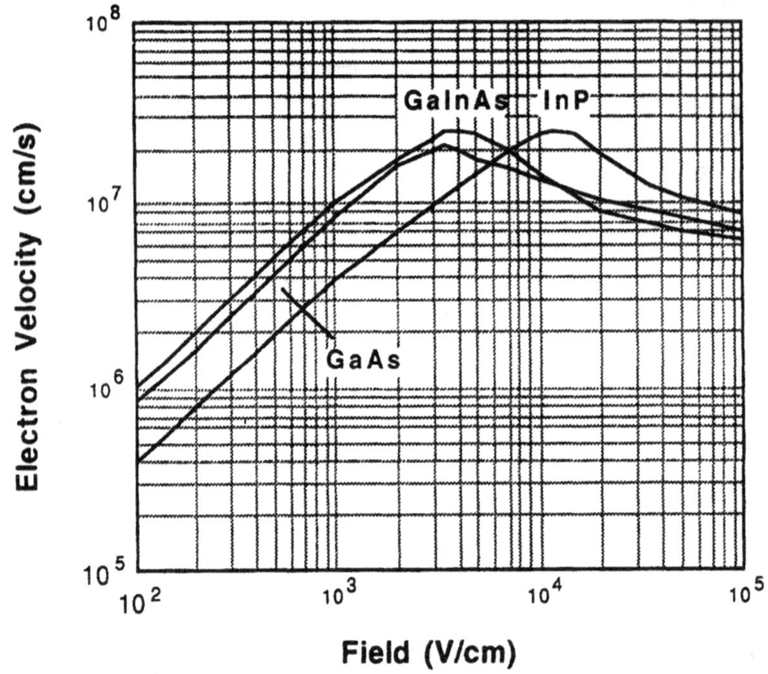


FIGURE 1 Velocity/field characteristics.

As well as trading off capacitance and transit time, the collector thickness also controls the breakdown voltage of the transistor. Plotting the $f_T$ against the emitter/collector breakdown voltage ($BV_{CEO}$) gives a roughly inverse relationship (FIGURE 2). Single heterojunction InP HBTs follow a similar trend to GaAs HBTs, but InP DHBTs, which have an InP collector layer, have roughly double the breakdown voltage for a given $f_T$. This is because of the higher electron velocity at moderate fields in InP compared with GaAs, so the collector layer can be made thicker while giving the same transit time. The thicker collector layer also reduces the collector/base capacitance which improves the power gain.

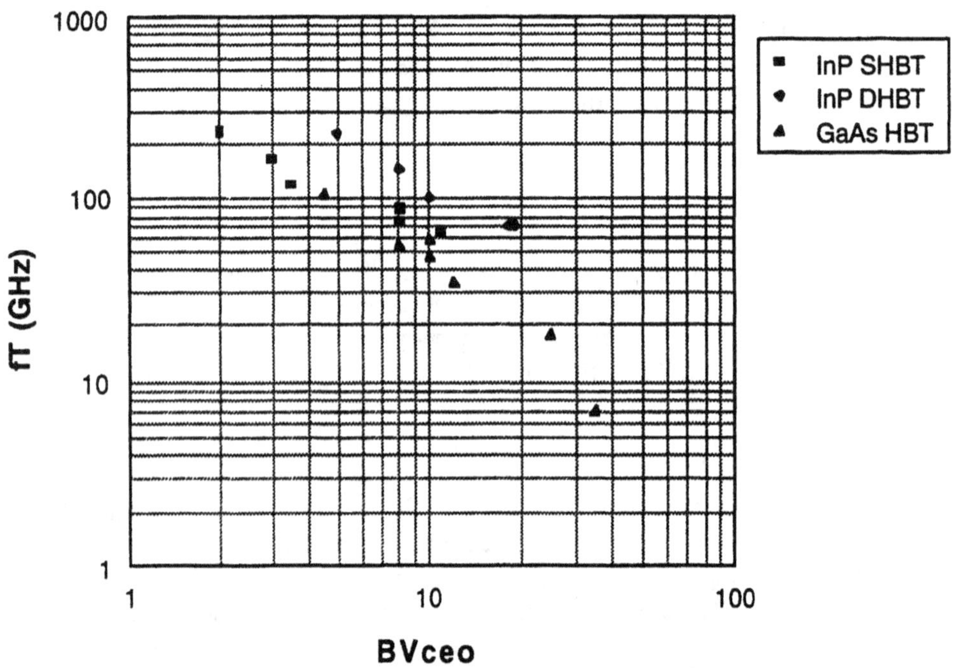


FIGURE 2 Relationship between breakdown and transition frequency.

These characteristics are attractive for power transistors, with the further advantage that InP has about 50% greater thermal conductivity than GaAs. Bipolar transistors lend themselves to running in class B or class C, as they are normally-off devices, which gives efficient operation. They have demonstrated power-added efficiencies of 90% in class C at 2 GHz with 1.2 W output, and 71% power-added efficiency in class B at 9 GHz with 2 W output [5].

### B3 Phototransistors

The possibility of getting gain from a photo detector is attractive as it reduces the demands on the following amplifier. Incorporating a photo diode into a transistor is therefore a useful combination. Optical gains of 1000 have been obtained [14] from InP phototransistors operating at 1.3 μm. The electrical current gain of these transistors was 1600. One drawback to using phototransistors is that the device structure for high optical efficiency may limit the speed of the transistor. A compromise has to be made with the thickness of the photodetector layer (principally the GaInAs collector) whether the structure is a PIN diode made from the base/collector junction, or a complete phototransistor, because the optical absorption, which determines the quantum efficiency of the detector, increases with thickness while the frequency response decreases [15]. This compromise is particularly significant for OEIC applications because it is only the photodetector which has to absorb light but all the other transistors have to have the same layer structure - this is discussed in Section D1.

## C ELECTRONIC CIRCUITS

### C1 Digital Circuits

The combination of a high transition frequency, low parasitic capacitance (resulting from the semi-insulating substrate), and a base/emitter voltage about half that of GaAs HBTs make InP HBTs an attractive choice for high speed digital circuit applications. A further advantage of HBTs is that they can directly exploit the wealth of circuits designed for silicon BJTs. Current mode logic (CML), with or without emitter followers forming emitter coupled logic (ECL), gives high speed and economic implementation of many circuit functions. Static frequency dividers have been operated at 44 GHz on InP substrates [16] and at 48 GHz using the transferred substrate process [17] to reduce parasitic capacitance.

As circuits become larger it can be difficult to distribute the high speed signals. The InP HBT has the advantage of a semi-insulating substrate, and the speed advantage is well illustrated for more complex circuits by a 12 bit accumulator for a direct digital frequency synthesiser [18]. This circuit, containing 928 transistors, has been clocked at 7 GHz.

### C2 Microwave Circuits

The semi-insulating substrate of InP lends itself to monolithic microwave integrated circuits (MMICs) because spiral inductors and transmission lines can have acceptably low loss. Distributed amplifiers (also known as travelling wave amplifiers) have been fabricated with 50 GHz bandwidth [19], and by further reducing parasitic capacitance with the transferred substrate process [20] 80 GHz bandwidth amplifiers have been made. The low operating voltage of InP HBT circuits has allowed a novel stacked bias approach to be used for a balanced amplifier with both stages in series running from a 5 V supply [21].

The advantages of InP HBTs as individual power devices have been discussed in Section B2. The low operating voltage of InP HBTs is also attractive for use in mobile telephones and an MMIC power amplifier with 2 W output at 1.9 GHz has been operated from a 3 V supply [22]. The individual output cells in the power amplifier still gave 68% power added efficiency even with a 2.7 V supply.

A key advantage of the InP HBT which has already been mentioned is its similarity to the silicon bipolar transistor so that a large repertoire of circuit function designs is already available. One of the most useful is the Gilbert cell multiplier, which can be used as a double balanced mixer, a gain control stage, a frequency doubler or an up-converter, with this last function demonstrated in InP HBTs [23] giving near unity conversion gain from a 3 GHz IF up to an output in the 5 - 12 GHz band.

The low surface recombination velocity of InP HBTs and the inherently buried active region result in good low frequency noise characteristics. This in turn leads to reduced close to carrier phase noise of InP HBT oscillators, that have been shown to be 10 dB quieter than similar oscillators using GaAs HBTs [24]. Low noise millimetre wave oscillators make good use of the properties of the InP HBT with a 62 GHz VCO having a phase noise of 78 dB below carrier at 100 kHz offset [25]. InP HBTs have also been used to make a fully integrated 94 GHz VCO [26].

Another advantage of bipolar transistors is the well-controlled threshold voltage. This makes InP HBTs useful for high-speed precision analogue functions, such as the decision circuits needed for high data rate fibre optic links. These functions are covered in Section C3 and Section D.

### C3 Data Converters

Bipolar transistors are well suited to analogue to digital converters due to the well controlled threshold voltage, and InP HBTs have enabled analogue to digital converters to be implemented at high conversion rates. As with many of the other HBT circuits the configurations used are very similar to silicon bipolar converters and use an all parallel or flash architecture to achieve 4 bit conversion at $2.4 \times 10^9$ samples per second (2.4 Gsps) [27]. This circuit, which contains 1038 transistors, illustrates the rapid advancement of InP HBT integrated circuit technology. Single bit data conversion, needed in fibre optic communications systems for example, has been achieved at 40 Gsps [11]. The corresponding conversion from digital to analogue uses simpler circuits, and a 4 bit converter running at 2.4 Gsps was included on the ADC mentioned above [27].

### C4 Mixed Functions

One of the most interesting developments has been the integration of both InP HBTs and InP HEMTs onto the same die [28]. Although inevitably involving a more complex process, this has enabled the best features to be obtained from each device, as demonstrated by an HBT DRO running at 23 GHz followed by a pair of HEMT frequency doublers on the same MMIC producing an output at 94 GHz [28]. This circuit takes advantage of both the low phase noise of the HBT oscillator and the high frequency gain of the HEMT amplifier, operated non-linearly to give the multiplier action.

A simpler but highly effective approach has been to include the bias regulation for a HEMT MMIC as part of an accompanying InP HBT circuit [29], taking advantage of the compact, high-accuracy analogue functions possible with HBTs.

## D OPTOELECTRONIC INTEGRATION

### D1 Receivers

The vertical structure of the HBT lends it to integration with photodiodes and the relatively coarse lithography needed in the HBT can cope more easily with the resulting topography. Receiver designs follow silicon bipolar practice using the transimpedance configuration. Using the base/collector of an HBT as a PIN photodiode a fully integrated receiver has been fabricated with a 46 dB Ω transimpedance gain and a 20 GHz 3 dB bandwidth for use up to 24 Gbit/s [30]. The advantages of a fully integrated receiver are shown by a 16 channel receiver [31] for use in wavelength division multiplexed fibre optic systems as only one chip has to be aligned rather than 16. Each channel had about 11 GHz bandwidth and by using a semi-insulating substrate the cross-talk between channels was kept down to -35 dB.

The well controlled threshold voltage and high frequency gain of InP HBTs makes them suited to the analogue support circuits for fibre optic receivers. A decision circuit for data recovery has been tested for operation up to 30 Gbit/s [32]. An impressive level of analogue integration is demonstrated by a complete data regenerator IC [33] which includes the clock recovery phase-lock-loop, the decision circuit and an output driver all operating up to 20 Gbit/s.

Most fibre optic systems are for high data rate digital transmission but there is increasing interest in utilising the very low loss of optical fibre to carry analogue microwave signals. An integrated PIN-HBT receiver has been fabricated for this application having 2 A/W responsivity and a 3 dB bandwidth of 18 GHz [34]. This application also makes use of the good linearity of the InP HBT with a spurious free dynamic range of 115 dB.

### D2 Transmitters

The good output drive capabilities and transconductance of the HBT makes it the natural choice as a laser driver. An InP HBT driver circuit for a separate laser chip (not on the same OEIC) has been operated at 10 Gbit/s [35].

However the full optoelectronic integration of high performance HBTs and lasers has proved to be difficult. The best approach so far has been to perform selective area growths for each device. Transistors have been integrated with a laser diode and the laser has been modulated at 1.6 GHz [36]. An alternative approach has been to grow an HBT structure directly on top of a multi-quantum well laser in a fully integrated device [37]. The laser had a 6 mA threshold which was almost identical to a control device without the overgrowth, while the HBT had a current gain of 50. As fibre optic data rates increase the enhanced performance would seem to favour on-chip driver circuits, but the lack of interconnection parasitics such as bond wire inductance has to be balanced against the compromises needed in HBT performance to accomplish the integration.

### D3 Optical Switches

As with the receiver circuits, the integration of HBTs and waveguide switches is somewhat easier due to the similarity of the structures. The need for integrated transistors is for supply of the relatively high drive voltages needed by waveguide optical switches. A 2×2 optical cross point switch has been fabricated from a modified HBT structure [38]; the switch had an isolation ratio of 14:1 whilst the transistor had a gain of 218.

## E FUTURE DEVELOPMENTS

As the InP HBT moves out of research and development and into production [18,22], it is necessary to weigh up the fundamental advantages of the InP HBT against likely applications for this technology. An inherent advantage of the InP HBT is the low close-to-carrier noise in oscillators which complements the low millimetre wave noise figure of InP HEMTs. Whether as a fully integrated HBT/HEMT MMIC or as separate chips in which each technology is used to best advantage, this is a promising combination for millimetre wave applications. One of the main drivers for the application of GaAs MMICs has been the growth of wireless communications. Operation from a battery places a premium on high-efficiency power amplifiers running off low-voltage supplies, which favours the InP HBT and particularly the InP DHBT because of its low saturation voltage.

The processing of compound semiconductor devices is less well developed than that of silicon, and InP based processing is in turn less advanced than that of GaAs. An advantage of GaAs is the availability of high quality 4 inch diameter wafers, but InP is catching up rapidly with the progression to 3 inch wafer processing. A key question for a production technology is that of reliability. Initial published results [39] have shown lifetimes of more than $10^7$ hours at a junction temperature of 125°C and this is supported by the long lifetimes of 1.55 μm lasers for fibre optic systems One of the main disadvantages of processing InP HBTs is the inability to render InP or GaInAs semi-insulating by ion implantation damage [2] using protons for example. This means that the transistors and diodes have to be isolated by mesa etching which takes up space and gives an uneven topography for subsequent processing. A possible solution has been to grow $n^+$ regions selectively [40] which also reduces parasitic capacitance. Another alternative is the transferred substrate process which, although more complex, further reduces capacitance by supporting the devices on a polymer film [17,20].

One of the most exciting areas is the integration of high performance HBTs with optoelectronic devices, opening up uses such as low-noise integrated photo-receivers and high-speed on-chip laser drivers. Given the recent rapid pace of development of InP HBT technology, it can be expected to be applied in an ever wider range of electronic and optoelectronic integrated circuits.

## REFERENCES


[1] J.A. Turner [ in *Properties of GaAs* (INSPEC, IEE, London, UK, 1996) p.820-3 ]

[2] S.J. Pearton, U.K. Chakrabarti [ in *Indium Phosphide and Related Materials* (Artech House, 1992) ch.7 ]

[3] K.W. Kobayashi et al [ *Microw. J. (USA)* vol.41 (June 1998) p.22-38 ]

[4] J.F. Jensen, L.M. Burns, W.E. Stanchina [ in *InP HBTs: Growth, Processing and Applications* (Artech House, 1995) p.265-315 ]

[5] C. Nguyen, T. Liu, M. Chen, R. Virk [ *Proc. Int. Conf. on InP and Related Materials* Cape Cod, MA, USA, 11-15 May 1997 (IEEE, New York, USA, 1997) p.15-9 ]

[6] M.A. Littlejohn, K.W. Kim, H. Tian [ in *Properties of Lattice-Matched and Strained InGaAs* (INSPEC, IEE, London, UK, 1993) p.107-16 ]

[7] [ *Properties, Processing and Applications of Indium Phosphide* (INSPEC, IEE, London, UK, 1999) ]

[8] Y. Matsuoka, S. Yamahata, K. Kurishma, H. Ito [ *Jpn. J. Appl. Phys. (Japan)* vol.35 no.11 (1996) p.5646-54 ]

[9] H.H. Weider [ in *Properties of Lattice-Matched and Strained InGaAs* (INSPEC, IEE, London, UK, 1993) p.137-44 ]

[10] R.N. Nottenburg, Y.K. Chen, M.B. Panish, R. Hamm, D.A. Humphrey [ *IEEE Electron Device Lett. (USA)* vol.EDL-9 no.10 (1988) p.524-6 ]

[11] M. Hafizi, J.F. Jensen [ *IEEE Radio Frequency IC Symposium* (IEEE, 1997) p.87-90 ]

[12] L.F. Eastman [ in *Properties of Lattice-Matched and Strained InGaAs* (INSPEC, IEE, London, UK, 1993) p.103-6 ]

[13] Q. Lee et al [ *IEEE Electron Device Lett. (USA)* vol.19 no.3 (1998) p.77-9 ]

[14] S. Chandrasekar, J.C. Campbell, A.G. Dentai, C.H. Joyner, G.J. Qua, O. Sugiura [ *Electron. Lett. (UK)* vol.24 no.6 (1988) p.319 ]

[15] E. Sano, M. Yoneyama, S. Yamahata, Y. Matsuoka [ *IEEE Trans. Electron Devices (USA)* vol.ED-43 no.11 (1996) p.1826-32 ]

[16] H. Masuda et al [ *GaAs IC Symposium Digest 1997* (IEEE, New York, USA, 1997) p.139-42 ]

[17] R. Pullela et al [ *Electron. Lett. (UK)* vol.34 no.16 (1998) p.1580-1 ]

[18] W. Stanchina et al [ *GaAs IC Symposium Digest 1995* (IEEE, New York, USA, 1995) p.31-4 ]

[19] K.W. Kobayashi et al [ *IEEE Microw. Guid. Wave Lett. (USA)* vol.7 no.10 (1997) p.353-5 ]

[20] B. Agarwal, R. Pullela, Q. Lee, D. Mensa, J. Guthrie, M.J.W. Rodwell [ *IEEE MTT-S Int. Microw. Symp. Dig. (USA)* (1998) p.529-31 ]

[21] K.W. Kobayashi et al [ *IEEE MTT-S Int. Microw. Symp. Dig. (USA)* (1998) p.267-70 ]

[22] D.C. Streit et al [ *GaAs IC Symposium Digest 1997* (IEEE, New York, USA, 1997) p.135-8 ]
[23] K.W. Kobayashi et al [ *GaAs IC Symposium Digest 1993* (IEEE, New York, USA, 1993) p.219-22 ]
[24] J. Cowles et al [ *IEEE MTT-S Int. Microw. Symp. Dig. (USA)* (1995) p.689-92 ]
[25] H. Wang et al [ *IEEE Microw. Guid. Wave Lett. (USA)* vol.5 no.11 (1995) p.388-90 ]
[26] H. Wang et al [ *IEEE Radio Frequency IC Symposium* (IEEE, 1997) p.91-4 ]
[27] L. Tran, S. Southwell, J. Velebit, A. Oki, D. Streit, B. Oyama [ *GaAs IC Symposium Digest 1993* (IEEE, New York, USA, 1993) p.159-62 ]
[28] H. Wang et al [ *RFIC Symposium Digest 1998* (IEEE, New York, USA, 1998) p.275-8 ]
[29] K.W. Kobayashi [ *Microw. J. (USA)* vol.40 no.12 (December 1997) p.64-72 ]
[30] K. Yang, A.L. Gutierrez-Aitken, X. Zhang, G.Z. Haddad, P. Bhattacharya [ *J. Lighwave Technol. (USA)* vol.14 no.8 (1996) p.1831-8 ]
[31] K.C. Syao, K. Yang, X. Zhang, G.I. Hahhad, P. Bhattacharya [ *Electron. Lett. (UK)* vol.33 no.1 (1997) p.82-3 ]
[32] L. Samoska et al [ *IEEE MTT-S Symposium Digest 1998* (IEEE, New York, USA, 1998) p.1843-6 ]
[33] E. Sano, K. Kurishima, S. Yamahata [ *Electron. Lett. (UK)* vol.33 no.2 (1997) p.159 ]
[34] D. Yap, R.H. Walden, Y.-M. Xie, Y.K. Brown, J. Vivilecchia, A.C. Yec [ *IEEE Radio Frequency IC Symposium* (IEEE, 1997) p.59-62 ]
[35] M. Bann et al [ *Electron. Lett. (UK)* vol.27 no.3 (1991) p.278-80 ]
[36] S. Chandrasekhar et al [ *Proc. Optical Fibers Communications Conf.* San Francisco, CA, USA, 22-26 Jan. 1990 (reported in *J. Lightwave Technol. (USA)* vol.8 (September 1990)), post deadline paper PD-27 ]
[37] U. Eriksson, P. Evaldsson, J. Wallin, B. Stalnacke, S. Lourdudoss, B. Willen [ *IEE Proc., Optoelectron. (UK)* vol.143 no.1 (1996) p.107-9 ]
[38] N. Shaw, M.J. Wale, I. Bennion, P.J. Topham, A.J. Moseley [ *Proc. SPIE (USA)* vol.1314 (1990) p.68-70 ]
[39] M. Hafizi, W.E. Stanchina, F. Williams, J.F. Jenson [ *IEEE Trans. Microw. Theory Tech. (USA)* vol.43 no.12 (1995) p.3048-54 ]
[40] J.I. Song, H.R. Frei, J.R. Hayes, R. Bhat, H.M. Cox [ *IEEE Electron Device Lett. (USA)* vol.15 no.4 (1994) p.123-5 ]

# 1.5 InP-based HEMT low-noise and power amplifiers

D. Streit and Y. Chen

June 1998

## A INTRODUCTION

The transition from GaAs-AlGaAs high electron mobility transistors to pseudomorphic GaAs-InGaAs-AlGaAs HEMTs a decade ago resulted in significant improvements in device capability for both low-noise and power applications. This was due primarily to the increased electron velocity associated with the smaller electron effective mass in $In_xGa_{1-x}As$ compared to GaAs. However, the InAs lattice parameter of about 0.606 nm is considerably larger than the GaAs lattice constant of 0.565 nm, and due to strain effects the compositional limit of pseudomorphic $In_xGa_{1-x}As$ on GaAs substrates is limited to about x = 0.30. One method to achieve higher InAs content in the InGaAs channel is to use an InP substrate with a lattice constant of 0.587 nm. This newer generation of HEMT devices uses $In_{0.53}Ga_{0.47}As$ channels lattice-matched to InP substrates. InP-based pseudomorphic HEMTs with $In_xGa_{1-x}As$ channel compositions of up to about x = 0.80 have achieved improvements in performance capability for both low-noise and power amplifier applications compared to GaAs-based pseudomorphic HEMTs.

## B PROFILE DESIGN

The fundamental design goal for InP-based HEMTs is to achieve the best electron transport properties possible within the reliability, breakdown and leakage requirements of the application. Low-noise and power amplifier designs thus have different optimised profiles due to their different bias points and mode of operation. Low-noise HEMTs are usually designed with a high-InAs content pseudomorphic InGaAs channel for optimum cutoff frequency, but at the expense of breakdown voltage. Power HEMTs are usually designed for higher breakdown voltage using either lattice-matched $In_{0.53}Ga_{0.47}As$ channels or composite InGaAs-InP channels.

FIGURE 1 shows a typical InP-based HEMT profile and the contribution of the various profile elements to the circuit schematic. For low-noise applications the noise figure may be approximated by

$$NF_{min} \approx 1 + 2\pi K_F f\, C_{gs}\left(\frac{R_s + R_g}{g_m}\right)^{1/2}$$

where $K_F$ is Fukui's material and process factor, f is the frequency of operation, $C_{gs}$ is the gate-source capacitance, and $g_m$ is the transconductance, also defined as

$$g_m = \frac{C_{gs}}{\tau_d} = \frac{\varepsilon v_s}{(d + \Delta d)}$$

where $\tau_d$ is the electron transit time under the gate, d is the distance from the gate to the heterojunction, and $\Delta d$ is the distance from the heterojunction to the centroid of the two-dimensional electron gas. The most important parameters for a low noise device are high $g_m$, low $C_{gs}/vg_m$, and low gate and source resistances $R_g$ and $R_s$.

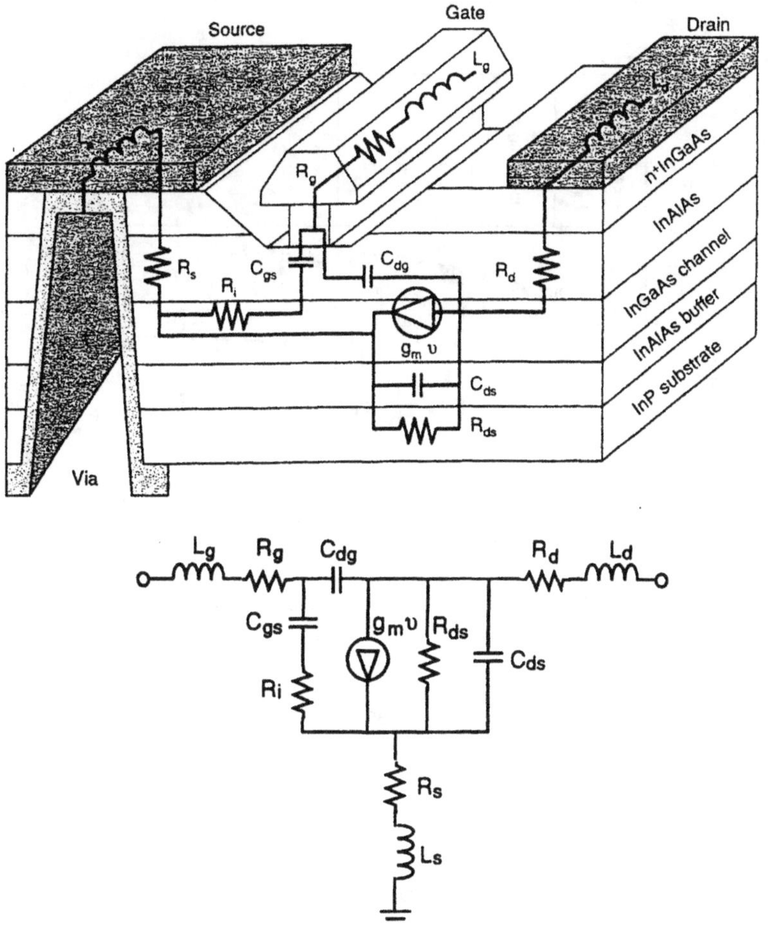


FIGURE 1 InGaAs-InP HEMT profile with schematic representation of the HEMT equivalent circuit.

Optimising the InGaAs channel design can maximise the electron velocity $v_s$ and transconductance. Higher InAs content using pseudomorphic channels increases the electron velocity, but also requires thinner channels to accommodate the increased strain associated with the increased lattice parameter. FIGURE 2 shows the band diagram and calculated electron concentration as a function of the compositional profile for a composite channel InP-InGaAs channel. The performance of power devices can be increased using this approach even with higher InAs content in the InGaAs portion of the channel, without compromising the breakdown of the device. Increasing InAs mole fraction allows a decrease in the channel sheet resistance through a combination of increased sheet charge and increased electron mobility, as shown in FIGURE 3.

The effect of InAs composition in the channel on device cutoff frequency is shown in FIGURE 4. The unity current gain cutoff frequency is related to device parameters as

$$f_T = \frac{v_s}{2\pi L_g}$$

where $v_s$ is the electron saturation velocity and $L_g$ is the effective gate length. Increasing the InAs content in the InGaAs channel increases $v_s$; increasing the electron saturation velocity in turn directly increases $f_T$. Keeping the gate length constant allows a comparison of the $In_xGa_{1-x}As$ devices, with $f_T$ increasing from about 100 GHz for x = 0 to over 300 GHz for x = 0.8.

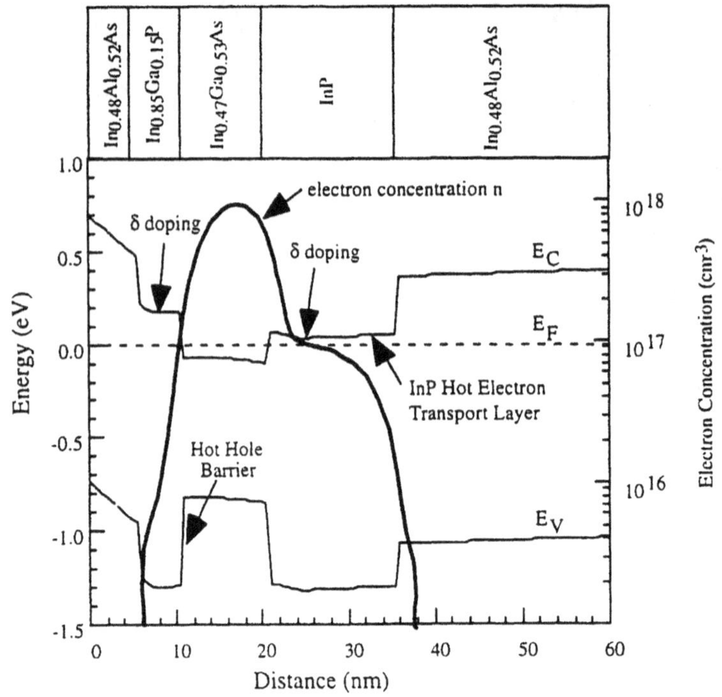


FIGURE. 2 Band diagram and calculated electron distribution for a composite-channel InGaAs-InP HEMT with InGaP hole-blocking layer.

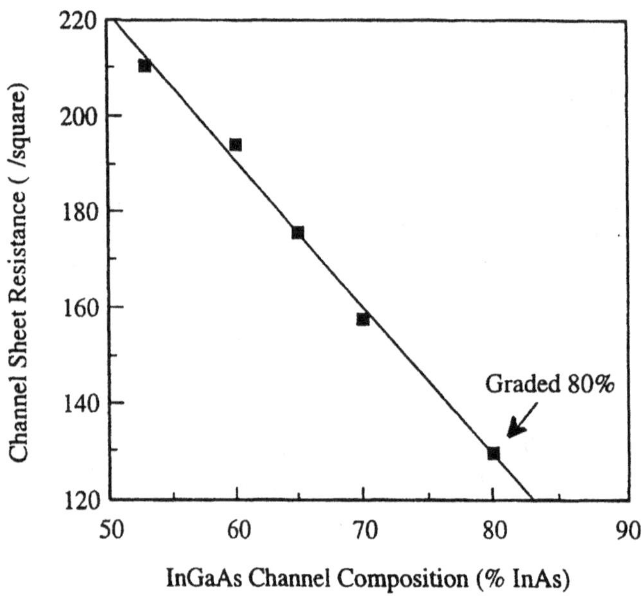


FIGURE 3 Channel resistance of InGaAs-InP HEMTs as a function of InAs mole fraction.

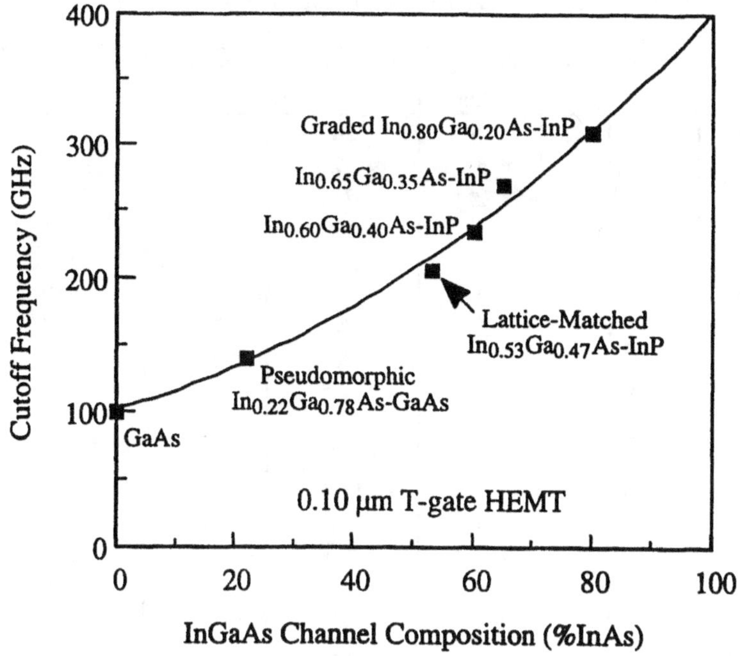


FIGURE 4 HEMT cutoff frequency as a function of InAs mole fraction.

## C InP-BASED LOW-NOISE HEMTS

TABLE 1 shows a compilation of results for InP-based low-noise amplifiers from 2 to 190 GHz.

TABLE 1 InP HEMT MMIC LNA performance.

| Frequency (GHz) | Gate length | Number of stages | Gain (dB) | NF (dB) | Month/year | Ref |
|---|---|---|---|---|---|---|
| 2.25 - 2.5 | 0.15 | 3 | 35 | 0.5 | 8/93 | [1] |
| 7 - 11 | 0.15 | 2 | 21 - 22 | 1.2 | 6/93 | [2] |
| 8 - 12 | | 1* | 8 - 10 | | 6/98 | [3] |
| 10 | | 1* | | 1.6 | 6/98 | [3] |
| 12 | | 1 | 15 | 0.78 | 6/91 | [4] |
| 12 | 0.15 | 1 | 16 | 1.3 | 4/91 | [5] |
| Ka-band | 0.1 | | 19 | 1.9 | 6/98 | [6] |
| 27 - 37 | | 2* | 10 | 3 | 1/92 | [7] |
| 34 - 40 | 0.15 | 2 | 10.6 - 11.6 | 2.4 - 2.8 | 8/92 | [8] |
| 35 | 0.15 | 1 | 10.9 | | 4/91 | [5] |
| 35.5 | | 1 | 13 | | 6/91 | [4] |
| 40 | 0.15 | 2 | 10.6 | 2.7 | 8/92 | [8] |
| Q-band | 0.1 | 1 | 10 | 1.6 | 6/96 | [9] |
| Q-band | 0.1 | 2 | 20 | 1.8 | 6/96 | [9] |
| 42 - 50 | | 3 | 26 - 31 | | 6/96 | [10] |
| 43 - 46 | 0.1 | 2 | 25 | 2.3 | 9/93 | [11] |
| 43.3 - 45.7 | | 3 | | 1.8 | 6/96 | [10] |
| 44.5 | 0.1 | 1 | 8 | 2.5 | 9/95 | [12] |
| 50 | 0.1 | | | 2.6 | 10/94 | [13] |
| 55 | | 1 | 8 | | 6/91 | [4] |
| 56 - 60 | | 2 | 15.25 | 4.2 | 6/92 | [14] |

*Hybrid amplifiers.

TABLE 1 continued

| Frequency (GHz) | Gate length | Number of stages | Gain (dB) | NF (dB) | Month/year | Ref |
|---|---|---|---|---|---|---|
| 56 - 64 | 0.1 | 3 | 24 | <3 | 10/93 | [15] |
| 56 - 64 | 0.1 | 3 | 24 - 25.5 | 3 | 12/93 | [16] |
| 58 | 0.15 | 2 | 21 | 3 | 3/90 | [17] |
| 60 | 0.15 | 1 | 8.4 | | 4/91 | [5] |
| 60 - 65 | 0.25 | 3 | 22.0 | 3 | 12/88 | [18] |
| 60 - 65 | 0.1 | 1 | 9.5 | 1.3 | 9/89 | [19] |
| 60 - 65 | 0.1 | 1 | 8 | 1.5 | 9/89 | [19] |
| V-band | | 4 | 28 | 2.3 | 6/96 | [10] |
| 61 | 0.2 | 3 | 19.5 | 2.6 | 6/89 | [20] |
| 62 | 0.15 - 0.25 | | | 1.7 | 10/89 | [21] |
| 62 | 0.1 | 3 | 24 - 25.5 | 2.7 | 12/93 | [16] |
| 62 | 0.1 | 2 | 11.8 | 4.3 | 6/96 | [22] |
| 63 | 0.2 | 1 | 8.7 | 0.8 | 11/88 | [23] |
| 63 | 0.2 | 1 | 6.5 | 1.8 | 12/88 | [24] |
| 63.5 | 0.1 | 1 | | 0.8 - 1.4 | 3/89 | [25] |
| 75 - 110 | 0.1 | | 23 | | 10/93 | [26] |
| 80 - 100 | 0.1 | 4 | 27 | 5 | 12/93 | [27] |
| 88 - 96 | 0.15 | 2 | 11.5 | 3.2 | 5/90 | [28] |
| 88 - 96 | 0.15 | 3 | 17.3 | 3.3 | 5/90 | [28] |
| 88 - 96 | 0.15 | 3* | 17.3 | 3.3 | 3/90 | [17] |
| 90 - 140 | | 6 | 15+ - 3 | | 6/97 | [29] |
| 90.4 | 0.25 | 2 | 10.2 | 4.5 | 6/89 | [30] |
| 91 - 96 | 0.1 | 3 | 17.5 | 3.2 - 3.5 | 5/19 | [31] |
| 92 | 0.1 | 2* | 14.2 | 2.6 | 6/92 | [32] |
| 93 | 0.15 | 3 | 16.5 | 3 | 6/91 | [33] |
| 93 | 0.15 | 2* | 16.5 | 3 | 4/91 | [34] |
| 93.4 | 0.15 | 2* | 16 | 3 | 8/91 | [35] |
| 94 | 0.1 | 3 | 20 | 3.3 | 10/96 | [36] |
| 94 | 0.15 - 0.25 | | | 3.3 | 10/89 | [21] |
| 94 | 0.1 | 3 | 20 | 3.3 | 5/95 | [37] |
| 94 | | 1 | 10 | 3.5 | 8/96 | [38] |
| 94 | 0.1 | 1 | | 2.3 | 10/96 | [36] |
| 96 | 0.1 | 1 | 7 | 2.6 | 12/93 | [27] |
| 100 | 0.1 | 3 | 19 | 4.3 | 6/93 | [39] |
| 110 | | 6 | | 7 | 6/97 | [29] |
| 155 | 0.1 | 3 | 12 | | 5/97 | [40] |
| 190 | 0.1 | 2 | 7.2 | | 9/98 | [41] |

*Hybrid amplifiers.

## D InP-BASED POWER HEMTS

TABLE 2 shows a compilation of results for InP-based power amplifiers from 2 to 95 GHz.

TABLE 2 InP-based HEMT power amplifier performance.

| Frequency | Gate length (μm) | Type | Power (mW) | Power density (mW/mm) | Gain (dB) | PAE (%) | Month/year | Ref |
|---|---|---|---|---|---|---|---|---|
| C-band | | MMIC | 200 - 300 | | | 67 | 7/93 | [42] |
| 4 | 0.5 | DEV | 225 | 450 | 14.25 | 59 | 11/93 | [43] |
| 7 | | DEV(d) | 270 | 900 | | 76 | 5/97 | [44] |
| 10 | 0.15 | DEV(c) | 350 | 780 | 12 | 60 | 5/94 | [45] |
| 12 | | DEV | | 730 | | 50 | 6/91 | [46] |
| 12 | | DEV | | 960 | | 40 | 6/91 | [46] |
| 12 | | DEV | | 470 | 11.3 | 59 | 6/91 | [46] |
| 18 | 0.15 | DEV(d) | 338 | 750 | 11.9 | 53 | 9/97 | [47] |
| 18 | 0.15 | DEV(d) | 531 | 590 | 10.5 | 53 | 9/97 | [47] |
| 20 | | MMIC | 516 | 645 | 7.1 | 47.1 | 5/93 | [48] |
| 20 | | MMIC | 560 | 700 | | | 5/93 | [48] |
| 20 | 0.15 | DEV(d) | 280 | 620 | | 46 | 4/95 | [49] |
| 30 | | DEV(a) | | 1450 | 6.2 | 24 | 5/92 | [50] |
| 44 | 0.15 | MMIC | 400 | | | 33 | 10/93 | [51] |
| 44 | 0.15 | MMIC | 250 | | | 33 | 11/93 | [52] |
| 44 | 0.2 | DEV | 225 | 375 | 5 | 39 | 3/95 | [53] |
| 44.5 | 0.15 | MMIC | 630 | | 7 | 31 | 8/94 | [54] |
| 57 | 0.22 | DEV | 150 | 333 | | 20 | 4/93 | [55] |
| 57 | 0.22 | DEV | 200 | 444 | | 17 | 4/93 | [55] |
| 59 | 0.15 | DEV | 155 | 346 | 4.9 | 30.1 | 6/93 | [56] |
| 59 | 0.15 | DEV(b) | 288 | 321 | 3.6 | 20.4 | 6/93 | [56] |
| 60 | 0.15 | DEV | | | | 41 | 4/90 | [57] |
| 60 | 0.15 | DEV | 17.5 | 350 | 7.2 | 41 | 8/91 | [58] |
| 60 | 0.1 - 0.25 | DEV | | 300 | 8.6 | 49 | 12/92 | [59] |
| 60 | 0.1 - 0.25 | DEV | | 520 | 5.9 | 33 | 12/92 | [59] |
| 60 | 0.15 | DEV | | 370 | 5.2 | 28.3 | 1/98 | [60] |
| 60 | 0.15 | DEV | | 355 | 6.2 | 12 | 2/98 | [61] |
| 60 | 0.15 | DEV | 158 | 316 | 6 | 37 | 8/98 | [62] |
| 94 | 0.15 | DEV | 10 | 200 | 4.9 | 26 | 8/91 | [58] |
| 94 | 0.15 | DEV | 15 | 300 | | 21 | 3/94 | [63] |
| 94 | 0.1 | DEV | 58 | 290 | 6.4 | 33 | 7/95 | [64] |
| 94 | 0.15 | MMIC | 40 | 125 | | 20 | 10/96 | [65] |
| 94 | 0.15 | MMIC | 130 | 203 | 4 | 13 | 5/97 | [66] |
| 95 | 0.15 | MMIC | 427 | 333 | 8.2 | 19 | 8/98 | [67] |

(a) InP channel; (b) Power combining of two devices; (c) AlInAsP Schottky layer; (d) InP/InGaAs composite channel HEMT.

## REFERENCES


[1] S.E. Rosenbaum et al [ *IEEE Microw. Guid. Wave Lett. (USA)* vol.3 no.8 (1993) p.265 ]

[2] S.E. Rosenbaum et al [ *IEEE MTT-S Int. Microw. Symp. Dig. (USA)* vol.2 (1993) p.1103 ]

[3] L. Pettersson, J.P. Starski, H. Zirath [ *IEEE MTT-S Int. Microw. Symp. Dig. (USA)* vol.3 (1998) p.1683 ]

[4] S.E. Rosenbaum et al [ *IEEE MTT-S Int. Microw. Symp. Dig. (USA)* vol.2 (1991) p.815 ]

[5] C.S. Chou et al [ *Proc. SPIE - Int. Soc. Opt. Eng. (USA)* vol.1475 (1991) p.151 ]

[6] D.L. Ingram et al [ *IEEE MTT-S Int. Microw. Symp. Dig. (USA)* vol.1 (1998) p.301 ]

[7] E.L. Holzman [ *Microw. Opt. Technol. Lett. (USA)* vol.5 no.1 (1992) p.1 ]

[8] Y. Umeda, T. Enoki, K. Arai, Y. Ishii [ *Int. Conf. on Solid State Devices and Materials* Tsukuba, Japan, 26-28 Aug. 1992, p.573 ]

[9] M.V. Aust et al [ *IEEE 1996 Microwave and Millimeter-Wave Monolithic Circuits Symposium Digest of Papers* San Francisco, CA, USA, 16-18 June 1996, Ed. I.J. Bahl (1996) p.89 ]
[10] Long Tran et al [ *IEEE MTT-S Int. Microw. Symp. Dig. (USA)* vol.1 (1996) p.9 ]
[11] D.C.W. Lo et al [ *IEEE Microw. Guid. Wave Lett. (USA)* vol.3 no.9 (1993) p.299-301 ]
[12] Huei Wang et al [ *IEEE Microw. Guid. Wave Lett. (USA)* vol.5 no.9 (1995) p.281 ]
[13] T. Enoki, T. Kobayashi, Y. Ishii [ *IEEE Gallium Arsenide Integrated Circuit Symposium 16th Annual Technical Digest* Philadelphia, PA, USA, 16-19 Oct. 1994 (1994) p.337 ]
[14] R.T. Webster, A.J. Slobodnik Jr., G.A. Roberts [ *IEEE Microw. Guid. Wave Lett. (USA)* vol.2 no.6 (1992) p.236 ]
[15] K.W. Chang et al [ *15th Annual GaAs IC Symposium Technical Digest* San Jose, CA, 10-13 Oct. 1993 (1993) p.211 ]
[16] R. Lai et al [ *IEEE Microw. Guid. Wave Lett. (USA)* vol.3 no.12 (1993) p.447-9 ]
[17] P.C. Chao et al [ *Proc. SPIE - Int. Soc. Opt. Eng. (USA)* vol.1288 (1990) p.2 ]
[18] P. Ho et al [ *Int. Electron Devices Meet. Tech. Dig. (USA)* (1988) p.184-6 ]
[19] U.K. Mishra et al [ *IEEE Trans. Microw. Theory Tech. (USA)* vol.37 no.9 (1989) p.1279 ]
[20] S. Vaughn et al [ *IEEE MTT-S Int. Microw. Symp. Dig. (USA)* vol.2 (1989) p.801 ]
[21] A.W. Swanson et al [ *1989 IEEE Military Communications Conference* Boston, MA, USA, 15-18 Oct. (1989) vol.3 p.744 ]
[22] Y. Umeda et al [ *IEEE 1996 Microwave and Millimeter-Wave Monolithic Circuits Symposium Digest of Papers* San Francisco, CA, USA, 16-18 June 1996, Ed. I.J. Bahl, p.115 ]
[23] U.K. Mishra, A.S. Brown [ *10th Annual GaAs IC Symposium Technical Digest* Nashville, TN, USA, 6-9 Nov. 1988 (1988) p.97-100 ]
[24] U.K. Mishra et al [ *IEEE Trans. Electron Devices (USA)* vol.35 no.12 (1988) p.2441 ]
[25] U.K. Mishra [ *Proceedings of the OSA Topical Meeting* Salt Lake City, UT, USA, 8-10 March 1989, Eds. T.C.L.G. Sollner, D.M. Bloom (Washington DC, 1989) vol.4 p.260 ]
[26] H. Wang, R. Lai, S.T. Chen, J. Berenz [ *IEEE Microw. Guid. Wave Lett. (USA)* vol.3 no.10 (1993) p.381 ]
[27] H. Wang et al [ *Int. Electron Devices Meet. Tech. Dig. (USA)* (1993) p.239 ]
[28] K.H.G. Duh et al [ *IEEE MTT-S Int. Microw. Symp. Dig. (USA)* vol.1 (1990) p.595 ]
[29] S. Weinreb, P.C. Chao, W. Copp [ *IEEE MTT-S Int. Microw. Symp. Dig. (USA)* vol.3 (1997) p.1279 ]
[30] K.H.G. Duh et al [ *IEEE MTT-S Int. Microw. Symp. Dig. (USA)* (1989) p.805 ]
[31] K.H.G. Duh et al [ *IEEE Microw. Guid. Wave Lett. (USA)* vol.1 no.5 (1991) p.114 ]
[32] P.D. Chow et al [ *IEEE MTT-S Int. Microw. Symp. Dig. (USA)* vol.2 (1992) p.807 ]
[33] P.D. Chow et al [ *IEEE MTT-S Int. Microw. Symp. Dig. (USA)* vol.3 (1991) p.1041 ]
[34] P.D. Chow et al [ *Proc. SPIE - Int. Soc. Opt. Eng. (USA)* vol.1475 (1991) p.48 ]
[35] D.C. Streit, K.L. Tan, P.-H. Liu, P.D. Chow [ *Proc. IEEE/Cornell Conf. on Advanced Concepts in High Speed Semiconductor Devices and Circuits* Ithaca, NY, 5-7 Aug. 1991, p.455 ]
[36] H. Wang et al [ *IEE Proc. H, Microw. Antennas Propag. (UK)* vol.143 no.5 (1996) p.361 ]
[37] G.I. Ng et al [ *IEEE 1995 Microwave and Millimeter-Wave Monolithic Circuits Symposium Digest of Papers* Orlando, FL, USA, 15-16 May 1995, Ed. I.J. Bahl, p.63 ]
[38] T. Gaier et al [ *Proc. SPIE - Int. Soc. Opt. Eng. (USA)* vol.2842 (1996) p.46 ]
[39] H. Wang et al [ *IEEE MTT-S Int. Microw. Symp. Dig. (USA)* vol.2 (1993) p.519 ]
[40] R. Lai et al [ *1997 Int. Conf. on Indium Phosphide and Related Materials* Cape Cod, MA, 11-15 May 1997, p.241 ]
[41] R. Lai et al [ *IEEE Microw. Guid. Wave Lett. (USA)* to be published (1998) ]
[42] L.E. Larson et al [ *Electron. Lett. (UK)* vol.29 no.15 (1993) p.1324 ]
[43] J.J. Brown et al [ *IEEE Trans. Electron Devices (USA)* vol.40 no.11 (1993) p.2111 ]
[44] J.B. Shealy et al [ *1997 Int. Conf. on Indium Phosphide and Related Materials* Cape Cod, MA, 11-15 May 1997, p.20 ]

[45] L.M. Jelloian et al [ *IEEE Electron Device Lett. (USA)* vol.15 no.5 (1994) p.172 ]

[46] M. Matloubian et al [ *IEEE MTT-S Int. Microw. Symp. Dig. (USA)* vol.2 (1991) p.721 ]

[47] J.B. Shealy et al [ *IEEE Microw. Guid. Wave Lett. (USA)* vol.7 no.9 (1997) p.261 ]

[48] M. Matloubian et al [ *IEEE Microw. Guid. Wave Lett. (USA)* vol.3 no.5 (1993) p.142 ]

[49] M. Matloubian et al [ *Electron. Lett. (UK)* vol.31 no.9 (1995) p.761 ]

[50] O. Aina et al [ *IEEE Electron Device Lett. (USA)* vol.13 no.5 (1992) p.300 ]

[51] A. Kurdoghlian et al [ *15th Annual GaAs IC Symposium Technical Digest* San Jose, CA, USA, 10-13 Oct. 1993, p.375 ]

[52] W. Lam et al [ *IEEE Microw. Guid. Wave Lett. (USA)* vol.3 no.11 (1993) p.420 ]

[53] K.Y. Hur et al [ *Electron. Lett. (UK)* vol.31 no.7 (1995) p.585 ]

[54] F. Lam et al [ *IEEE Microw. Guid. Wave Lett. (USA)* vol.4 no.8 (1994) p.277 ]

[55] M. Matloubian et al [ *IEEE Electron Device Lett. (USA)* vol.14 no.4 (1993) p.188 ]

[56] M. Matloubian et al [ *IEEE MTT-S Int. Microw. Symp. Dig. (USA)* vol.2 (1993) p.535 ]

[57] P.M. Smith et al [ *2nd Int. Conf. on Indium Phosphide and Related Materials* Denver, CO, 23-25 April 1990, p.39 ]

[58] M.Y. Kao, P.M. Smith, P.C. Chao, P. Ho [ *Proc. IEEE/Cornell Conf. on Advanced Concepts in High Speed Semiconductor Devices and Circuits* Ithaca, NY, 5-7 Aug. 1991, p.469 ]

[59] P. Ho et al [ *Semiconductor Heterostructures for Photonic and Electronic Applications Symposium* Boston, MA, USA, 30 Nov.-4 Dec. 1992, Eds. C.W. Tu, D.C. Houghton, R.T. Tung (Materials Research Society, Pittsburgh, PA, USA, 1993) p.251 ]

[60] S. Piotrowicz et al [ *IEEE Microw. Guid. Wave Lett. (USA)* vol.8 no.1 (1998) p.10 ]

[61] P. Chevalier, X. Wallart, B. Bonte, R. Fauquembergue [ *Electron. Lett. (UK)* vol.34 no.4 (1998) p.409 ]

[62] R. Grundbacher et al [ to be published ]

[63] K.C. Hwang et al [ *Proc. 1994 IEEE 6th Int. Conf. on Indium Phosphide and Related Materials* Santa Barbara, CA, USA, 27-31 March 1994, p.18 ]

[64] P.M. Smith et al [ *IEEE Microw. Guid. Wave Lett. (USA)* vol.5 no.7 (1995) p.230 ]

[65] R. Lai et al [ *IEEE Microw. Guid. Wave Lett. (USA)* vol.6 no.10 (1996) p.366 ]

[66] Y.C. Chen et al [ *1997 Int. Conf. on Indium Phosphide and Related Materials* Cape Cod, MA, 11-15 May 1997, p.509 ]

[67] Y.C. Chen et al [ *IEEE Microw. Guid. Wave Lett. (USA)* to be published (1998) ]

# 1.6 InP in integrated optics

M. Horita, S. Tanaka and Y. Matsushima

November 1998

## A INTRODUCTION

Wavelength-division multiplexed (WDM) optical transmission systems have been intensively studied and begun to be put to practical use, because they are suitable for constructing optical networks as well as increasing capacity of an optical fibre transmission line. Wavelength filters are key components in WDM systems as demultiplexers (DEMUXs) and optical add and drop multiplexers (OADMs). Tunable filters are particularly desired, because they can promote flexibility of system design. Many tunable filters based on InP waveguides have been proposed and studied, because they are compact, potentially tunable and suitable for monolithic integration with lasers, detectors and so on. Most of the wavelength filters consist of optical waveguides and gratings formed near the waveguide. They can select signals of wavelengths which correspond to the Bragg wavelength which is determined by the grating pitch and the effective refractive index of the waveguide. Major characteristics required for practical use are listed below:

- low insertion loss;
- high isolation;
- low crosstalk;
- low polarisation sensitivity;
- wavelength selectivity appropriate for application;
- large tuning range;
- stability of the above filtering characteristics during tuning operation.

This Datareview concerns tunable wavelength filters using InP waveguides, focusing on their bandwidth and tuning range.

## B WAVELENGTH FILTERS USING WAVEGUIDE RESONATORS

Wavelength tunable filters using a $\lambda/4$-shifted passive waveguide grating resonator were demonstrated [1]. The full width at half-maximum (FWHM) of the resonance width was 0.5 nm, and the width of the stop band was 2.5 nm. A wavelength shift of 4.2 nm toward shorter wavelength was obtained at a current injection of 100 mA due to the plasma effect by carrier injection. However, a decrease in transmissivity was observed because the waveguide loss increased at a rate of 1 $cm^{-1}$/mA due to the free carrier absorption. Two channel wavelength signal switching by a current control was demonstrated using two signals with 0.7 nm displacement. The crosstalk between the channels was -6 dB.

Electrically tunable Fabry-Perot filters using InGaAsP/InP rib waveguides were reported [2]. The free spectral range (FSR) was 3.07 nm for an etalon with a cavity length of 102 μm. Tuning through one FSR was possible for a bias current of 2.75 mA. An FSR of 5.9 nm and FWHM of 0.3 nm for a finesse of 20 were obtained for an etalon with a cavity length of 54 μm.

## C WAVELENGTH FILTERS USING REFLECTION GRATINGS

Wavelength filters can be made by an optical waveguide and a reflection grating. Tunable filtering operation was demonstrated in tunable multi-section DFB lasers [3-5] and DBR lasers [6-8]. Although the structures are simple, a 3 dB coupler or an optical circulator is necessary to make use of the reflected signal.

The DBR laser structure is preferable in terms of tuning range. A continuous tuning range of 1.7 nm and a total tuning range of 5 nm were obtained with an FWHM bandwidth of 5 GHz in a three-section DBR laser [6]. BER measurement of a 1 Gbit/s signal was performed for selected wavelengths, and less than 2 dB power penalty was observed by cross-talk from the perturbing channel at 15 GHz separation.

A polarisation independent tunable filter/amplifier was demonstrated by introducing a square-shaped core waveguide and a polarisation independent gain medium [7]. A maximum tuning range of 9.5 nm and an FWHM of 0.9 nm were shown. The displacement between the peak wavelengths for two polarisations was less than 0.1 nm. Electrically tunable channel selection of 5 Gbit/s WDM signals with three channels was demonstrated by using the device [8].

A four channel tunable optical notch filter was presented by using reflection gratings with four different pitches [9]. Each channel can be selected by controlling a current injected to the region with corresponding grating. The channel spacing was 4 nm, and the operation current was 2 mA. A polarisation-independent and zero-loss device was also demonstrated by introducing a specially designed waveguide structure and gain medium [10]. The difference between the peak wavelengths for two polarisations was less than 0.2 nm. Fibre to fibre gain of 2 dB was obtained, although a polarisation dependence of 3 dB was observed. Crosstalk between channels was less than -16 dB. The grating tuning current was 20 mA.

## D WAVELENGTH FILTERS WITH TRACK-CHANGING FUNCTION

Tunable filters based on coupled waveguide structure are especially attractive for application to the OADM, for instance, because they have a track changing function in addition to the wavelength selectivity. The coupler filters can be further classified into co-directional couplers [11-16] and contra-directional couplers. A wider tuning range is expected with the co-directional couplers, while a narrower bandwidth is expected with contra-directional couplers if the coupling length is the same.

A tuning range of up to 37 nm was reported with a bandwidth of 5.5 nm in vertically stacked co-directional couplers [12]. The minimum bandwidth for a co-directional coupler filter was 1.25 nm for a 5 mm-long filter [13]. Polarisation-independent wavelength selectivity was demonstrated by introducing a double periodic grating. The difference between the peak wavelengths for two polarisations was about 0.5 nm in a fabricated device. A tunable receiver with a 30 nm wavelength tuning range was reported by integrating a preamplifier, a two-stage grating-assisted co-directional coupler filter and a photodetector [16]. The FWHM of the receiver was 4 nm.

For a contra-directional coupler scheme, a grating filter integrated with a directional coupler was presented for a wavelength selective router [17]. The FWHM was 7 nm. A phase-synchronised sampled grating was employed to suppress sidemode intensity, and crosstalk between two channels, 6 nm away from each other, was improved to -13 dB. Insertion loss was 15 dB. BER was measured for 2.5 Gbit/s NRZ signals, and power-penalty-free wavelength routing was demonstrated. The switching time was 1 ns.

A vertically and contra-directionally coupled semiconductor waveguide type OADM (VECCS-OADM) loaded with a striped thin-film heater was presented as an electrically tunable and narrow-band OADM device [18]. A schematic view of the device in the InGaAsP/InP material system is depicted in FIGURE 1. The width of the grating was trimmed along the waveguides in order to improve the sidemode suppression ratio (SMSR), and an SMSR of 18 dB was attained [19]. Variation of the centre wavelength is plotted against the current to the striped thin-film heater in FIGURE 2 [20]. The wavelength shift was proportional to the applied electrical power, and was up to 11 nm at I = 120 mA. The inset shows the spectra around the Bragg wavelength at I = 70 mA. Total insertion loss from the input fibre to the output fibre was 9 dB. The FWHM of the dropped spectrum was as narrow as 0.24 nm. Crosstalk of the dropped light from the adjacent channel, 0.8 nm away, was -25 dB. Variations of

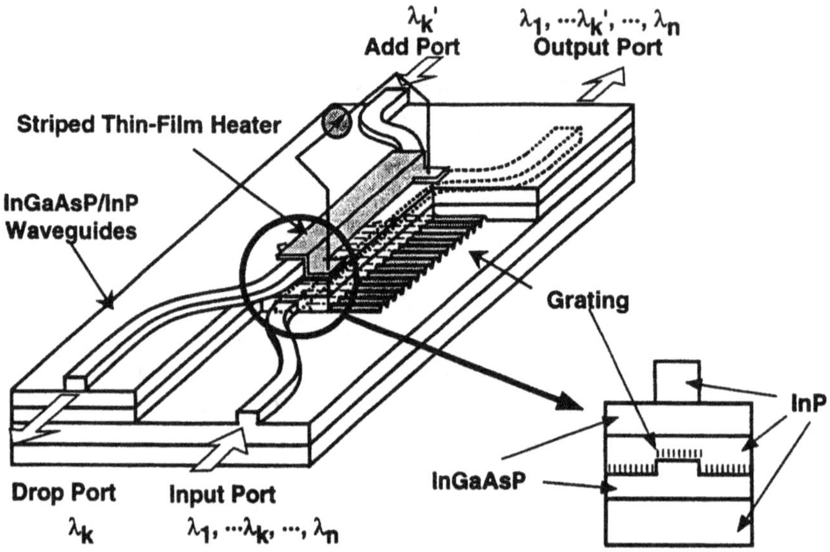


FIGURE 1 Schematic structure of the VECCS-OADM with a trimmed grating and a striped thin-film heater.

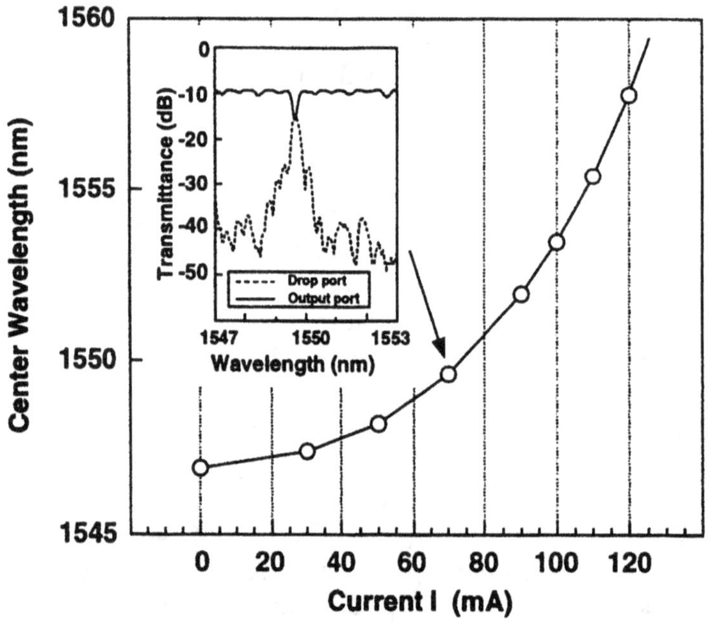


FIGURE 2 Centre wavelength versus current induced in the striped thin-film heater. The inset shows the spectra around the Bragg wavelength at I = 70 mA.

major optical characteristics are summarised in FIGURE 3. Stability of filtering characteristics, such as the insertion loss, the crosstalk and the FWHM, was confirmed over the tuning range. This stability is a strong advantage of this tuning scheme over the plasma effect, in which the degradation of the insertion loss was an essential problem. Power-penalty-free tunable demultiplexing of 5 Gbit/s PRBS signals was demonstrated.

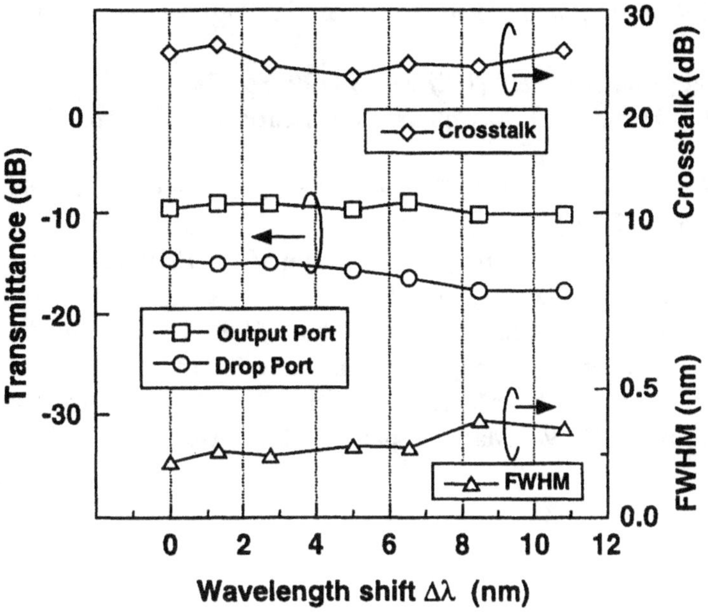


FIGURE 3 Variation of the insertion loss, the crosstalk and the FWHM of the VECCS-OADM during the tuning.

## E CONCLUSION

Tunable wavelength filters using InP waveguides were reviewed, focusing on their bandwidth and tuning range. Tunability can be added to wavelength filters by using InP waveguides. Devices with narrow bandwidth and wide tuning range were attained, and the polarisation dependence was reduced and the insertion loss can be compensated by integrating gain medium in some devices. However, it seems to be very difficult to reduce the polarisation dependent loss over a wide wavelength range and over the operating temperature. Technical innovation in device structure, fabrication technology and materials will be necessary to overcome the difficulties.

## REFERENCES

[1] T. Numai [ *Jpn. J. Appl. Phys. (Japan)* vol.30 (1991) p.2519-25 ]

[2] A. Dentai, J. Stone, E.C. Burrows, C.A. Burrus, L.W. Stulz, M. Zirngibl [ *IEEE Photonics Technol. Lett. (USA)* vol.6 (1994) p.629-31 ]

[3] K. Magari, H. Kawaguchi, K. Oe, Y. Nakano, M. Fukuda [ *Appl. Phys. Lett. (USA)* vol.51 (1987) p.1974-6 ]

[4] S. Suzuki et al [ *J. Lightwave Technol. (USA)* vol.LT-8 (1990) p.660-6 ]

[5] M.J. Chawki, R. Auffret, L. Berthou [ *Electron. Lett. (UK)* vol.26 (1990) p.1146-8 ]

[6] O. Sahlén, M. Öberg, S. Nilsson [ *Electron. Lett. (UK)* vol.27 (1991) p.578-80 ]

[7] U. Koren et al [ *Electron. Lett. (UK)* vol.31 (1995) p.2171-2 ]

[8] G. Raybon et al [ *IEEE Photonics Technol. Lett. (USA)* vol.9 (1997) p.40-2 ]

[9] J.-P. Weber, B. Stoltz, M. Dasler, B. Koek [ *IEEE Photonics Technol. Lett. (USA)* vol.6 (1994) p.77-9 ]

[10] J.-P. Weber, B. Stoltz, M. Dasler [ *Electron. Lett. (UK)* vol.31 (1995) p.220-1 ]

[11] R.C. Alferness et al [ *Appl. Phys. Lett. (USA)* vol.60 (1992) p.980-2 ]

[12] L.L. Buhl et al [ *Electron. Lett. (UK)* vol.29 (1993) p.81-2 ]

[13] F. Heismann et al [ *Appl. Phys. Lett. (USA)* vol.64 (1994) p.2335-7 ]

[14] S.-K. Han, R.V. Ramsaswamy, R.F. Tavlykaev [ *J. Lightwave Technol. (USA)* vol.14 (1996) p.77-83 ]

[15] C.Y. Park et al [ *Electron. Lett. (UK)* vol.33 (1997) p.773-4 ]

[16] Y.-H. Jan, M.E. Heinbuch, L.A. Coldren, S.P. DenBaars [ *Electron. Lett. (UK)* vol.32 (1996) p.1697-8 ]

[17] Y. Shibata, S. Oku, Y. Kondo, T. Tamamura, M. Naganuma [ *J. Lightwave Technol. (USA)* vol.14 (1996) p.1027-32 ]

[18] M. Horita, S. Tanaka, H. Furuya, Y. Matsushima [ in *PDP Tech. Digest OECC'97* Seoul, Korea, July 1997, paper PDP1-2, p.4-5 ]

[19] M. Horita, S. Tanaka, Y. Matsushima [ in *Proc. 10th Int. Conf. on Indium Phosphide and Related Materials* Tsukuba, Japan, 1998, TuA1-3, p.22-5 ]

[20] M. Horita, S. Tanaka, Y. Matsushima [ in *Proc. 24th European Conf. on Optical Communications (ECOC'98)* Madrid, Spain, 1998, MoD04, p.115-6 ]

# 1.7 InP as a solar cell material

S.G. Bailey

June 1999

## A INTRODUCTION

Indium phosphide solar cells have been of interest to the space community because of their radiation resistance properties although there has been only limited utilisation of these cells in space. This review encompasses the development of indium phosphide space solar cells and discusses the performance of these cells in the space environment. There is currently limited commercial availability of InP solar cells, although there may be a future market for radiation resistant solar cells for satellites orbiting in high radiation environments.

Most of the early research on InP solar cells was concerned with their possible use in the terrestrial environment [1] due to their higher solar energy conversion efficiency compared to silicon. The higher cost of InP substrates has so far precluded extensive development of these cells for terrestrial use. Since no radiation damage data were collected for this application there was little subsequent interest in developing this cell for use in space. However, in 1984 it was demonstrated in Japan that InP solar cells were significantly more resistant to electron irradiation than either GaAs or Si [2,3], and, later, InP cells were found to also be superior under proton irradiation [4]. It was observed that, in addition to being more radiation resistant, InP could be annealed at low temperature and under the influence of light [5]. The major obstacles encountered at this relatively early stage of development were the lower efficiencies obtained and the higher cost of InP wafers. A good review of the earlier historical development of InP cells can be found in a review by Coutts and Yamaguchi [6].

## B SOLAR CELL RESEARCH AND DEVELOPMENT

In FIGURE 1 it can be noted that InP with a bandgap energy of 1.34 eV is close to the optimum for the conversion of the Air Mass Zero (AM0) solar spectrum.

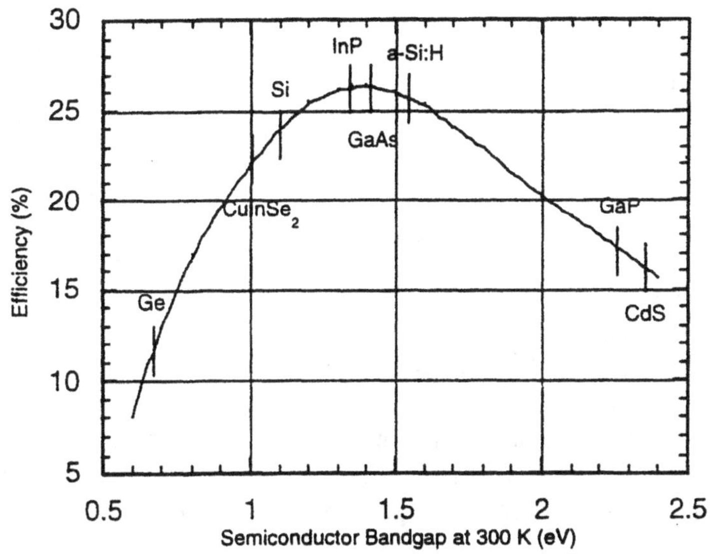


FIGURE 1 Solar cell efficiency limit as a function of the bandgap of the cell material for AM0.

The early InP solar cells were processed using closed tube diffusion [7] and achieved efficiencies of only 13%. By the late 1980s increased experience and refinement of processing including the advent of other techniques such as liquid phase epitaxy and metal organic chemical vapour deposition (MOCVD) resulted in higher efficiencies [8-10]. A world record 19.1% efficient InP cell [11], made by MOCVD, has an emitter, with a total thickness of about 25 nm, which is made up of two layers: a moderately heavily doped region ($3 \times 10^{18}$ $cm^{-3}$: Si) near the junction and a very heavily doped region ($3 \times 10^{19}$ $cm^{-3}$: Si) at the surface. A p-type InP substrate ($5 \times 10^{18}$ $cm^{-3}$: Zn) with a heavily doped buffer layer followed by a lightly doped p-type base layer form the base region of the cell. An n-type $In_{0.53}Ga_{0.47}As$ layer capped the cell to facilitate creation of good front contacts of first chromium for adhesion, then primarily silver with a thin gold layer for chemical protection. The InGaAs was then removed from the active area of the cell. Gold-zinc was used for the back contact. The cell characteristics as measured at NASA Lewis (now Glenn) Research Center were as follows: area = 4 $cm^2$; temperature = 25°C; short circuit current = 145.37 mA; open circuit voltage = 0.876 V; maximum power = 104.93; fill factor = 82.4%; efficiency = 19.1%.

Although the homoepitaxial n/p cell holds the efficiency record there are several good reasons to explore the p/n structure. When growing InP cells on Ge or Si the p/n configuration eliminates the need for a tunnel junction to counteract the natural blocking diode formed by the interdiffusion at the III-V group IV interface. The p/n configuration is also much less sensitive to the effects of a poor emitter surface recombination velocity permitting deeper junctions and significantly higher blue cell response demonstrating better radiation resistance. The record for a p/n InP cell is 17.6% [12]. The cell structure begins with an $n^+$ InP substrate with MOCVD grown ~0.35 μm $n^+$ InP back surface field, then 1.5 - 4.0 μm n InP base (~$10^{17}$ $cm^{-3}$: Zn) followed by a 0.1 - 0.2 μm p InP emitter (~$10^{17}$ $cm^{-3}$: Si) and a 0.05 - 0.1 μm $p^+$ layer (~$10^{18}$ $cm^{-3}$: Si). An InGaAs layer as above was utilised under the Au-Ge front contacts. The cell characteristics as measured at NASA Lewis Research Center were as follows: area = 0.36 $cm^2$; temperature = 25°C; short circuit current = 12.52 mA; open circuit voltage = 0.854 V; fill factor = 81.1%; efficiency = 17.6%.

Heterojunction InP cells have also been demonstrated. The best efficiency recorded for a CdS/InP cell is 15.3% [13]. ITO/InP cells have yielded much higher efficiencies than were predicted by standard heterojunction theory. Most of the ITO/InP cells were produced by sputter deposition of the ITO layer and efficiencies over 15% have been achieved [14]. The highest efficiency ITO/InP cell (17%) was an $n^+$-p-$p^+$ structure with the p-type InP layer grown by APMOVPE [15] and the ITO by DC magnetron sputtering. In most ITO/InP cell designs, the ITO layer is 50 - 70 nm in thickness.

Although the cost of InP substrates has decreased during the past decade, they remain more expensive than silicon or germanium wafers. Both the cost and fragility of InP substrates has led to the development of n/p InP cells on both GaAs and Si and p/n InP on Ge. The lattice mismatch for InP on Ge or GaAs is 3.7% and on Si is 7.5%. Due to the large differences in lattice constant and thermal expansion coefficients between InP and these substrates high mechanical stresses are produced in the epitaxial InP which leads to the generation of a high density of defects. The defects include dislocations, stacking faults and even microcracks. The record efficiency for InP/GaAs is 13.7% [16], for InP/Si 13.0% [17] and for InP/Ge 9.1% [18]. It has also been demonstrated that an InP epitaxial layer can be removed from an InP substrate by a preferentially etched epitaxial liftoff (PEEL) technique [19,20]. The release layer is lattice matched InGaAs which is selectively etched with $HF:H_2O_2:H_2O$ (1:1:10). Apiezon black wax is used to provide compression to the InP film to facilitate peeling of the film. The film and substrate remain in good condition after the PEEL process and the substrate may be reused. A process for cleavage of a lateral epitaxial film for transfer to another substrate (CLEFT) [21] has also successfully removed InP structures from an InP substrate which can also be reused.

InP is also a component of a thin base (0.25 - 0.8 μm) $InAs_xP_{1-x}/InP$ multiquantum well p-i-n cell fabricated by chemical beam epitaxy. The best reported efficiency is 8.2% [22]. Another example of heteroepitaxial devices on InP substrates is a high performance monolithic interconnected module (MIM) for thermophotovoltaic (TPV) applications [23]. A MIM device consists of fifteen InGaAs photovoltaic cells series-connected on a single 1 × 1 cm semi-insulating InP substrate. Both lattice-matched and lattice mismatched InGaAs/InP devices were fabricated with bandgaps of 0.74 eV and 0.60 eV. The spectral utilisation factor, defined as the above-bandgap absorbed energy divided by the total absorbed energy, for a MIM 0.6 eV device is >80% [24].

There are three additional components to consider for high efficiency solar cell fabrication: texturing, contacts and antireflection coatings. Both V-groove [25] and maskless texturing [26] have been utilised in InP solar cell designs with similar enhancements of efficiency compared to textured silicon solar cells. Good ohmic contacts are essential for obtaining the highest cell efficiencies. The majority of cell structures use gold-zinc as the contact for the p-type material and as noted above [11] use chromium-silver-gold as the n-type contact. It has been noted [27] that device degradation at elevated temperatures can be prevented by incorporation of indium or gallium into the gold and the formation of a thin gold-phosphorus layer at the metal/InP interface. The most common antireflection coatings for InP are industry standards of magnesium fluoride and zinc sulphide dual coatings with the coating thickness dependent on whether a coverglass will be used or on the thickness and optical constants of the ITO layer in ITO/InP cells [28].

## C CELL PERFORMANCE UNDER IRRADIATION

A comparison of InP (n/p) [29] with an initial efficiency of 16%, InGaP [30], GaAs/Ge [31], and 200 μm thick silicon [31] shows that InP (see FIGURE 2) is the most resistant to 1 MeV electron irradiation. A similar plot for exposure to a proton energy range from 2 to 50 MeV [32,33] again points to the radiation resistance superiority of InP. There are some differences in the degradation dependent on the cell design (diffused homojunction [34], ion implanted and ITO/InP [35,32,33]). Typical laboratory values for the percentage degradation for 10 MeV protons at a fluence of $1 \times 10^{12}$ $cm^{-2}$ are 10% for InP, 20 - 25% for GaAs and 35 - 40% for silicon. A comparison of p/n with a lower beginning of life efficiency to a high efficiency n/p InP cell shows that n/p cells degrade at a slightly

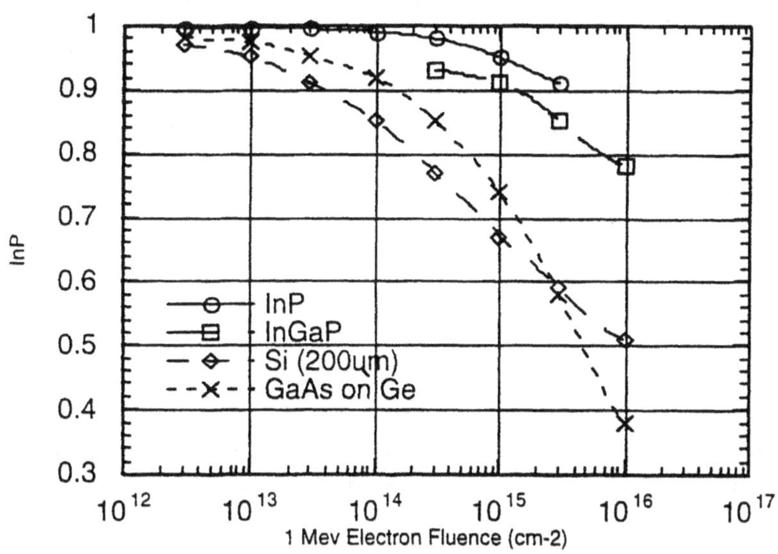


FIGURE 2 Normalised efficiency after 1 MeV electron irradiation.

slower rate initially than p/n. However, the p/n cells eventually survive to a higher proton fluence [36]. Quantum efficiency measurements show that, initially, the cells degrade primarily due to radiation-induced reduction in the minority carrier diffusion length in the base region. In $n^+/p$ InP/Si solar cells carrier removal induced by heavy proton irradiation causes the depletion region to expand. Due to enhanced carrier collection in the depletion region the short circuit current increases above the pre-irradiation level. However, the expansion of the depletion region also induces a large increase in the junction dark current which causes the open circuit voltage to rapidly degrade. These competing effects cause a brief 'plateau' in the cell maximum power just prior to junction failure. Eventually carrier removal destroys the junction and the cell fails [37].

Complete recovery after a heavy radiation dosage has been observed in relatively-lightly-doped InP after heating to 90°C [2]. Annealing at room temperature has also been observed due to photo-injection of minority carriers [38,39]. Light bias was found to enhance the diffusion length in the emitter region. The annealing thermal and illumination characteristics of InP are significant in the operation of InP cells on satellites.

## D SPACE FLIGHT PERFORMANCE

The only InP powered satellite, MUSES-A, was launched into a lunar orbit in January 1990 [40]. 1300 diffused junction ($2 \times 1$ cm$^2$) InP cells with efficiencies of 16 - 17% were used. Unfortunately only a few months of data were obtained, showing no degradation of the InP cells. Several other flight experiments have provided data from space. Two InP experiments were mounted on the LIPS III satellite launched in 1987 into a nearly circular orbit of altitude 1100 km and an inclination slightly in excess of 60° [41]. The experiments were two four-cell modules containing InP homojunction cells and ITO/InP cells, respectively. The ITO/InP cells showed no degradation in short circuit current after 900 days in orbit [33]. The short circuit current of a cell in the homojunction module had an increase in short circuit current of ~4% after 971 days in orbit. This can be attributed to the enhanced carrier collection discussed above. In 1989 two InP cells were launched into an elliptical orbit at 75° aboard the Japanese scientific satellite, EXOS-D [42]. These were $n^+p$ diffused junction cells without coverglass. The cells showed severe short-circuit current reduction after only one week (equivalent 1 MeV electron fluence of $2 \times 10^{16}$ cm$^{-2}$). In 1994 the Photovoltaic Array Space Power Plus experiment was launched on the APEX satellite in an elliptical orbit with an initial apogee of 2552 km and a perigee of 363 km at an inclination of 70° [43]. The experiments ran for 373 days. The indium phosphide module on PASP Plus consisted of ten 2 cm × 2 cm 305 μm thickness $n^+/p$ cells made by Spire Corp with 152 μm coverglass. The initial measured performance of the InP module was 16.9% AM0 efficiency with an initial $P_{max}$ of 22.23 mW/cm$^2$. In the proton dominated orbit InP degraded to about 92% of its original value while GaAs and silicon dropped to 87% and 82% respectively. For high radiation orbits InP should be superior to other cell types.

## E CONCLUSION

The major impediments to the future use of InP for space power applications have been the relatively high cost of the cell on an InP substrate and the lower efficiency of the InP cell on other substrates. There is a niche market however for radiation resistant solar cells. It is anticipated that future efforts will focus on enhancing the efficiency of the InP cell on Ge or Si substrates.

**REFERENCES**

[1] I. Weinberg, D. Brinker [ *Proc. 21st Intersociety Energy Conversion Engineering Conf.* San Diego, CA, USA, 25-29 Aug. 1986 (American Chem. Soc., 1986) p.1431 ]
[2] M. Yamaguchi, C. Uemura, A. Yamamoto [ *J. Appl. Phys. (USA)* vol.55 (1984) p.1429-35 ]
[3] M. Yamaguchi, C. Uemura, A. Yamamoto, A. Shibukawa [ *Jpn. J. Appl. Phys. (Japan)* vol.23 (1984) p.302-6 ]
[4] I. Weinberg, C. Swartz, R. Hart Jr. [ *Proc. 18th IEEE Photovoltaic Specialists Conf.* Las Vegas, NV, USA, 21-25 Oct. 1985 (IEEE, New York, USA, 1985) p.1722-4 ]
[5] M. Yamaguchi, Y. Itoh, K. Ando [ *Appl. Phys. Lett. (USA)* vol.45 (1984) p.206-11 ]
[6] T. Coutts, M. Yamaguchi [ *Curr. Top. Photovoltaics (UK)* vol.3 (1989) p.29-234 ]
[7] I. Weinberg et al [ *Sol. Cells (Switzerland)* vol.22 (1987) p.213 ]
[8] K. Parat, S. Bothra, J. Borrego, S. Ghandhi [ *Solid-State Electron. (UK)* vol.30 (1987) p.283 ]
[9] K. Choi, C. Shen [ *J. Appl. Phys. (USA)* vol.63 (1988) p.1198 ]
[10] C. Keavney, M. Spitzer [ *Appl. Phys. Lett. (USA)* vol.52 (1988) p.1439 ]
[11] C. Keavney, V. Haven, S. Vernon [ *Proc. 21st IEEE Photovoltaic Specialists Conf.* Kissimmee, FL, USA, 21-25 May 1990 (IEEE, New York, USA, 1990) p.141-4 of vol.1 ]
[12] R. Hoffman et al [ *Proc. 26th IEEE Photovoltaic Specialists Conf.* Anaheim, CA, USA, 29 Sept.-3 Oct. 1997 (IEEE, New York, USA, 1997) p.815 ]
[13] F. Himrane, N. Pearsall, R. Hill [ *Proc. 18th IEEE Photovoltaic Specialists Conf.* Las Vegas, NV, USA, 21-25 Oct. 1985 (IEEE, New York, USA, 1985) p.338 ]
[14] T. Coutts, S. Naseem [ *Appl. Phys. Lett. (USA)* vol.46 no.2 (1985) p.164 ]
[15] X. Li, M. Wanlass, T. Gessert, K. Emery, T. Coutts [ *Appl. Phys. Lett. (USA)* vol.54 no.26 (1989) p.2674 ]
[16] M. Wanlass, T. Coutts, J. Ward, K. Emery [ *Proc. 22nd IEEE Photovoltaic Specialists Conf.* Las Vegas, NV, USA, 7-11 Oct. 1991 (IEEE, New York, USA, 1991) p.159-65 of vol.1 ]
[17] S. Wojtczuk et al [ *IEEE 1st World Conf. on Photovoltaic Energy Conversion - WCPEC* Waikoloa, HI, USA, 5-9 Dec. 1994 (IEEE, New York, USA, 1994) p.1705-8 of vol.2 ]
[18] S. Wojtczuk, S. Vernon, D. Burke [ *SPRAT XIII* (1994) p.91 ]
[19] S. Bailey, D. Wilt, F. DeAngelo, E. Clark [ *Proc. 23rd IEEE Photovoltaic Specialists Conf.* Louisville, KY, USA, 10-14 May 1993 (IEEE, New York, USA, 1993) p.783-5 ]
[20] S. Bailey, D. Wilt, F. DeAngelo [ US Patent no.5,641,381 (1997) ]
[21] R. McClelland, C. Bozler, J. Fan [ *Appl. Phys. Lett. (USA)* vol.37 (1980) p.560 ]
[22] A. Freundlich, M. Vilela, C. Monier, L. Aguilar, F. Newman, I. Serdiukova [ *Proc. 26th IEEE Photovoltaic Specialists Conf.* Anaheim, CA, USA, 29 Sept.-3 Oct. 1997 (IEEE, New York, USA, 1997) p.911 ]
[23] N. Fatemi, R. Hoffman, J. Stan, V. Weizer, P. Jenkins [ *IEEE 2nd World Conf. on Photovoltaic Energy Conversion - WPEC* (IEEE, New York, USA, 1998) p.345 ]
[24] M. Clevenger, C. Murray, D. Riley [ MRS (1997) ]
[25] S. Bailey, N. Fatemi, G. Landis, D. Brinker, M. Faur [ *Proc. 2nd Int. Conf. on Indium Phosphide and Related Materials* Denver, CO, USA, 23-25 April 1990 (IEEE, New York, USA, 1990) p.73 ]
[26] S. Bailey, M. Young, N. Fatemi, G. Landis [ *Proc. 7th Int. Photovoltaic Science and Engineering Conf.* (1993) p.617 ]
[27] N. Fatemi, V. Weizer [ *J. Electron. Mater. (USA)* vol.20 no.10 (1991) p.875 ]
[28] N. Pearsall et al [ *Proc. 22nd IEEE Photovoltaic Specialists Conf.* Las Vegas, NV, USA, 7-11 Oct. 1991 (IEEE, New York, USA, 1991) p.425 ]
[29] I. Weinberg [ *Sol. Cells (Switzerland)* vol.31 (1991) p.331 ]
[30] T. Takamoto, E. Ikeda, H. Kurita, M. Yamaguchi [ *14th European Photovoltaic Solar Energy Conf.* (1997) p.970 ]
[31] B. Anspaugh [ *Solar Cell Radiation Handbook* Addendum 1 (1989) ]

[32] N. Pearsall, C. Goodbody, O. Oparaku, A. Dollery, R. Hill [ *Proc. 20th IEEE Photovoltaic Specialists Conf.* Las Vegas, NV, USA, 26-30 Sept. 1988 (IEEE, New York, USA, 1988) p.898-902 of vol.2 ]

[33] N. Pearsall, C. Goodbody, N. Robson, I. Forbes, R. Hill [ *Proc. 21st IEEE Photovoltaic Specialists Conf.* Kissimmee, FL, USA, 21-25 May 1990 (IEEE, New York, USA, 1990) p.1172 ]

[34] I. Weinberg, C. Swartz, R. Hart, T. Coutts [ *Proc. 20th IEEE Photovoltaic Specialists Conf.* Las Vegas, NV, USA, 26-30 Sept. 1988 (IEEE, New York, USA, 1988) p.893 ]

[35] G. Augustine, A. Smith, A. Rohatgi, C. Keavney [ *Proc. 20th IEEE Photovoltaic Specialists Conf.* Las Vegas, NV, USA, 26-30 Sept. 1988 (IEEE, New York, USA, 1988) p.903 ]

[36] S. Messenger, R. Walters, G. Summers, R. Hoffman [ *IEEE 2nd World Conf. on Photovoltaic Energy Conversion - WPEC* (IEEE, New York, USA, 1998) p.3726 ]

[37] S. Messenger et al [ *Proc. 26th IEEE Photovoltaic Specialists Conf.* Anaheim, CA, USA, 29 Sept.-3 Oct. 1997 (IEEE, New York, USA, 1997) p.995 ]

[38] M. Yamaguchi, K. Ando, A. Yamamoto, C. Uemura [ *Appl. Phys. Lett. (USA)* vol.58 (1985) p.568 ]

[39] M. Yamaguchi, K. Ando, A. Yamamoto, C. Uemura [ *Appl. Phys. Lett. (USA)* vol.44 (1984) p.432 ]

[40] M. Yamaguchi et al [ *Proc. 21st IEEE Photovoltaic Specialists Conf.* Kissimmee, FL, USA, 21-25 May 1990 (IEEE, New York, USA, 1990) p.1198 ]

[41] J. Severns et al [ *Proc. 20th IEEE Photovoltaic Specialists Conf.* Las Vegas, NV, USA, 26-30 Sept. 1988 (IEEE, New York, USA, 1988) p.801 ]

[42] M. Yamaguchi et al [ *Proc. 22nd IEEE Photovoltaic Specialists Conf.* Las Vegas, NV, USA, 7-11 Oct. 1991 (IEEE, New York, USA, 1991) p.1576 ]

[43] H. Curtis, D. Marvin [ *SPRAT XIV* (1995) p.268 ]

CHAPTER 2

# MECHANICAL, THERMAL, PIEZOELECTRIC AND ELECTRO-OPTIC PROPERTIES

# 2.1 Density of InP

S. Adachi

April 1998

## A INTRODUCTION

The crystal density d is one of the most important material parameters. It can be given by

$$d = \frac{4M}{a^3 N_A} = \frac{d_M M}{N_A} \quad (1)$$

where a, M, $N_A$ and $d_M$ are the lattice parameter, molecular weight of the crystal, Avogadro's constant ($6.0221367 \times 10^{23}$ $mol^{-1}$) and molecular density, respectively. The molecular weight M of InP is 145.7938 g.

## B CALCULATED VALUES

If an accurate lattice parameter is available, the calculation of d then gives a good reliable value. Data on the lattice parameter of InP suggest that a = 586.90 pm [1] so that at 300 K

$$d_M = 4/a^3 = 1.9786 \times 10^{22}\ cm^{-3} \quad \text{and} \quad d = 4.7902\ g/cm^3$$

TABLE 1 lists the values of d for temperatures between 4.2 and 600 K, as calculated from EQN (1) with a from [1]. The X-ray d data for T ≥ 700 K from Brice [2] are also listed in this table.

TABLE 1 Values of d at various temperatures.

| T (K) | d ($g/cm^3$) | T (K) | d ($g/cm^3$) |
|---|---|---|---|
| 4.2 | 4.7982 | 214 | 4.7945 |
| 12 | 4.7982 | 232 | 4.7936 |
| 20 | 4.7982 | 251 | 4.7926 |
| 29 | 4.7983 | 272 | 4.7914 |
| 39 | 4.7984 | 286 | 4.7909 |
| 49 | 4.7984 | 291 | 4.7904 |
| 59 | 4.7985 | 300 | 4.7902 |
| 70 | 4.7985 | 400 | 4.7836 |
| 80 | 4.7985 | 500 | 4.7770 |
| 90 | 4.7985 | 600 | 4.7704 |
| 100 | 4.7984 | 700 | 4.763 |
| 109 | 4.7984 | 800 | 4.756 |
| 119 | 4.7982 | 900 | 4.749 |
| 127 | 4.7979 | 1000 | 4.742 |
| 136 | 4.7977 | 1100 | 4.733 |
| 156 | 4.7971 | 1200 | 4.723 |
| 175 | 4.7963 | 1300 | 4.712 |
| 195 | 4.7954 | | |

## C MEASURED DATA

TABLE 2 lists the measured values of d taken from [3,4]. We obtain, from EQN (1), that d = 4.81 g/cm$^3$ at 25°C [3] corresponds to a value of a = 586.1 pm. Note that this value is much smaller than those in [1] (a = 586.827 - 587.01 pm). The data of Glazov et al [4] showed that InP becomes denser on melting. This densification of the structure implies an increase in the coordination number, consistent with a transition to the electronic configuration corresponding to metallic bonding of the atoms in the melt [4].

TABLE 2 Measured values of d.

| T (°C) | d (g/cm$^3$) | Ref |
|---|---|---|
| 25 | 4.81 | [3] |
| 800 | 4.655 | [4] |
| 950 | 4.645 | [4] |
| 1050 | 4.64 | [4] |
| 1060 | 5.07 | [4] |
| 1080 | 5.03 | [4] |
| 1100 | 4.975 | [4] |
| 1125 | 4.925 | [4] |
| 1150 | 4.855 | [4] |

## D CONCLUSION

We have discussed the crystal density of InP. No precise experimental data have been found on this material parameter. Thus the best that can be done is to quote an X-ray density which is given by EQN (1).

## REFERENCES


[1] S. Adachi [ Datareview in this book: *2.2 Lattice parameter of InP* ]

[2] J.C. Brice [ in *Properties of Indium Phosphide* EMIS Datareviews Series No.6 (INSPEC, IEE, London, UK, 1991) ch.1 p.3-4 ]

[3] U. Piesbergen [ *Z. Nat.forsch. A (Germany)* vol.18 (1963) p.141-7 ]

[4] V.M. Glazov, K. Davletov, A.Ya. Nashel'skii, M.M. Mamedov [ *Russ. J. Phys. Chem. (UK)* vol.51 (1977) p.1496-9 ]

# 2.2 Lattice parameter of InP

S. Adachi

April 1998

## A INTRODUCTION

The lattice constant a in the zinc-blende structure is defined by the length of the face-centred cube edge. It is relatively heavily dependent on both temperature and pressure. The temperature dependence of a is explained by the thermal expansion coefficient. The constant a is related to hydrostatic pressure by Murnaghan's equation [1]. The lattice constant is also a function of the purity, perfection and stoichiometry (see [2]).

## B ROOM-TEMPERATURE DATA

Experimental evaluations of a for undoped (or low-doped) InP have been carried out by several authors [3-10]. TABLE 1 summarises these results. From this table, we obtain a mean value of a = 586.90 pm with a standard deviation of 62 fm.

TABLE 1 Values of a at 300 K.

| a (pm) | Ref |
|---|---|
| 586.875 ± 0.01 | Giesecke and Pfister [3] |
| 587.01 | Kudman and Paff [4] |
| 586.83 | Bisaro et al [5] |
| 586.93 | Sugii et al [6] |
| 586.827 | Sirota et al [7] |
| 586.945 ± 0.01 | Deus et al [8] |
| 586.90 | Haruna et al [9] |
| 586.8647 | Bak-Misiuk et al [10] |

## C TEMPERATURE DEPENDENCE

Values of a as a function of temperature (T) have been reported below room temperature by Sirota et al [7], Deus et al [8] and Haruna et al [9] and above room temperature by Kudman and Paff [4] and Bisaro et al [5].

Assuming the relationship between a and T to be linear, the formula for thermal expansion $\alpha$ is given by

$$a(T) = a_0(1 + \alpha\Delta T) \quad (1)$$

where $a_0$ is the lattice constant at temperature $T_0$ and $\Delta T = T - T_0$. Inserting the value of $\alpha = 4.55 \times 10^{-6}\ K^{-1}$ [11] into EQN (1), we obtain

$$a(T) = 586.90 + 0.0027\Delta T \text{ pm} \quad (2)$$

with $\Delta T = T - 300$ K ($a_0 = 586.90$ pm at $T_0 = 300$ K). This relation is valid for $300 < T < 600$ K.

TABLE 2 lists the values of a for $4.2 \le T \le 600$ K. The data are taken from Haruna et al [9] for $4.2 \le T \le 300$ K and from EQN (2) for $300 \le T \le 600$ K. They have uncertainties which range from 0.03 pm at 4.2 K to 0.10 pm at 600 K [12]. The a values in TABLE 2 are plotted in FIGURE 1.

TABLE 2 Values of a at various temperatures.

| T (K) | a (pm) | T (K) | a (pm) |
|---|---|---|---|
| 4.2 | 586.5740 | 136 | 586.5948 |
| 12 | 586.5744 | 156 | 586.619 |
| 20 | 586.5738 | 175 | 586.651 |
| 29 | 586.5720 | 195 | 586.687 |
| 39 | 586.5684 | 214 | 586.725 |
| 49 | 586.5655 | 232 | 586.763 |
| 59 | 586.5632 | 251 | 586.805 |
| 70 | 586.5616 | 272 | 586.851 |
| 80 | 586.5610 | 286 | 586.873 |
| 90 | 586.5622 | 291 | 586.891 |
| 100 | 586.5646 | 300 | 586.90 |
| 109 | 586.5682 | 400 | 587.17 |
| 119 | 586.5738 | 500 | 587.44 |
| 127 | 586.5850 | 600 | 587.71 |

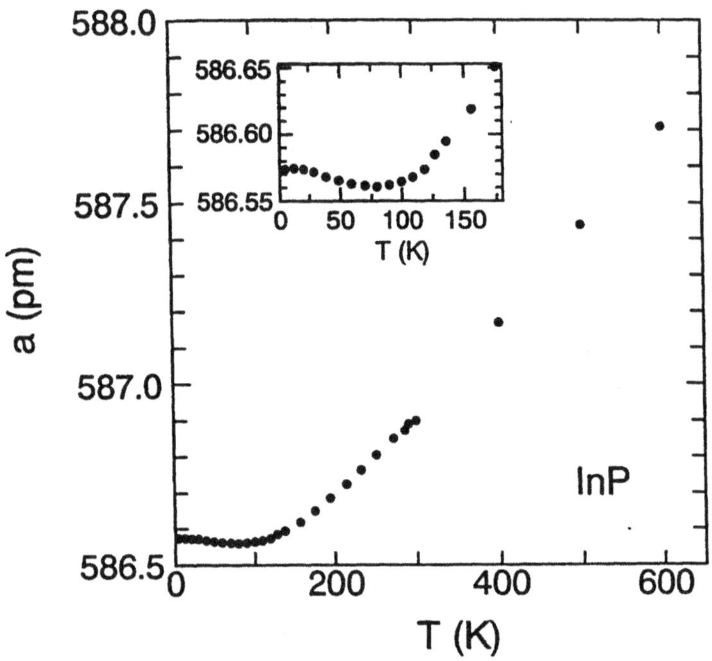


FIGURE 1 Lattice parameter a for InP. Values are taken from Haruna et al [9] for $4.2 \le T \le 300$ K and from EQN (2) for $300 \le T \le 600$ K.

## D DOPING EFFECT

The lattice parameters of solids can expand or contract when impurity atoms are incorporated in the solids. An example is the dilation, or expansion, of the InP lattice induced by Te doping. An increase in the InP lattice parameter of ~0.15% was observed at a Te concentration of $\sim 10^{20}$ $cm^{-3}$ [13]. TABLE 3 summarises the lattice parameter changes observed in Zn-, Si-, Sn-, S- and Te-doped InP [6,13-16]. Vegard's law was found to be valid in Te-doped InP [13], but not in Si- and Sn-doped InP [14].

TABLE 3 Doping effect on the InP lattice parameter.

| Dopant | Dilation (D) or Contraction (C) | Ref |
|---|---|---|
| Zn | C | [6] |
| Si | D | [14] |
| Sn | D | [6] |
| | D | [14] |
| Sn + Ga + As | D | [15] |
| S | C | [6] |
| | D | [15,16] |
| Te | D | [13] |

## E CONCLUSION

The available experimental data on the lattice parameter of InP have been presented. The parameter of undoped (or low-doped) InP is about 586.90 pm at 300 K. Much more work is needed: in particular, precision measurements at high temperatures (≥600 K), quantitative estimates of the dopant effects and evaluation of the variations with stoichiometry and dislocation density.

## REFERENCES


[1] S. Adachi [ *Physical Properties of III-V Semiconductor Compounds: InP, InAs, GaAs, GaP, InGaAs, and InGaAsP* (Wiley-Interscience, New York, 1992) ]

[2] J.C. Brice [ in *Properties of Gallium Arsenide* EMIS Datareviews Series No.2 (INSPEC, IEE, London, UK, 1990) ch.1 p.7-12 ]

[3] G. Giesecke, H. Pfister [ *Acta Crystallogr. (Denmark)* vol.11 (1958) p.369-71 ]

[4] I. Kudman, R.J. Paff [ *J. Appl. Phys. (USA)* vol.43 (1972) p.3760-2 ]

[5] R. Bisaro, P. Merenda, T.P. Pearsall [ *Appl. Phys. Lett. (USA)* vol.34 (1979) p.100-2 ]

[6] K. Sugii, H. Koizumi, E. Kubota [ *J. Electron. Mater. (USA)* vol.12 (1983) p.701-12 ]

[7] A.N.N. Sirota, A.M. Antyukhov, A.A. Sidorov [ *Sov. Phys.-Dokl. (USA)* vol.29 (1984) p.662-5 ]

[8] P. Deus, H.A. Schneider, U. Voland, K. Stiehler [ *Phys. Status Solidi A (Germany)* vol.103 (1987) p.443-7 ]

[9] K. Haruna, H. Maeta, K. Ohashi, T. Koike [ *J. Phys. C, Solid State Phys. (UK)* vol.20 (1987) p.5275-9 ]

[10] J. Bak-Misiuk, J. Wolf, U. Pietsch [ *Phys. Status Solidi A (Germany)* vol.118 (1990) p.209-17 ]

[11] S. Adachi [ Datareview in this book: *2.6 Thermal expansion, thermal conductivity and specific heat of InP* ]

[12] J.C. Brice [ in *Properties of Indium Phosphide* EMIS Datareviews Series No.6 (INSPEC, IEE, London, UK, 1991) ch.1 p.5-7 ]

[13] F.G. Kellert, K.T. Chan, J.E. Turner, V.M. Robbins [ *J. Electron. Mater. (USA)* vol.19 (1990) p.1425-8 ]

[14] C. Ferrari, P. Franzosi [ *J. Appl. Phys. (USA)* vol.79 (1996) p.6890-4 ]

[15] A. Knauer, J. Karusslich, R. Kittner, R. Staske, A. Barwolff [ *Cryst. Res. Technol. (Germany)* vol.25 (1990) p.441-50 ]

[16] J.P. Farges, C. Schiller, W.J. Bartels [ *J. Cryst. Growth (Netherlands)* vol.83 (1987) p.159-66 ]

# 2.3 Elastic constants and acoustic properties of InP

S. Adachi

April 1998

## A INTRODUCTION

The bulk modulus of a semiconductor crystal is of fundamental physical interest and also of practical importance in the interpretation of high-pressure experimental data as it provides the connection between pressure and lattice constant through Murnaghan's equation [1]. We discuss in this Datareview the elastic and acoustic properties of InP.

## B ELASTIC CONSTANTS

### B1 Elastic Stiffness Constants

Adiabatic elastic constants can be determined most accurately from ultrasound velocity measurements. The difference between the adiabatic and isothermal elastic constants was found to be small in most III-V semiconductors [2].

InP crystallises in the zinc-blende structure, and so the elastic stiffness tensor [c] has three independent components, $c_{11}$, $c_{12}$ and $c_{44}$ (second-order elastic constants). There have been several experimental determinations of the elastic stiffness constants $c_{ij}$ using bulk InP single crystals [3-6]. We summarise in TABLE 1 the results obtained in these studies.

TABLE 1 Room-temperature stiffness constants for bulk InP (in GPa). USV = ultrasound velocity measurement, PDC = phonon dispersion curve analysis.

| $c_{11}$ | $c_{12}$ | $c_{44}$ | Method | Ref |
|---|---|---|---|---|
| 102.2 | 57.6 | 46.0 | USV | Hickernell and Gayton [3] |
| 101.1 | 56.1 | 45.6 | USV | Nichols et al [4] |
| 103.4 | 58.31 | 41.13 | USV | Gerlich and Wolf [5] |
| 105 ± 2 | 65 ± 2 | 44 ± 1 | PDC | Borcherds and Kunc [6] |

The data in TABLE 1 show variations in $c_{11}$, $c_{12}$ and $c_{44}$ of 1%, 4% and 11%, respectively. The scatter in $c_{44}$ is surprising since this should be the best determined of the three constants. Average values may be taken from all the velocity measurements [3-5] and errors estimated from the scatter:

$$c_{11} = 102.2 \pm 0.9 \quad c_{12} = 57.3 \pm 0.9 \quad c_{44} = 44.2 \pm 2.2 \quad \text{GPa} \tag{1}$$

The average value of $c_{44}$ is close to the value of 44 ± 1.0 GPa from phonon dispersion curve analysis [6], while there is ~15% (~3%) disagreement in $c_{12}$ ($c_{11}$).

The elastic constants of an epitaxial InP film have been successfully determined using a Brillouin scattering technique [7]. The epitaxial film, about 1 μm thick, was grown on InP substrate by chemical beam epitaxy. A set of the $c_{ij}$ values obtained in this study is

$$c_{11} = 101 \pm 2 \quad c_{12} = 56.2 \pm 1.5 \quad c_{44} = 44.4 \pm 1.8 \quad \text{GPa} \tag{2}$$

No clear difference can thus be recognised between the bulk and epitaxial film values.

## B2 Elastic Compliance Constants

The elastic compliance tensor [s], which has the same form as [c], is connected reciprocally with [c] through Hook's relation. Explicit expressions for the components $s_{ij}$ in terms of $c_{ij}$ can be given in the following equations:

$$s_{11} = (c_{11} + c_{12})(c_{11} - c_{12})^{-1}(c_{11} + 2c_{12})^{-1} \tag{3a}$$
$$s_{12} = -c_{12}(c_{11} - c_{12})^{-1}(c_{11} + 2c_{12})^{-1} \tag{3b}$$
$$s_{44} = 1/c_{44} \tag{3c}$$

The elastic stiffness constants are discussed in subsection B1. The best data come from work on ultrasound velocity measurements [3-5]. TABLE 2 gives the derived values of $s_{ij}$. The best values are obtained by averaging these data [2].

TABLE 2 Room-temperature compliance constants for bulk InP ($10^{-12}$ $m^2/N$).

| $s_{11}$ | $s_{12}$ | $s_{44}$ | Ref |
|---|---|---|---|
| 16.5 | -5.9 | 21.7 | Hickernell and Gayton [3] |
| 16.4 | -5.8 | 21.9 | Nichols et al [4] |
| 16.3 | -5.9 | 24.3 | Gerlich and Wolf [5] |
| 16.4 ± 0.1 | -5.9 ± 0.1 | 22.6 ± 1.2 | Best values |

## B3 Temperature and Pressure Coefficients

The elastic stiffness constants $c_{ij}$ in most III-V semiconductors decrease with increasing temperature (T) or increase with increasing pressure (p) [8]. Using the data of Gerlich and Wolf [5] at 273 and 298 K, Jordan [9] estimated the temperature coefficient of $c_{ij}$ for InP. The pressure dependence of $c_{ij}$ for InP has been studied by Nichols et al [4] for hydrostatic pressures up to 0.4 GPa; they found that the stiffness constants vary linearly up to the highest pressure reached (~0.4 GPa). TABLE 3 summarises these studies.

TABLE 3 Temperature and hydrostatic-pressure coefficients of $c_{ij}$ for InP.

| | $c_{11}$ | $c_{12}$ | $c_{44}$ | Ref |
|---|---|---|---|---|
| $dc_{ij}/T$ (MPa/K) | -14.0 | -8.3 | -4.0 | Jordan [9] |
| $dc_{ij}/p$ | 4.17 | 4.80 | 0.36 | Nichols et al [4] |

## B4 Third-Order Elastic Stiffness Constants

For the cubic point groups, the third-order elastic stiffness tensor has six independent components: $c_{111}$ ($= c_{222} = c_{333}$), $c_{112}$ ($= c_{223} = c_{133} = c_{113} = c_{122} = c_{233}$), $c_{123}$, $c_{144}$ ($= c_{255} = c_{366}$), $c_{166}$ ($= c_{155} = c_{244} = c_{344} = c_{266} = c_{355}$) and $c_{456}$. Elastic stiffness constant versus pressure data allow the determination of the third-order elastic constants. Nichols et al [4] obtained these values from their measured pressure-dependent data for InP, as listed in TABLE 4.

TABLE 4 Third-order elastic stiffness constants for InP (in GPa).

| $c_{111}$ | $c_{112}$ | $c_{123}$ | $c_{144}$ | $c_{166}$ | $c_{456}$ |
|---|---|---|---|---|---|
| -860 | -185 | -510 | -650 | 160 | -4.2 |

## C YOUNG'S MODULUS, POISSON'S RATIO, ETC.

Young's modulus Y and Poisson's ratio P are not isotropic in the cubic, zinc-blende crystals. In TABLE 5 we list the functional expressions of Y and P for directions within the three important (100), (110) and (111) planes in the cubic, zinc-blende lattice.

TABLE 5 Functional expressions for Young's modulus (Y) and Poisson's ratio (P) within the three crystallographic planes (100), (110) and (111) in the cubic, zinc-blende lattice. $S = [s_{11} - s_{12} - (1/2)s_{44}]$. m is the direction for a longitudinal stress and n is the direction for a transverse strain (orthogonal to the direction m).

| Y, P | Functional expression |
|---|---|
| Y (100) plane | |
| [001] direction | $1/s_{11}$ |
| [011] direction | $1/[s_{11} - (1/2)S]$ |
| Y (110) plane | |
| [001] direction | $1/s_{11}$ |
| [111] direction | $1/[s_{11} - (2/3)S]$ |
| Y (111) plane | $1/[s_{11} - (1/2)S]$ |
| P (100) plane | |
| m = [010], n = [001] | $-s_{12}/s_{11}$ |
| m = [011], n = $[0\bar{1}1]$ | $-[s_{12} + (1/2)S]/[s_{11} - (1/2)S]$ |
| P (110) plane | |
| m = [001], n = $[1\bar{1}0]$ | $-s_{12}/s_{11}$ |
| m = $[1\bar{1}1]$, n = $[11\bar{2}]$ | $-[s_{12} + (1/3)S]/[s_{11} - (2/3)S]$ |
| P (111) plane | $-[s_{12} + (1/6)S]/[s_{11} - (1/2)S]$ |

In TABLE 6 we list the numerical values of Y and P for directions within the three important planes (100), (110) and (111) as calculated using the best values of $s_{ij}$ listed in TABLE 2.

TABLE 6 Young's modulus (Y) and Poisson's ratio (P) within the three important planes (100), (110) and (111) for InP at 300 K. Y is in GPa, m is the direction for a longitudinal stress and n is the direction for a transverse strain (orthogonal to the direction m).

| Y, P | Value |
|---|---|
| Y (100) plane | |
| [001] direction | 61 |
| [011] direction | 92 |
| Y (110) plane | |
| [001] direction | 61 |
| [111] direction | 110 |
| Y (111) plane | 92 |
| P (100) plane | |
| m = [010], n = [001] | 0.36 |
| m = [011], n = $[0\bar{1}1]$ | 0.037 |
| P (110) plane | |
| m = [001], n = $[1\bar{1}0]$ | 0.36 |
| m = [111], n = $[11\bar{2}]$ | 0.25 |
| P (111) plane | 0.37 |

The bulk (B) and shear moduli ($C_s$) are, respectively, given by

$$B = (c_{11} + 2c_{12})/3 \tag{4}$$

$$C_s = (c_{11} - c_{12})/2 \tag{5}$$

The isotropy factor A can be given by

$$A = (c_{11} - c_{12})/2c_{44} \tag{6}$$

Note that A = 1.0 for a perfectly isotropic material.

The compressibility $C_o$ is now given by

$$C_o = [(c_{11} + 2c_{12})/3]^{-1} \tag{7}$$

The Cauchy ($C_a$) and Born ($B_o$) ratios can be defined by

$$C_a = c_{12}/c_{44} \tag{8}$$

$$B_o = (c_{11} + c_{12})^2/[4c_{11}(c_{11} - c_{44})] \tag{9}$$

We list in TABLE 7 the calculated results for B, $C_s$, $C_o$, $C_a$ and $B_o$ of InP using the best values of $c_{ij}$ in EQN (1).

TABLE 7 Various elastic moduli for InP at 300 K.

| Elastic modulus | Value |
|---|---|
| B (GPa) | 72.3 |
| $C_s$ (GPa) | 22.5 |
| A | 0.51 |
| $C_o$ ($GPa^{-1}$) | 0.0138 |
| $C_a$ | 1.30 |
| $B_o$ | 1.07 |

## D SOUND VELOCITIES

If the crystal density d and stiffness constant $c_{ij}$ of solids are known, one can easily calculate the bulk sound velocity v from the general relation:

$$v = (c_{ij}/d)^{1/2} \tag{10}$$

If we neglect non-linear terms in the equation of motion, pure longitudinal sound waves may propagate in the [100], [110] and [111] directions. Pure transverse waves may propagate not only in these directions, but also in the directions [hkl] (h = k; l arbitrary) and [hk0] (h and k arbitrary). We list in TABLE 8 expressions for the elastic constants $c_{ij}$ which control the sound velocities (see EQN (10)) propagating along the three major directions [100], [110] and [111] in a cubic, zinc-blende lattice.

TABLE 9 lists the calculated sound velocities propagating along the [100], [110] and [111] directions in InP. In the calculation, we used the crystal density d = 4.79 g/cm$^3$ [10].

TABLE 8 Propagation direction, direction or plane of polarisation, $c_{ij}$ (see EQN (10)) and sound wave mode (longitudinal acoustic (LA) and transverse acoustic waves (TA1 and TA2)) in cubic, zinc-blende lattice.

| Propagation direction | Direction or plane of polarisation | $c_{ij}$ | Wave mode |
|---|---|---|---|
| [100] | <100> | $c_{11}$ | LA |
| [100] | (100) | $c_{44}$ | TA1, TA2 |
| [110] | [110] | $(c_{11} + c_{12} + 2c_{44})/2$ | LA |
| [110] | (110) | $(c_{11} - c_{12})/2$ | TA1 |
| [110] | (001) | $c_{44}$ | TA2 |
| [111] | [111] | $(c_{11} + 2c_{12} + 4c_{44})/3$ | LA |
| [111] | (111) | $(c_{11} - c_{12} + c_{44})/3$ | TA1, TA2 |

TABLE 9 Calculated sound velocities in InP at 300 K.

| Propagation direction | Wave mode | Sound velocity ($10^5$ cm/s) |
|---|---|---|
| [100] | LA | 4.62 |
| [100] | TA1, TA2 | 3.04 |
| [110] | LA | 5.09 |
| [110] | TA1 | 2.16 |
| [110] | TA2 | 3.04 |
| [111] | LA | 5.23 |
| [111] | TA1, TA2 | 2.49 |

## E CONCLUSION

We have discussed the elastic and acoustic properties of InP. The ultrasound technique is the most accurate method for the determination of the elastic stiffness constants, but the measured data show variations in $c_{11}$, $c_{12}$ and $c_{44}$ of 1%, 4% and 11%, respectively. Such variations are probably due to experimental error, not to different crystals used by different authors. Various elastic moduli such as Young's modulus, Poisson's ratio, and the bulk modulus of InP have been obtained. Numerical information about the sound velocities propagating along the three major directions [100], [110] and [111] in InP has also been presented. These data can be used in a general discussion of the elastic and acoustic properties of InP.

## REFERENCES


[1] S. Adachi [ *Physical Properties of III-V Semiconductor Compounds: InP, InAs, GaAs, GaP, InGaAs, and InGaAsP* (Wiley-Interscience, New York, 1992) ]

[2] A.D. Prins, D.J. Dunstan [ in *Properties of Indium Phosphide* EMIS Datareviews Series No.6 (INSPEC, IEE, London, UK, 1991) ch.1 p.8-9 ]

[3] F.S. Hickernell, W.R. Gayton [ *J. Appl. Phys. (USA)* vol.37 (1966) p.462 ]

[4] D.N. Nichols, D.S. Rimai, R.J. Sladek [ *Solid State Commun. (USA)* vol.36 (1980) p.667-9 ]

[5] D. Gerlich, M. Wolf [ in *High Pressure Science and Technology*, vol.1, Eds B. Vodar, Ph. Marteau (Pergamon, Oxford, 1980) p.506-11 ]

[6] P.H. Borcherds, K. Kunc [ *J. Phys. C, Solid State Phys. (UK)* vol.11 (1978) p.4145-55 ]

[7] G. Garlotti, D. Fioretto, L. Palmieri, G. Socino, A. Verdini, C. Rigo [ *J. Phys., Condens. Matter (UK)* vol.8 (1996) p.2265-72 ]

[8] S. Adachi [ *GaAs and Related Materials: Bulk Semiconducting and Superlattice Properties* (World Scientific, Singapore, 1994) ]

[9] A.S. Jordan [ *J. Cryst. Growth (Netherlands)* vol.71 (1985) p.559-65 ]

[10] S. Adachi [ Datareview in this book: *2.1 Density of InP* ]

# 2.4 Piezoelectric constants of InP

S. Adachi

April 1998

## A INTRODUCTION

The III-V compounds crystallising in the zinc-blende structure are the simplest crystals lacking a centre of symmetry and, hence, capable of exhibiting piezoelectric and related effects depending on polar symmetry. The zinc-blende crystals, like InP and GaAs, possess only one piezoelectric tensor component ($e_{14} = e_{1,23}$ or $d_{14} = d_{1,23}$) [1].

## B PIEZOELECTRIC STRESS CONSTANT

Piezoelectricity is the generation of electronic polarisation by application of stress to a crystal lacking a centre of symmetry. It is important especially for the analysis of transport properties in semiconductors [2]. In TABLE 1 we list the experimental and theoretical piezoelectric stress constant $e_{14}$ of InP [3-8].

TABLE 1 Piezoelectric stress constant $e_{14}$ of InP.

| $e_{14}$ (C/m$^2$) | | Ref |
|---|---|---|
| Measured | Calculated | |
| \|0.040\| ± 0.002 | | Boyle and Sladek [3] |
| -0.083 | | Rottner et al [4] |
| | -0.035 | Phillips and Van Vechten [5] |
| | -0.115 | Hubner [6] |
| | -0.093 | Miura et al [7] |
| | +0.056 | Gironcoli et al [8] |

Arlt and Quadflieg [9] determined piezoelectric constants for a variety of semiconductors by piezoelectric Hall-effect measurements and have proposed the macroscopic origins of piezoelectricity as being due to effects such as ionic polarisation, strain-dependent ionicity and electronic polarisation. They did not, however, carry out measurements on InP. Boyle and Sladek [3] measured the attenuation of ultrasonic longitudinal waves travelling in the [110] direction in Cr-doped, n-type InP at 300 K, and determined the piezoelectric constant to be $|e_{14}| = 0.040 \pm 0.002$ C/m$^2$ at 1 bar. Nichols and Sladek [10] also subjected InP to 2 kbar of hydrostatic pressure and found that the pressure did not change the piezoelectric constant significantly ($|e_{14}| = 0.042$ C/m$^2$ at 2 kbar). More recently, Rottner et al [4] have obtained the value of $e_{14} = -0.083$ C/m$^2$ ($d_{14} = -1.8 \times 10^{-12}$ m/V) at 300 K. The InP samples used were grown by the liquid-encapsulated Czochralski method and were doped with Fe at concentrations of about $10^{17}$ cm$^{-3}$. The sign of the constant was determined by qualitative static measurements of the piezovoltage by compressing the crystal.

The piezoelectric scattering of electrons in semiconductors is usually not of major practical importance, except in high-quality crystals, because of impurity scattering. At low temperatures, however, lattice scattering of electrons is known to be dominated by the piezoelectric interaction, which causes elastic scattering due to the relatively low-energy acoustic phonons. Rode [11] has used the value of $|e_{14}| = 0.035$ C/m$^2$ to analyse electron-transport properties of InP. This value corresponds to that predicted by Phillips and Van Vechten [5] based on their ionicity model ($e_{14} = -0.035$ C/m$^2$).

Many theoretical works [5-8,12-18] have been performed to understand piezoelectric properties of crystals. The plots of experimental $e_{14}$ versus Phillip's ionicity $f_i$ for some of the III-V and II-VI compounds showed that $e_{14}$ passes through zero at $f_i \sim 0.5$ while undergoing a reversal in sign (smaller $f_i$ values of III-V compounds possess negative $e_{14}$ numbers, but larger $f_i$ values of II-VI compounds possess positive ones) [19]. Hubner [6] has predicted the $e_{14}$ values for a number of the III-V, II-VI and I-VII compounds and found them to be in good numerical agreement (with correct sign) with experimental data. He gave the value of $e_{14} = -0.115$ C/m$^2$ for InP. Similar considerations by Miura [7] gave $e_{14} = -0.093$ C/m$^2$ for this material.

More recently, Gironcoli et al [8] have presented an ab initio approach to piezoelectricity in semiconductors. The piezoelectric tensor was given in their model by the stress induced by a homogeneous electric field, and the perturbation was treated self-consistently by linear response, thus avoiding both supercells and numerical differentiation. They calculated $e_{14}$ of nine III-V binary semiconductors (Al,Ga,In)(P,As,Sb) and found that, contrary to common belief, two III-V semiconductors AlP and InP have positive $e_{14}$ values. The $e_{14}$ value of InP obtained was +0.056 C/m$^2$ (see TABLE 1).

## C PIEZOELECTRIC STRAIN AND ELECTROMECHANICAL COUPLING CONSTANTS

The piezoelectric strain component $d_{k,ij}$ is connected reciprocally with the stress component $e_{k,mn}$ through

$$d_{k,ij} = \sum_{m,n} s_{ijmn} e_{k,mn} \tag{1}$$

where $s_{ijmn}$ is the elastic compliance constant. In the case of the zinc-blende crystals, the piezoelectric tensor [d] can be expressed only by the component $d_{14}$:

$$d_{14} = s_{44} e_{14} \tag{2}$$

It is known [20] that injected ultrasonic waves can be amplified in piezoelectric semiconductors by the application of a sufficiently high electric field because of the strong interaction with mobile charge carriers. Domains of intense acoustic flux, showing a broad band of frequencies in the low-GHz range, can also be produced in piezoelectric semiconductors, such as GaAs, by acoustoelectric amplification of phonons from the thermal equilibrium spectrum [21]. The gain of such acoustoelectric interactions can be explained in terms of the material parameter $K^2$, the so-called electromechanical coupling constant. The carrier mobilities in semiconductors are also strongly affected by this material parameter [2]. In the zinc-blende type crystals, the maximum coupling of the transverse acoustic waves occurs along the (110) direction. The coupling constant in this case can be written as

$$K_{14}^2 = e_{14}^2 / (\varepsilon_s \varepsilon_0 c_{44}) \tag{3}$$

where $e_{14}$ is the piezoelectric stress constant, $\varepsilon_s\varepsilon_0$ is the electronic permittivity for constant strain and $c_{44}$ is the elastic stiffness constant. Using the values of $e_{14}$ [3,4], $c_{44} = 4.60 \times 10^{10}$ N/m$^2$ ($s_{44} = 1/c_{44}$) [22] and $\varepsilon_s = 12.4$ [23], we can calculate $d_{14}$ and $K_{14}^2$ for InP. The results are listed in TABLE 2.

TABLE 2 Piezoelectric strain constant $d_{14}$ and electromechanical coupling constant $K_{14}^2$ for InP as calculated from the values of $e_{14}$ in [3,4].

| $d_{14}$ ($10^{-13}$ m/V) | $K_{14}^2$ | Ref |
|---|---|---|
| $\lvert 8.70 \rvert$ | $3.17 \times 10^{-4}$ | Boyle and Sladek [3] |
| -18.0 | $1.36 \times 10^{-3}$ | Rottner et al [4] |

## D CONCLUSION

The available experimental and theoretical data on the piezoelectric constants of InP have been reviewed. We hope that highlighting the scarcity of such data will encourage further work in the field.

## REFERENCES


[1] J.F. Nye [ *Physical Properties of Crystals* (Clarendon, Oxford, 1972) ]
[2] D.L. Rode [ *Semicond. Semimet. (USA)* vol.10 (1975) ]
[3] W.F. Boyle, R.J. Sladek [ *Solid State Commun. (USA)* vol.16 (1975) p.323-6 ]
[4] K. Rottner, R. Helbig, G. Muller [ *Appl. Phys. Lett. (USA)* vol.62 (1993) p.352-3 ]
[5] J.C. Phillips, J.A. Van Vechten [ *Phys. Rev. Lett. (USA)* vol.23 (1969) p.1115-7 ]
[6] K. Hubner [ *Phys. Status Solidi B (Germany)* vol.57 (1973) p.627-34 ]
[7] M. Miura, H. Murata, Y. Shiro, K. Iishi [ *J. Phys. Chem. Solids (UK)* vol.42 (1981) p.931-6 ]
[8] S.D. Gironcoli, S. Baroni, R. Resta [ *Phys. Rev. Lett. (USA)* vol.62 (1989) p.2853-6 ]
[9] G. Arlt, P. Quadflieg [ *Phys. Status Solidi (Germany)* vol.25 (1968) p.323-30 ]
[10] D.N. Nichols, R.J. Sladek [ *Phys. Rev. B (USA)* vol.24 (1981) p.6161-2 ]
[11] D.L. Rode [ *Phys. Rev. B (USA)* vol.2 (1970) p.1012-24 ]
[12] R.M. Martin [ *Phys. Rev. B (USA)* vol.5 (1972) p.1607-13 ]
[13] R.M. Martin, K. Kunc [ *Phys. Rev. B (USA)* vol.24 (1981) p.2081-8 ]
[14] J.B. McKitterick [ *Phys. Rev. B (USA)* vol.28 (1983) p.7384-6 ]
[15] M.R. Press, D.E. Ellis [ *Phys. Rev. B (USA)* vol.38 (1988) p.3102-11 ]
[16] A. Dal Corso, M. Posternak, R. Resta, A. Baldereschi [ *Phys. Rev. B (USA)* vol.50 (1994) p.10715-21 ]
[17] A.P. Mirgorodsky, M.B. Smirnov, E. Abdelmounim, T. Merle, P.E. Quintard [ *Phys. Rev. B (USA)* vol.52 (1995) p.3993-4000 ]
[18] T. Kamiya [ *Jpn. J. Appl. Phys. (Japan)* vol.35 (1996) p.4421-6 ]
[19] S. Adachi [ *J. Appl. Phys. (USA)* vol.58 (1985) p.R1-R29 ]
[20] N.I. Meyer, M.H. Jorgensen [ *Advances in Solid State Physics* Ed. O. Madelung (Pergamon, Vieweg, 1970) ]
[21] D.L. Spears [ *Phys. Rev. B (USA)* vol.2 (1970) p.1931-51 ]
[22] F.S. Hickernell, W.R. Gayton [ *J. Appl. Phys. (USA)* vol.37 (1966) p.462 ]
[23] S. Adachi [ *J. Appl. Phys. (USA)* vol.53 (1982) p.8775-92 ]

# 2.5 Electro-optic constants of InP

S. Adachi

April 1998

## A INTRODUCTION

A crystal that is optically isotropic becomes birefringent when it is subject to an external electric field (E): this phenomenon is known as the electro-optic effect [1]. The effect linearly proportional to E is called the linear electro-optic (Pockels) effect. The effect affords a convenient and widely used means of controlling the intensity or phase of optical radiation in crystals. This property has been extensively explored and used for a wide variety of optoelectronic devices.

## B GENERAL CONSIDERATIONS

The III-V compounds crystallising in the zinc-blende structure are the simplest crystals lacking a centre of symmetry and, hence, capable of exhibiting linear electro-optic and related effects. In order to describe the linear electro-optic effect rigorously, we must take into account a possible piezoelectrically induced elasto-optic (photoelastic) contribution to the impermeability change [2,3]. The electro-optic constant in this case can be given by

$$r_{ij,k}^{T} = r_{ij,k}^{S} + r_{ij,k}^{P} \qquad (1)$$

where $r_{ij,k}^{T}$ is the free value determined at constant stress (e.g. by making a measurement at low frequencies well below acoustic resonances of the sample), $r_{ij,k}^{S}$ is the clamped value determined at constant strain (e.g. at frequencies well above the acoustic resonances of the sample) and $r_{ij,k}^{P}$ is the elasto-optic contribution (piezoelectrically induced) to the impermeability change.

The zinc-blende crystals, like InP and GaAs, possess only one electro-optic tensor component $r_{41} = r_{23,1}$ [1]. The elasto-optic effect can now be written in a zinc-blende family as

$$r_{41}^{P} = p_{44}s_{44}e_{14} \qquad (2)$$

where $p_{44}$ is the photoelastic constant, $s_{44}$ is the elastic compliance constant and $e_{14}$ is the piezoelectric constant. It is noted that for many III-V and II-VI semiconductors $r_{41}^{P}$ is much smaller than $r_{41}^{S}$ in a wavelength region below the band edge (i.e. $r_{ij,k}^{T} \sim r_{ij,k}^{S}$) [2] (see also TABLE 2 below).

## C EXPERIMENTAL DATA

The linear electro-optic constants of InP have been determined by Tada and Suzuki [4,5]. TABLE 1 summarises their obtained values. These data suggest that there should be little wavelength dependence within this experimental range ($\lambda$ = 1.06 - 1.50 $\mu$m).

TABLE 1 Linear electro-optic constants $r_{41}^T$ and $r_{41}^S$ for InP.

| Wavelength (μm) | $r_{41}^T$ ($\times 10^{-12}$ m/V) | $r_{41}^S$ ($\times 10^{-12}$ m/V) | Ref |
|---|---|---|---|
| 1.06 | -1.45 | | [4] |
| 1.064 | -1.32 | -1.34 | [5] |
| 1.208 | -1.49 | -1.54 | [5] |
| 1.306 | -1.53 | -1.59 | [5] |
| 1.35 | -1.3 | | [4] |
| 1.50 | -1.63 | -1.68 | [5] |

## D THEORETICAL VALUES

The electro-optic tensor $r_{ij,k}$ is a complex third-rank tensor. Since the imaginary part of the dielectric impermeability may be taken as zero in a wavelength region near and below the fundamental absorption edge, the tensor component can be recognised as a real physical constant with a good approximation.

A microscopic tensor theory for calculating the linear electro-optic constant in the transparent region of solids was proposed by DiDomenico and Wemple [6]. Garrett [7] also presented a one-dimensional anharmonic oscillator model to understand a number of non-linear optical phenomena. Garrett's model has been extended by Sugie and Tada [8] to the three-dimensional oscillator so that it is applicable to materials with arbitrary crystal structures. Adachi and Oe [2,3,9] also proposed a method for the calculation of the linear and quadratic electro-optic constants based on a simplified model of the interband transitions. More recently, Hernandez-Cabrera et al [10] have used a tight-binding scheme for computing the linear electro-optic constant $r_{41}$ in the transparent region of various zinc-blende semiconductors, including InP.

We show in TABLE 2 and FIGURE 1 the linear electro-optic constants $r_{41}^S$ and $r_{41}^P$ for InP as calculated from Adachi's model [2,3]. It is understood that the $|r_{41}^S|$ ($|r_{41}^P|$) value begins to increase steeply toward the band edge ($E_0 \sim 1.35$ eV). It is also clear that $|r_{41}^P|$ is much smaller than $|r_{41}^S|$, hence $r_{ij,k}^T \sim r_{ij,k}^S$.

TABLE 2 Calculated $r_{41}^S$ and $r_{41}^P$ values for InP.

| Photon energy (eV) | Wavelength (μm) | $r_{41}^S$ ($\times 10^{-12}$ m/V) | $r_{41}^P$ ($\times 10^{-14}$ m/V) |
|---|---|---|---|
| 0.50 | 2.48 | -1.32 | 7.41 |
| 0.60 | 2.07 | -1.33 | 7.14 |
| 0.70 | 1.77 | -1.33 | 6.79 |
| 0.80 | 1.55 | -1.34 | 6.34 |
| 0.90 | 1.38 | -1.35 | 5.75 |
| 1.00 | 1.24 | -1.37 | 4.97 |
| 1.10 | 1.13 | -1.42 | 3.84 |
| 1.20 | 1.03 | -1.53 | 2.02 |
| 1.30 | 0.953 | -1.98 | -2.31 |
| 1.31 | 0.946 | -2.12 | -3.28 |
| 1.32 | 0.939 | -2.33 | -4.58 |
| 1.33 | 0.932 | -2.68 | -6.54 |
| 1.34 | 0.925 | -3.51 | -10.4 |

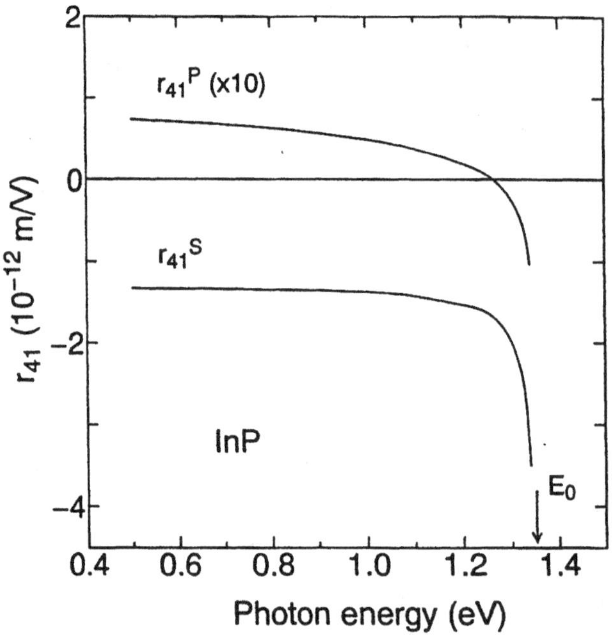


FIGURE 1 Calculated lineshapes of $r_{41}^S$ and $r_{41}^P$ for InP. The vertical arrow indicates the position of the fundamental absorption edge ($E_0 \sim 1.35$ eV).

## E FRANZ-KELDYSH EFFECT

As the incident photon energy approaches the $E_0$ edge, the change in refractive index associated with the Franz-Keldysh (FK) effect will begin to dominate over the linear electro-optic effect. The FK effect is a field-induced extension of the low-energy absorption tail. The corresponding change in the refractive index is positive at below-$E_0$ energies and is non-linear with an applied field E. The FK effect in InP has been studied both theoretically [11] and experimentally [12,13].

## F CONCLUSION

We have reviewed here the linear electro-optic effect in InP. The clamped electro-optic constant should be $r_{41}^S \sim (-1.45 \pm 0.2) \times 10^{-12}$ m/V for incident photon energies far from the fundamental absorption edge $E_0$. There are no measured values near the $E_0$-edge resonance. Theoretical predictions show that $|r_{41}^S|$ ($|r_{41}^T|$) begins to increase steeply toward the $E_0$ edge.

## REFERENCES

[1] J.F. Nye [*Physical Properties of Crystals* (Clarendon, Oxford, 1964) ]
[2] S. Adachi [*J. Appl. Phys. (USA)* vol.56 (1984) p.74-80 ]
[3] S. Adachi [*J. Appl. Phys. (USA)* vol.56 (1984) p.1499-504 ]
[4] K. Tada, N. Suzuki [*Jpn. J. Appl. Phys. (Japan)* vol.19 (1980) p.2295-6 ]
[5] N. Suzuki, K. Tada [*Jpn. J. Appl. Phys. (Japan)* vol.23 (1984) p.291-5 ]
[6] M. DiDomenico Jr., S.H. Wemple [*J. Appl. Phys. (USA)* vol.40 (1969) p.720-34 ]
[7] C.G.B. Garrett [*IEEE J. Quantum Electron. (USA)* vol.QE-4 (1968) p.70-84 ]
[8] M. Sugie, K. Tada [*Jpn. J. Appl. Phys. (Japan)* vol.15 (1976) p.257-63 ]
[9] S. Adachi [*J. Appl. Phys. (USA)* vol.72 (1992) p.3702-4 ]

[illegible]-Cabrera, C. Tejedor, F. Meseguer [ *J. Appl. Phys. (USA)* vol.58 (1985) p.4666-9 ]

[illegible] Bennett, R.A. Soref [ *IEEE J. Quantum Electron. (USA)* vol.QE-23 (1987) p.2159-66 ]

[12] S. Aytac, A. Schlachetzki [ *J. Opt. Commun. (Germany)* vol.6 (1985) p.82-8 ]

[13] T.E. Van Eck, L.M. Walpita, W.S.C. Chang, H.H. Wieder [ *Appl. Phys. Lett. (USA)* vol.48 (1986) p.451-3 ]

# 2.6 Thermal expansion, thermal conductivity and specific heat of InP

S. Adachi

April 1998

## A INTRODUCTION

Investigation of the thermal properties of solids is an old topic which arises in strong connection with the fundamental physical properties of the solids. We review here the thermal expansion coefficient, thermal conductivity and specific heat of InP.

## B THERMAL EXPANSION COEFFICIENT

The thermal expansion coefficient is a second-rank symmetric tensor ([α]) relating the temperature T and the second-rank strain tensor [e] by

$$[e] = [\alpha]T \tag{1}$$

The thermal expansion coefficient is usually determined by measuring the temperature dependence of the lattice constant. Experimental evaluations of the coefficient for InP have been made below room temperature by Sirota et al [1], Deus et al [2] and Haruna et al [3] and above room temperature by Bernstein and Beals [4], Kudman and Paff [5], Bisaro et al [6], Glazov et al [7], Sugii et al [8] and Bak-Misiuk et al [9]. We show in TABLE 1 and FIGURE 1 the α values for InP for temperatures below 300 K [3,10]. It is understood from these data that α decreases from its positive value on cooling, passing through zero for T ~ 80 K. The negative character of α is recognised for temperatures between 80 and 15 K. Kagaya and Soma [11] predicted theoretically a negative thermal expansion below 75 K using lattice dynamic and homogeneous deformation methods.

TABLE 1 Thermal expansion coefficient α of InP below 300 K [3,10].

| T (K) | α ($10^{-6}$/K) | T (K) | α ($10^{-6}$/K) | T (K) | α ($10^{-6}$/K) |
|---|---|---|---|---|---|
| 10 | 0.20 | 90 | 0.32 | 200 | 3.38 |
| 15 | 0.00 | 100 | 0.68 | 210 | 3.53 |
| 20 | -0.23 | 110 | 1.05 | 220 | 3.69 |
| 25 | -0.33 | 120 | 1.31 | 230 | 3.84 |
| 30 | -0.48 | 130 | 1.60 | 240 | 3.97 |
| 35 | -0.53 | 140 | 1.94 | 250 | 4.08 |
| 40 | -0.53 | 150 | 2.28 | 260 | 4.24 |
| 45 | -0.48 | 160 | 2.54 | 270 | 4.35 |
| 50 | -0.44 | 170 | 2.78 | 280 | 4.40 |
| 60 | -0.35 | 180 | 3.01 | 290 | 4.49 |
| 70 | -0.16 | 190 | 3.21 | 300 | 4.56 |
| 80 | 0.03 | | | | |

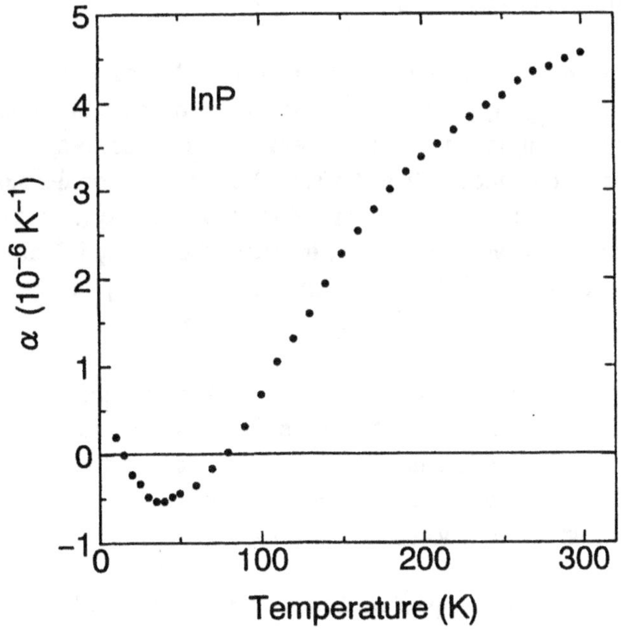


FIGURE 1 Thermal expansion coefficient α for InP below 300 K.
The experimental data are taken from Haruna et al [3,10].

At high temperatures (>280 K) the thermal expansion coefficient can be assumed to have a constant value. TABLE 2 lists the values found in this range [4-8].

TABLE 2 Thermal expansion coefficient α of InP evaluated by linear approximation.

| $\alpha$ ($10^{-6}$/K) | T (K) | Ref |
|---|---|---|
| 4.6 | 300 - 573 | Bernstein and Beals [4] |
| 4.75 ± 0.10 | 288 - 923 | Kudman and Paff [5] |
| 4.56 ± 0.10 | 298 - 673 | Bisaro et al [6] |
| 3.97 | 300 - 773 | Glazov et al [7] |
| 4.85 ± 0.10 | 293 - 633 | Sugii et al [8] |

A mean value of $4.55 \times 10^{-6}$ $K^{-1}$ is obtained from TABLE 2.

Above 773 K Glazov et al [7] found that α departs from linearity in the sense of increased thermal expansion. Their α values are listed in TABLE 3.

TABLE 3 Thermal expansion coefficient of InP above 550 K [7].

| T (K) | $\alpha$ ($10^{-6}$/K) |
|---|---|
| 550 | 4.5 |
| 650 | 5.4 |
| 750 | 6.2 |
| 850 | 7.0 |
| 950 | 7.8 |

At high temperatures the thermal expansion coefficient depends on dopants and their concentration [12]. For the dopants studied [8], the expansion coefficient is increased and, in some cases, decreased.

## C THERMAL CONDUCTIVITY

Thermal conductivity (K), or thermal resistivity ($K^{-1}$), results essentially from interactions between phonons and from the scattering of phonons by crystalline imperfections. Knowledge of the thermal conductivity of semiconductors forms an important part in the design of power-dissipating devices, such as diodes, transistors and optoelectronic devices (laser and light-emitting diodes). Numerical thermal-conductivity data are also necessary in calculating the figure of merit for thermoelectric devices (e.g. Peltier devices). A useful description of the theoretical and practical aspects of the thermal conductivity in semiconductors is given by Maycock [13], Holland [14], Bhandari and Rowe [15] and Srivastava [16] who review some of the work done on III-V compounds, including InP and its related alloy semiconductors.

An exact calculation of the lattice thermal conductivity for the III-V compounds is possible in principle, but lack of knowledge of various parameters (e.g. anharmonic forces and lattice vibration spectra) and the difficulty of obtaining an exact solution of phonon-phonon interactions are formidable barriers to progress. Experimental evaluations of K for undoped (or low-doped) InP have been carried out by several authors [17-19]. TABLE 4 lists the room-temperature K values reported for InP.

TABLE 4 Room-temperature thermal conductivity K for InP.

| K (W/cm/K) | Ref |
|---|---|
| 0.68 | Kudman and Steigmeier [17] |
| 0.70 | Aliev et al [18] |
| 0.88 | Both et al [19] |

In most semiconductors, we observe experimentally that the thermal conductivity of a pure single crystal is zero at T = 0 K and rises approximately exponentially to a maximum near 10 - 20 K, falls somewhat faster than $T^{-1}$, and then varies approximately as $T^{-1}$ to the melting temperature. The K values listed in TABLE 5 are taken from Aliev et al [18] (T = 2.5 - 220 K) and Kudman and Steigmeier [17] (T = 300 - 800 K; [13]). The samples used in these studies were coarse-grained polycrystals with carrier densities of $n = 2 \times 10^{16}$ cm$^{-3}$ [18] and $n = 6.3 \times 10^{16}$ cm$^{-3}$ [17] at 77 K. The data in TABLE 5 are plotted in FIGURE 2. We can see that the thermal conductivity K for InP shows a maximum at T ~ 17 K.

TABLE 5 Thermal conductivity K for InP at various temperatures [17,18].

| T (K) | K (W/cm/K) | T (K) | K (W/cm/K) | T (K) | K (W/cm/K) | T (K) | K (W/cm/K) |
|---|---|---|---|---|---|---|---|
| 2.5 | 0.47 | 15.0 | 28.8 | 80 | 4.7 | 450 | 0.36 |
| 2.8 | 0.87 | 16.5 | 31.0 | 86 | 4.3 | 500 | 0.31 |
| 3.0 | 1.23 | 20.5 | 27.0 | 93 | 4.0 | 550 | 0.28 |
| 3.4 | 1.54 | 27 | 22.0 | 120 | 2.6 | 600 | 0.25 |
| 3.6 | 1.84 | 36 | 14.4 | 130 | 2.3 | 650 | 0.22 |
| 4.0 | 2.30 | 41 | 12.5 | 145 | 1.98 | 700 | 0.20 |
| 4.3 | 2.73 | 51 | 10.0 | 220 | 1.14 | 750 | 0.18 |
| 6.0 | 7.00 | 57 | 8.0 | 300 | 0.68 | 800 | 0.17 |
| 7.8 | 12.3 | 63 | 6.6 | 350 | 0.51 | | |
| 10.0 | 18.0 | 73 | 5.4 | 400 | 0.42 | | |

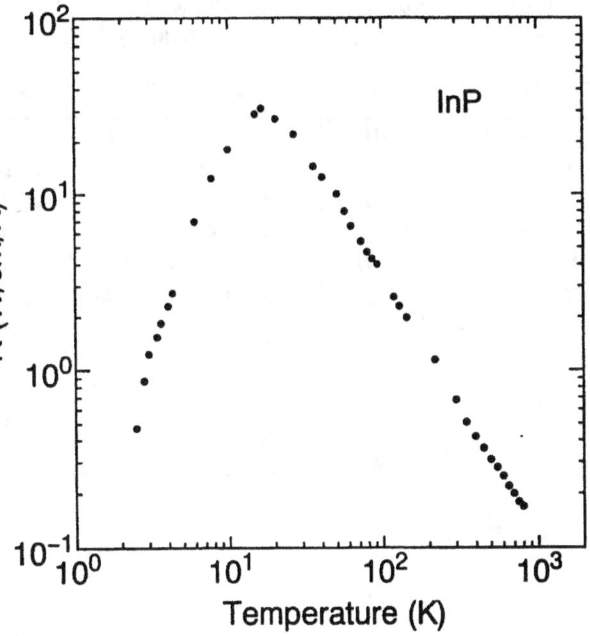


FIGURE 2 Thermal conductivity K for InP. The experimental data are taken from Aliev et al [18] (T = 2.5 - 220 K) and Kudman and Steigmeier [17] (T = 300 - 800 K; [13]).

In doped semiconductors, the total thermal conductivity can be generally given by the sum of the lattice ($K_L$) and electronic ($K_e$) contributions. In a metal, the electronic thermal conductivity $K_e$ and electrical conductivity $\sigma$ are related by the Wiedemann-Franz-Lorentz law:

$$K_e = L\sigma T \tag{2}$$

where L is the Lorentz number. In a semiconductor, a more complicated relationship exists between $K_e$ and $\sigma$ [14].

## D SPECIFIC HEAT

Many practical uses of InP demand a knowledge of the thermal properties over a wide range of temperatures. A major step forward in our knowledge concerns the thermal energy content of a solid. This leads us to one of the most essential thermal properties, the specific heat or heat capacity of a solid [20].

The specific heat of InP below 300 K has been determined by Piesbergen [21] and Sirota et al [22,23] using a calorimetric technique. These data agree and TABLE 6 gives the data of Sirota et al.

The accuracy of the measurements varied from about 3% in the range 5 - 20 K to 0.6% at 160 K. The Debye temperature versus T data were also presented in graphical form in [21-23].

Above 300 K Barin et al [24] recommended (in J/g/at/K):

$$C_p = 20.5 + 0.0073T \quad \text{for} \quad 298 < T < 910 \text{ K} \tag{3a}$$

$$C_p = 27.6 \quad \text{for} \quad 910 < T < 1335 \text{ K} \tag{3b}$$

TABLE 6 Specific heat at constant pressure ($C_p$) and at constant volume ($C_v$) for InP (in J/g/at/K) [22,23].

| T (K) | $C_p$ | $C_v$ | T (K) | $C_p$ | $C_v$ |
|---|---|---|---|---|---|
| 5 | 0.0104 | 0.0104 | 100 | 13.24 | 13.23 |
| 10 | 0.1364 | 0.1364 | 120 | 15.20 | 15.18 |
| 15 | 0.6735 | 0.6735 | 140 | 16.72 | 16.71 |
| 20 | 1.622 | 1.622 | 160 | 18.00 | 17.98 |
| 30 | 3.876 | 3.876 | 180 | 19.07 | 19.04 |
| 40 | 5.819 | 5.819 | 200 | 20.04 | 20.00 |
| 50 | 7.430 | 7.430 | 220 | 20.80 | 20.77 |
| 60 | 8.732 | 8.731 | 240 | 21.40 | 21.36 |
| 70 | 9.917 | 9.912 | 260 | 21.87 | 21.82 |
| 80 | 11.00 | 11.00 | 280 | 22.26 | 22.20 |
| 90 | 12.08 | 12.08 | 300 | 22.56 | 22.50 |

Yamaguchi et al [25] also recommended (in J/g/at/K):

$$C_p = 25.1 + 2.54 \times 10^{-3}T - 2.51 \times 10^{5}T^{-2} \quad \text{for} \quad 800 < T < 910\text{ K} \tag{4a}$$
$$C_p = 26.5 + 0.91 \times 10^{-3}T - 3.38 \times 10^{5}T^{-2} \quad \text{for} \quad 910 < T < 1340\text{ K} \tag{4b}$$

The melting point for stoichiometric InP reported by Yamaguchi et al [25] is 1340 ± 1 K. The $C_p$ values obtained from these studies are plotted in FIGURE 3.

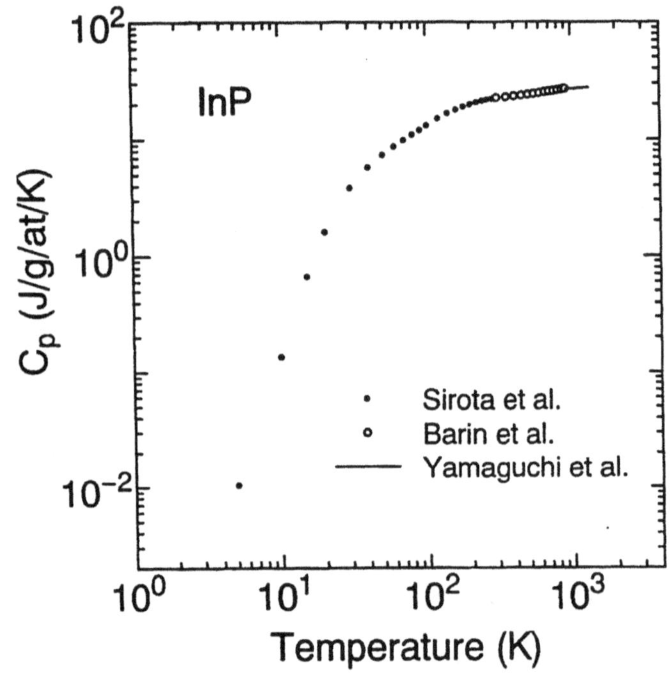


FIGURE 3 Specific heat $C_p$ for InP. The experimental data are taken below 300 K from Sirota et al (solid circles; [22,23]) and above 300 K from Barin et al (open circles; [24]) and Yamaguchi et al (solid line; [25]).

## E CONCLUSION

The available experimental data on the thermal expansion coefficient, thermal conductivity and specific heat of InP have been presented. These quantities are found to depend markedly on temperature. The data presented here can be used in a general discussion of the thermal properties of InP.

## REFERENCES


[1] A.N.N. Sirota, A.M. Antyukhov, A.A. Sidorov [ *Sov. Phys.-Dokl. (USA)* vol.29 (1984) p.662-5 ]

[2] P. Deus, H.A. Schneider, U. Voland, K. Stiehler [ *Phys. Status Solidi A (Germany)* vol.103 (1987) p.443-7 ]

[3] K. Haruna, H. Maeta, K. Ohashi, T. Koike [ *J. Phys. C, Solid State Phys. (UK)* vol.20 (1987) p.5275-9 ]

[4] L. Bernstein, R.J. Beals [ *J. Appl. Phys. (USA)* vol.32 (1961) p.122-3 ]

[5] I. Kudman, R.J. Paff [ *J. Appl. Phys. (USA)* vol.43 (1972) p.3760-2 ]

[6] R. Bisaro, P. Merenda, T.P. Pearsall [ *Appl. Phys. Lett. (USA)* vol.34 (1979) p.100-2 ]

[7] V.M. Glazov, K. Davletov, A.Ya. Nashel'skii, M.M. Mamedov [ *Russ. J. Phys. Chem. (UK)* vol.51 (1977) p.1496-9 ]

[8] K. Sugii, H. Koizumi, E. Kubota [ *J. Electron. Mater. (USA)* vol.12 (1983) p.701-12 ]

[9] J. Bak-Misiuk, J. Wolf, U. Pietsch [ *Phys. Status Solidi A (Germany)* vol.118 (1990) p.209-17 ]

[10] K. Haruna, H. Maeta [in *Properties of Indium Phosphide* EMIS Datareviews Series No.6 (INSPEC, IEE, London, UK, 1991) ch.1 p.16-7 ]

[11] H.-Matsuo Kagaya, T. Soma [ *Solid State Commun. (USA)* vol.58 (1986) p.479-82 ]

[12] U. Pietsch, J. Bak-Misiuk, V. Gottschalch [ *Phys. Status Solidi A (Germany)* vol.82 (1984) p.K137-40 ]

[13] P.D. Maycock [ *Solid-State Electron. (UK)* vol.10 (1967) p.161-8 ]

[14] M.G. Holland [ *Semicond. Semimet. (USA)* vol.2 (1966) p.3-31 ]

[15] C.M. Bhandari, D.M. Rowe [ *Thermal Conduction in Semiconductors* (Wiley, New York, 1988) ]

[16] G.P. Srivastava [ *The Physics of Phonons* (Adam Hilger, Bristol, 1990) ]

[17] I. Kudman, E.F. Steigmeier [ *Phys. Rev. (USA)* vol.133 (1964) p.A1665-7 ]

[18] S.A. Aliev, A.Ya. Nashel'skii, S.S. Shalyt [ *Sov. Phys.-Solid State (USA)* vol.7 (1965) p.1287-8 ]

[19] W. Both, V. Gottschalch, G. Wagner [ *Cryst. Res. Technol. (Germany)* vol.21 (1986) p.K85-7 ]

[20] U. Piesbergen [ *Semicond. Semimet. (USA)* vol.2 (1966) p.49-60 ]

[21] U. Piesbergen [ *Z. Nat.forsch. A (Germany)* vol.18 (1963) p.141-7 ]

[22] A.N.N. Sirota, A.M. Antyukhov, V.V. Novikov, V.A. Fedorov [ *Sov. Phys.-Dokl. (USA)* vol.26 (1981) p.701-3 ]

[23] A.N.N. Sirota, A.M. Antyukhov, V.V. Novikov, A.A. Sidorov [ *Sov. Phys.-Dokl. (USA)* vol.27 (1982) p.754-6 ]

[24] I. Barin, O. Knacke, O. Kubaschewski [ *Thermochemical Properties of Inorganic Substances* (Springer, Berlin, 1977) ]

[25] K. Yamaguchi, K. Itagaki, A. Yazawa [ *J. Jpn. Inst. Met. (Japan)* vol.53 (1989) p.764-70 ]

CHAPTER 3

# ELECTRONIC TRANSPORT PROPERTIES

# 3.1 Electronic transport properties of InP: general remarks

R.K. Ahrenkiel

June 1998

This chapter will detail the electrical transport properties of InP that are described in the literature. These properties include majority-carrier mobility as a function of carrier concentration and temperature. Closely associated with the latter topic are the bulk resistivities as a function of carrier concentration and temperature. These properties are then examined as a function of growth conditions including bulk wafer and the various types of epitaxial growth. The remainder of the chapter discusses minority-carrier properties beginning with the minority-carrier lifetime. Closely associated with the lifetime is the diffusion length for minority carriers. A topic associated with minority-carrier lifetime, the surface recombination velocity, concludes the chapter.

There is scant information on another topic, the minority carrier mobility and diffusivity, because measurement of these quantities is quite complex. By contrast, the majority-carrier mobility is routinely measured by the variation of the Hall effect, which is called the van der Pauw technique [1]. The most common technique for estimating the minority-carrier mobility, $\mu$, is to extrapolate from majority-carrier data using a common ionised impurity concentration and temperature. The diffusivity, D, is related to the minority carrier mobility for non-degenerate semiconductors by means of the Einstein relationship

$$D = kT/q\mu \tag{1}$$

Here, k is the Boltzmann constant, T is the absolute temperature and q is the electronic charge. At room temperature, kT/q is about 0.026 eV.

The chapter is constructed from relevant sections of the 1991 edition of Properties of Indium Phosphide [2] updated and revised by this author. There have been significant additions to the database of knowledge concerning majority carrier properties in this time period. There has also been considerable activity in the study of minority-carrier properties, particularly the minority carrier lifetime. Work on InP-based devices has increased very significantly during the last seven or eight years. A large class of III-V epitaxial devices are grown on InP substrates, lattice-matched $In_{0.53}As_{0.47}As$ being the primary example. These devices are primarily transistors, but include low-bandgap thermophotovoltaic cells. Photovoltaic cells, with applications to space power, have been an active research area using InP junction devices. Finally, high-speed switching devices using InP, doped or ion implanted with 'lifetime killing' impurities, have been a popular topic. This chapter will attempt to provide the latest transport parameters to support this research and development.

## REFERENCES


[1] D.K. Schroder [ *Semiconductor Material and Device Characterization* (John Wiley & Sons Inc., New York, USA, 1990) ]

[2] INSPEC [ *Properties of Indium Phosphide* (IEE, London, UK, 1991) ]

# 3.2 Majority-carrier mobility in InP: general remarks

R.K. Ahrenkiel

June 1998

The majority-carrier mobility (electron mobility in n-type and hole mobility in p-type material) in InP crystals and thin films grown by a variety of techniques is presented. The mobility is the carrier-drift velocity per unit of applied electric field. Using simple statistical theory, the quasiparticles (electrons and holes) undergo momentum-scattering processes by collisions with impurities, defects and lattice vibrations or phonons. One can ascribe a scattering time, $\tau$, to the inverse of the scattering probability of a given defect or process. Simple theory then produces the result for the mobility in terms of $\tau$ and the quasi particle effective mass m*

$$\mu = \frac{q\tau}{m^*} \tag{1}$$

Here, q is the electronic charge. When more than one scattering mechanism is present, the scattering probabilities are additive and the net scattering probability is given by:

$$\left\langle \frac{1}{\tau} \right\rangle = \sum_i \left\langle \frac{1}{\tau_i} \right\rangle \tag{2}$$

where $\tau_i$ is the scattering time for each mechanism. A variant of Mathiessen's rule, which proposes that the total mobility can be deduced from the mobility corresponding to each scattering mechanism, cannot be used unless all scattering relaxation times have the same energy dependence. Therefore in almost all cases:

$$\frac{1}{\mu} \neq \sum_i \frac{1}{\mu_i} \tag{3}$$

The use of Mathiessen's rule leads to significant errors in the total mobility of about 20% for samples with about $10^{16}$ carriers/cm$^3$. Fitting mobility data using an expression like EQN (3) leads to similar errors in parameter extraction. Appropriate modelling of the mobility from the component scattering mechanisms is accomplished by solution of the Boltzmann transport equation using an iterative method or by Monte Carlo simulation, in order to determine

$$\left\langle \frac{1}{\tau} \right\rangle$$

[1,2]. The principal scattering mechanisms are phonons, various types of impurities, and mechanical defects such as dislocations. As phonon scattering is strongly temperature dependent and becomes much weaker at low temperature, measurement as a function of temperature is a common method of separating these mechanisms.

## REFERENCES

[1] Y. Takeda [ in *GaInAsP Alloy Semiconductors* Ed. T.P. Pearsall (Chichester, John Wiley, 1982) ]

[2] M.A. Littlejohn [ in *GaInAsP Alloy Semiconductors* Ed. T.P. Pearsall (Chichester, John Wiley, 1982) ]

# 3.3 Electron and hole mobilities in InP

G.W. Iseler, D. Lancefield and R.K. Ahrenkiel

June 1998

## A OVERVIEW

The general properties of the electron mobility of InP as a function of temperature and doping density can be considered independently of the growth technique used. The growth techniques considered here give information about the electron mobility with respect to developments in the growth procedures, incorporation of background impurities and doping characteristics. The electron mobility, $\mu$, is the velocity per unit electric field, and at low fields is described by $\mu = q\langle\tau\rangle/m^*$, where q is the electronic charge, $\langle\tau\rangle$ is the average scattering time, and m* is the electron effective mass. The principal scattering mechanisms controlling the scattering time in InP are polar-optic and acoustic-phonon scattering at temperatures above about 100 K. Below these temperatures, ionised, neutral impurity, and piezoelectric scattering dominate the mobility. At a given doping concentration, impurity scattering becomes the more dominant mechanism as the temperature decreases. Thus, mobility measurements at 77 K in undoped InP are used as a measure of background impurity content of samples. Measurements of the electron mobility at 300 K and 77 K are used as a standard means of assessing InP as well as being an important consideration in device design. Walukiewicz et al [1] and Takeda and Sasaki [2] have tabulated the electron mobility calculated at temperatures of 300 K and 77 K for a wide range of doping densities and compensation ratios ($r = N_a/N_d$), but with various material parameters. Anderson et al [3] compared the Walukiewicz curves with their own data for samples doped with Si, S, Sn and Se over a range of doping densities from $10^{15}$ to $10^{19}$ $cm^{-3}$. Typical mobilities at 300 K range from 5000 $cm^2V^{-1}s^{-1}$ at carrier densities of $10^{15}$ $cm^{-3}$ to 1000 $cm^2V^{-1}s^{-1}$ at carrier densities of $10^{19}$ $cm^{-3}$. At 77 K, mobilities at similar doping levels were 70,000 $cm^2V^{-1}s^{-1}$ and 1000 $cm^2V^{-1}s^{-1}$, respectively. No significant differences were observed between dopants, although the experimental mobilities are significantly less than the calculated values for doping densities greater than $10^{18}$ $cm^{-3}$. One possible explanation for this discrepancy has been given by Yanchev et al [4] in terms of impurity correlation effects arising during growth. Taguchi and Yamada [5] have pointed out that piezoelectric scattering should also be included in the analysis of data taken at low temperatures. High-temperature mobilities, measured by a number of groups, appear to give anomalously low values despite having high mobilities at low temperature. A number of models for these effects have been proposed, including residual deep donors [6] and non-uniform depletion effects, but as yet no consensus has been reached. Today, high-purity samples are routinely available, grown by a number of techniques [7-9], with mobilities that exceed those given in the tables of Walukiewicz et al. For example, Boud et al [10] have reported extremely high mobilities of 300,000 $cm^2V^{-1}s^{-1}$ at 77 K. The maximum measured mobility was about 400,000 $cm^2V^{-1}s^{-1}$ at temperatures of about 40 K. Calculations of electron transport at high electric fields [11] indicate that some improvement in high field characteristics, principally due to the larger Γ-L and Γ-X separations, agrees well with experiment. Takagi and Sugano [12] have measured the electron mobility in accumulation and inversion-type metal-insulator-semiconductor field effect transistors (MISFETs) on SI and p-type substrates. The mobility was found to depend on the carrier density in the channel, the temperature, and the surface electric field. The mobility was less in the p-type substrates: 600 $cm^2V^{-1}s^{-1}$, as compared to 2000 $cm^2V^{-1}s^{-1}$ in SI substrates.

The mobility is a function of the crystal impurities that come from both the intentional dopants and the unwanted, residual impurities that were present in the starting material or incorporated during growth. Thus, the carrier mobility is a function of material purity and is sometimes used as a diagnostic. In particular, the mobility of undoped semiconductors, at low temperature such as 78 K, is often used for residual impurity evaluation. Nevertheless, a typical range of mobility values versus carrier concentration and temperature is useful from a material and device perspective. Such values are aids in evaluating the purity, and predicting the characteristics, of devices to be made from the material.

The carrier mobility is also decreased by scattering by extended mechanical defects such as dislocations. In this case, the theoretical analysis is somewhat more complex than for point defects. The effect of dislocations on the mobility of InP and GaAs grown on silicon has recently been analysed [13]. As both materials are not lattice-matched to silicon, dislocation densities exceeding $1 \times 10^8$ $cm^{-2}$ are generated in heteroepitaxial growth. These studies indicated practically no reduction in mobility at 300 K, but significant effects were observed at low temperature.

## A1 Electron Mobility in n-Type InP

If InP crystals did not contain electrically active impurities or native defects, they would be semi-insulating, with a calculated room temperature resistivity of order of magnitude $10^9$ ohm cm. Crystals grown without intentional doping are actually n-type because they contain shallow donor impurities, primarily Si and S [14-17], that are ionised at room temperature to produce conduction-band electrons. In such crystals grown by state-of-the-art techniques, the electron concentration is usually in the high $10^{14}$ to low $10^{16}$ $cm^{-3}$ range [13-16,18-23], and the room temperature electron mobility $\mu_n$ is between 3500 and 5000 $cm^2V^{-1}s^{-1}$. The intrinsic mobility is controlled by lattice scattering and is calculated to be 4800 $cm^2V^{-1}s^{-1}$ [24]. (Carrier concentrations are those defined by $1/qR_H$, where $R_H$ is the Hall constant and q is the electronic charge. The carrier mobilities are Hall mobilities defined by $R_H/\rho$, where $\rho$ is the resistivity.)

FIGURE 1 shows the value of electron mobility of InP over a wide range of carrier concentrations and at temperatures of 300 K and 77 K. For each value of $n < 1 \times 10^{17}$ $cm^{-3}$, $\mu_n$ was read from the curve of $\mu_n$ versus n calculated for uncompensated samples by Walukiewicz et al [1]. For this concentration range, the curve of Walukiewicz et al gives an excellent fit to the experimental data for the highest mobility samples at each value of electron concentration. For the higher concentrations, the values of $\mu_n$ in FIGURE 1 were read from an experimental curve based on reported data and our own unpublished results, because the curve of Walukiewicz et al appears to overestimate the values of $\mu_n$ for uncompensated samples. The values of $\mu_n$ at 77 K were read from an experimental curve for $\mu_n$ versus n based on the highest mobility values reported in the literature [24] and obtained in our

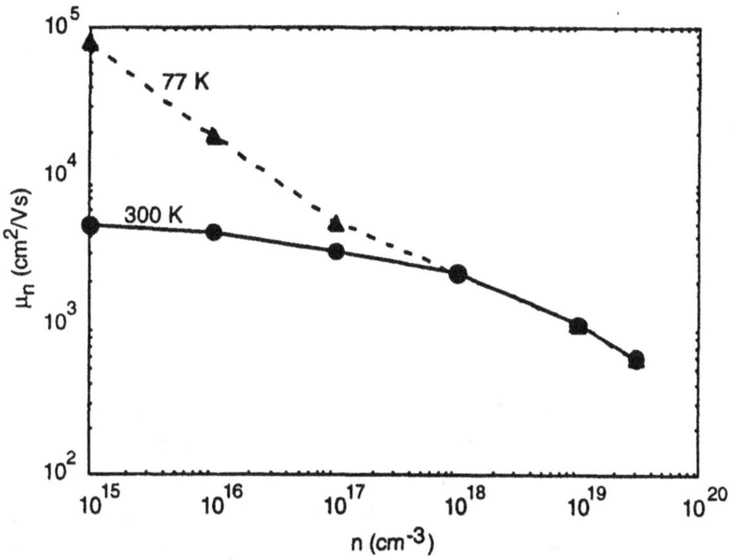


FIGURE 1 The electron mobility versus electron concentration at 77 K and 300 K ambient temperatures.

laboratory. (Note that almost all electrical transport data are reported in terms of n and $\mu_n$, although ρ and $R_H$ are the directly measured quantities.)

Note that for the lightly doped sample ($1 \times 10^{15}$ $cm^{-3}$), the 77 K mobility is calculated as 80,000 $cm^2V^{-1}s^{-1}$ because of the background donor scattering. The lightly doped regions of the curve are sensitive to both background doping levels and residual impurities. Larger values of electron mobility (77 K) will be referenced in this Datareview owing to lower doping levels and low residual impurity levels.

At carrier concentrations greater than $1 \times 10^{18}$ $cm^{-3}$, the mobility is nearly independent of temperature. In this range, the mobility is dominated by ionised impurity scattering from the dopant atoms, and phonon scattering is relatively insignificant.

### A2 Hole Mobility in p-Type InP

Bulk InP with p-type conductivity is obtained by doping with shallow acceptors such as Zn, Cd, Hg or Be, which substitute for In. Acceptors substituting for P have not been found in bulk InP [25].

FIGURE 2 shows the hole mobility in p-type InP with hole concentrations ranging from $1 \times 10^{15}$ $cm^{-3}$ to $8 \times 10^{19}$ $cm^{-3}$. The figure shows the same sample set at 300 K and 77 K. The values of hole mobility $\mu_p$ shown in FIGURE 2 range from approximately 150 $cm^2V^{-1}s^{-1}$ for lattice-scattering to 45 $cm^2V^{-1}s^{-1}$ for ionised impurity scattering in the most heavily doped samples at 300 K. These values were read from a curve for $\mu_p$ versus p based on reported data [26-29] and our own unpublished results.

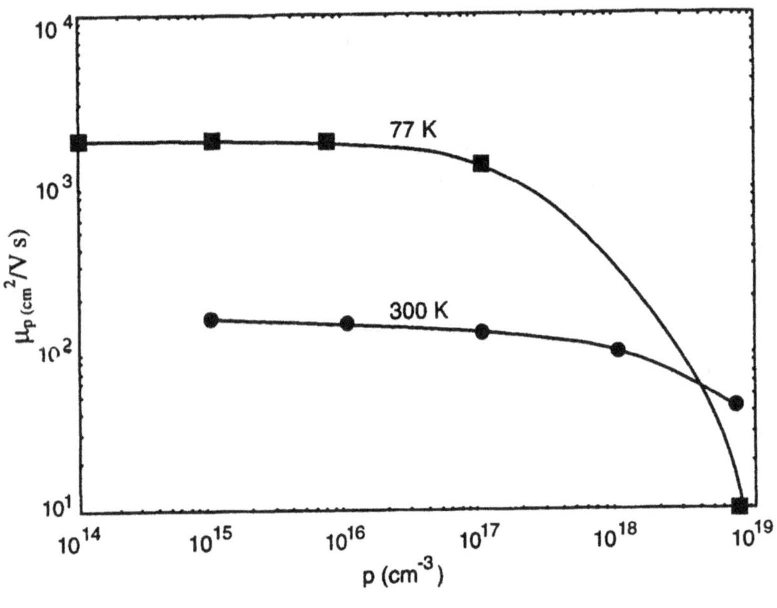


FIGURE 2 The hole mobility versus hole concentration at 77 K and 300 K.

The temperature dependence of mobility for p-type InP differs significantly from that for n-type material because the energy levels of shallow acceptors are somewhat deeper than those of shallow donors. For samples with $p < 1 \times 10^{18}$ $cm^{-3}$ at 300 K, acceptor deionisation decreases the free-hole concentration at 77 K. In FIGURE 2, the values of p decreased by factors of ten to fourteen when the temperature was reduced to 77 K, except for the most heavily doped sample. As shown in FIGURE 2, for $p < 1 \times 10^{17}$ $cm^{-3}$, $\mu_p$ increases by a factor of twelve to fourteen in cooling from 300 to 77 K. For the most highly doped sample, p changes little between 300 and 77 K. However $\mu_p$ decreases by about a factor of four in cooling to 77 K, presumably resulting from transport in the impurity band [30].

## B ELECTRON MOBILITY IN BULK InP

Bulk InP is important as substrate material for both InP-based devices and those based on lattice matched and pseudomorphically strained heterojunction structures for optoelectronic device applications. The primary points of importance are the ability to provide substrates that may be both conducting and semi-insulating (SI) with a large area and low dislocation density. In particular, the SI state in InP can be used to insulate different device components because of the larger bandgap as compared with Si. The material is grown by a number of variations on the basic techniques of liquid-encapsulated Czochralski (LEC) [31] and Bridgman [32]. Nominally undoped InP crystals are n-type with typical electron densities of $10^{15}$ $cm^{-3}$ with corresponding mobilities at room temperature in the range 4000 to 5000 $cm^2V^{-1}s^{-1}$ [33].

A number of approaches have been taken to achieve SI material. InP doped with Fe gives resistivities in the range of 2.1 to 4.6 × $10^7$ ohm cm with mobilities in the range 1700 to 3000 $cm^2V^{-1}s^{-1}$ [34], with the mobility profile describing a V-shape pattern across the diameter of a two-inch wafer. The authors calculate the neutral impurity density to be about $10^{16}$ $cm^{-3}$. By heat-treating the material in a phosphorus atmosphere, Hofmann et al [33] obtain resistivities of up to 2 × $10^7$ ohm cm with mobilities of 1400 $cm^2V^{-1}s^{-1}$, indicating that the mobility in their material was increased due to reduced impurity scattering. Brandt et al [35] reported resistivities in excess of $10^7$ ohm cm by Ti-doping. Although no mobility measurements were reported, the low diffusivity of Ti in InP may be technologically important. The authors claim this to be the first compensation mechanism in III-V materials based on a deep donor impurity.

High-purity polycrystalline InP required for the starting material in LEC is often provided by a Bridgman or gradient-freezing method [36]. The major contaminant is believed to be Si. Pak et al [32] showed that Si contamination could be reduced by modifying the temperature of the phosphorus region; they produced mobilities of approximately 30,000 $cm^2V^{-1}s^{-1}$ at 77 K for a carrier density of 2 × $10^{15}$ $cm^{-3}$. Kubota et al [37] found that synthesised, solute-diffusion grown starting material had mobilities at 77 K in the range 30,000 to 100,000 $cm^2V^{-1}s^{-1}$ with corresponding carrier densities in the range 3.0 to 0.4 × $10^{15}$ $cm^{-3}$. Kubota and Katsui [21], using a ceramic AlN crucible, obtained similar mobilities and carrier densities. The same authors [38] have increased the mobility of LEC-grown material by pre-encapsulating the starting material to reduce impurity incorporation such as Si and C from the carbon-heater and radiation-shield materials. They reported increases in mobility of at least 15% over material grown without pre-encapsulation.

Dowling et al [39] have reported an apparatus for the direct synthesis and growth of polycrystalline InP ingots of 10 to 11 kg. Typical mobilities were approximately 4500 $cm^2V^{-1}s^{-1}$ at room temperature and 22,000 $cm^2V^{-1}s^{-1}$ at 77 K.

## C ELECTRON MOBILITY IN LPE InP

Work by Ip et al [40] was the first to produce reasonably high mobility material grown by liquid phase epitaxy (LPE). They used long bake-out periods to reduce the Si background level in the In, and produced mobilities of approximately 67,000 $cm^2V^{-1}s^{-1}$ at 77 K. Astles et al [41] studied the mobility using a range of dopants producing carrier densities ranging from $10^{16}$ $cm^{-3}$ to 4 × $10^{19}$ $cm^{-3}$, and obtained mobilities of 4000 $cm^2V^{-1}s^{-1}$ and 600 $cm^2V^{-1}s^{-1}$, respectively. No significant difference in the mobility for different dopants was observed. Baumann et al [42] studied high levels of Si doping and found mobilities of 1000 $cm^2V^{-1}s^{-1}$ for n ~ 2 × $10^{19}$ $cm^{-3}$ at 77 K. No amphoteric effects were observed. Groves and Plonko [43] reported mobilities up to 70,000 $cm^2V^{-1}s^{-1}$ by adding $H_2O$ during the extended bake-out. Korber et al [44] have used Yb to remove residual impurities of S, Se, Si, Te and C. These elements form stable compounds with Yb, which are insoluble in In, and can thus be used to reduce the background-impurity concentration. They used Yb mole fractions between 9 × $10^{-5}$ and 2.5 × $10^{-4}$, which they find leads to a maximum impurity reduction without noticeable Yb incorporation into the grown layer. These films produced mobilities of 5700 and 90,000 $cm^2V^{-1}s^{-1}$ at room temperature and 77 K, respectively. Similarly Kondo et al [45] purified an InP solution using Co and obtained 77 K mobilities in the range of 100,000 to 110,000 $cm^2V^{-1}s^{-1}$. Beneking and Emeis [46] have reported on InP-doping with As or Ga. The inclusion of As and Ga expands and contracts the host lattice constant, respectively, and is found to reduce the density of dislocations and deep levels, resulting in an increased electron mobility and improved uniformity across the wafer.

## D ELECTRON MOBILITY OF VPE AND MOVPE InP

### D1 Vapour Phase Epitaxial (VPE) InP

The growth of InP by vapour phase epitaxy (VPE) and especially metal-organic vapour phase epitaxy (MOVPE) has become an area of increased activity over the last twenty years. In hydride VPE, high-purity material has been grown by McCollum et al [47]. They measured 77 K mobilities as high as 125,000 $cm^2V^{-1}s^{-1}$, and identified Si as the main residual impurity. Mobilities measured over a wide range of doping densities for chloride VPE have been analysed by Anderson et al [3]. Flow modulation has also been used with the hydride technique with layer growth rates of 25 Å/cycle. Mobilities at room temperature were 3650 $cm^2V^{-1}s^{-1}$ for an electron density of about $5 \times 10^{15}$ $cm^{-3}$ [48]. Kondo et al [49] grew InP by chloride VPE using group III metal-organic sources. Preliminary results gave mobilities of about 3500 $cm^2V^{-1}s^{-1}$ at room temperature, and about 12,000 $cm^2V^{-1}s^{-1}$ at 77 K for samples grown with no special attention to high purity growth. Teng et al [50] have grown InP on GaAs and report that the layer mobilities and doping profiles are similar to those grown on InP substrates.

### D2 Metal-Organic Vapour Phase Epitaxial (MOVPE) InP

Many published papers per year describing metal-organic vapour phase epitaxy (MOVPE) or metal-organic chemical vapour deposition (MOCVD) refer to electron mobility in InP, indicating a high level of interest in this material and the growth technique. These papers have been reviewed by Gerrard et al [51] and Ludowise [52]. We will consider a number of areas. A variety of different gas sources have been used in the growth of epitaxial InP. Probably the most common is trimethylindium (TMI) and phosphine [53]. However, safety concerns have provided an incentive to replace the highly toxic hydrides with less hazardous alkyl sources, such as isobutylphosphine and tertiarybutylphospine [54]. High-quality epitaxial layers have been grown with 300 K mobilities of about 3800 $cm^2V^{-1}s^{-1}$ and 77 K values of 12,000 $cm^2V^{-1}s^{-1}$ [55]. Solid red-phosphorus has also been used [56], resulting in $\mu \sim 3600$ $cm^2V^{-1}s^{-1}$ at 300 K. Trimethylindium-trimethylamine and dimethyl (3-dimethylaminopropyl) indium have been used as indium sources [57,58]. Both have produced mobilities close to 50,000 $cm^2V^{-1}s^{-1}$ at 77 K. High-purity material has been grown by a number of authors using solid TMI [59] and both low pressure [8] and atmospheric-pressure [60] MOVPE. Measurements at 77 K indicate mobilities in the range of 70,000 to 140,000 $cm^2V^{-1}s^{-1}$. Chen et al [60] give an interesting table comparing the highest mobilities obtained in InP grown by various techniques. Razeghi et al [61] have grown InP at low pressures using TMI; they produced residual doping levels of $3 \times 10^{13}$ $cm^{-3}$ and Hall mobilities at 300 K of 6000 $cm^2V^{-1}s^{-1}$ and at 50 K of 200,000 $cm^2V^{-1}s^{-1}$. Boud et al [10] reported persistent photogenerated carriers in InP, and measured a maximum mobility of about 400,000 $cm^2V^{-1}s^{-1}$ at 40 K. Silicon-doping (from disilane) [62] and Te-doping (from diethyltelluride) [63] have been studied. Under constant gas flow, Si-doped layers are uniform whereas Te-doped layers are tapered indicating a 'memory' effect. Mobilities vary from 10,000 to 1000 $cm^2V^{-1}s^{-1}$ for carrier concentrations from $10^{16}$ to $10^{19}$ $cm^{-3}$, respectively, at 77 K. Semi-insulating InP has been grown by Fe-doping with iron pentacarbonyl [64]. Growth of InP on GaAs has been reported by Horikawa et al [65], with mobilities of 15,000 $cm^2V^{-1}s^{-1}$ and n about $1.3 \times 10^{15}$ $cm^{-3}$ at 77 K. Growth on ZnSe-coated Si has been reported by Lee et al [66]. Mobilities of 3100 $cm^2V^{-1}s^{-1}$ at 300 K are the highest reported to date.

**REFERENCES**


[1] W. Walukiewicz et al [ *J. Appl. Phys. (USA)* vol.51 no.5 (1980) p.2659-68 ]

[2] Y. Takeda, A. Sasaki [ *Solid-State Electron. (UK)* vol.27 no.12 (1984) p.1127-9 ]

[3] D.A. Anderson, N. Apsley, P. Davies, P.L. Giles [ *J. Appl. Phys. (USA)* vol.58 no.8 (1985) p.3059-67 ]

[4] I.Y. Yanchev, B.G. Arnaudov, S.K. Evtimova [ *J. Phys. C (UK)* vol.12 no.19 (1979) p.765 ]

[5] A. Taguchi, S. Yamada [ *J. Appl. Phys. (USA)* vol.61 no.6 (1987) p.2412-5 ]

[6] M. Bezaquen, D. Walsh, K. Mazuruk [ *Phys. Rev. B (USA)* vol.36 no.8 (1987) p.4388-93 ]

[7] L.L. Taylor, D.A. Anderson [ *J. Cryst. Growth (Netherlands)* vol.64 no.1 (1983) p.55-9 ]

[8] M.A. DiForte-Poisson, C. Brylinski, J.P. Duchemin [ *Appl. Phys. Lett. (USA)* vol.46 no.5 (1985) p.476-8 ]

[9] E. Kuphal [ *J. Cryst. Growth (Netherlands)* vol.54 no.1 (1981) p.117-26 ]

[10] J.M. Boud, M.A. Fisher, D. Lancefield, A.R. Adams, E.J. Thrush, C.G. Cureton [ *Inst. Phys. Conf. Ser. (UK)* no.91 (1987) ch.8 p.801-4 ]

[11] J. Costa, A. Peczalski, M. Shur [ *J. Appl. Phys. (USA)* vol.66 no.2 (1989) p.674-9 ]

[12] S. Takagi, T. Sugano [ *J. Appl. Phys. (USA)* vol.62 no.6 (1987) p.2387-91 ]

[13] A. Bartels, E. Peiner, A. Schlachtezki [ *J. Appl. Phys. (USA)* vol.78 (1995) p.6141 ]

[14] S. Shinoyama, C. Uemura, A. Yamamoto, S. Tohno [ *J. Electron. Mater. (USA)* vol.10 no.5 (1981) p.941-56 ]

[15] M. Gauneau, R. Chaplain, A. Rupert, R. Coquille, Y. Toudic, G. Grandpierre [ *J. Cryst. Growth (Netherlands)* vol.76 no.1 (1986) p.128-34 ]

[16] G. Muller, J. Pfannenmuller, E. Tomzig, J. Volkl, F. Kohl [ *J. Cryst. Growth (Netherlands)* vol.64 no.1 (1983) p.37-9 ]

[17] P.J. Dean, M.S. Skolnick, B. Cockayne, W.R. MacEwan, G.W. Iseler [ *J. Cryst. Growth (Netherlands)* vol.67 no.3 (1984) p.1186-94 ]

[18] B. Cockayne, G.T. Brown, W.R. MacEwan [ *J. Cryst. Growth (Netherlands)* vol.54 no.1 (1981) p.9-15 ]

[19] W.A. Bonner [ *J. Cryst. Growth (Netherlands)* vol.54 no.1 (1981) p.22-31 ]

[20] T. Inoue, H. Shimakura, K. Kainosho, R. Hirano, O. Oda [ *J. Electrochem. Soc. (USA)* vol.137 no.4 (1990) p.1283-7 ]

[21] E. Kubota, A. Katsui [ *Jpn. J. Appl. Phys. 2 (Japan)* vol.24 no.2 (1985) p.L69-72 ]

[22] G.W. Iseler [ *J. Cryst. Growth (Netherlands)* vol.54 no.1 (1981) p.16-20 ]

[23] R.L. Henry, E.M. Swiggard [ *J. Electron. Mater. (USA)* vol.7 no.5 (1978) p.647-57 ]

[24] P.S. Whitney, K. Uwai [ *J. Appl. Phys. (USA)* vol.63 no.5 (1988) p.1585-90 ]

[25] S.J. Pearton, U.K. Chakrabarti, C.R. Abernathy, W.S. Hobson [ *Appl. Phys. Lett. (USA)* vol.55 no.19 (1989) p.2014-6 ]

[26] M. Morioka, K. Tada, S. Akai [ *Annu. Rev. Mater. Sci. (USA)* vol.17 (1987) p.75 ]

[27] R.E. Nahory, A.A. Ballman, H. Brown, M.R. Wilson [ *Inst. Phys. Conf. Ser. (UK)* no.65 (1983) p.7-14 ]

[28] B. Tuck, M.D. Zahari [ *J. Phys. D (UK)* vol.10 no.18 (1977) p.2473-9 ]

[29] K.J. Bachman, E. Buehler, B.I. Miller, J.H. McFee, F.A. Thiel [ *J. Cryst. Growth (Netherlands)* vol.39 no.1 (1977) p.137-50 ]

[30] N.F. Mott [ *Metal-Insulator Transitions, 2nd Edition* (Taylor & Francis, London, UK, 1990) p.146-50 ]

[31] G.W. Iseler [ *J. Electron. Mater. (USA)* vol.13 no.6 (1984) p.989-1011 ]

[32] K. Pak, M. Matsui, H. Fukuhara, T. Nishinaga, T. Nakamura, Y. Tasuda [ *J. Electrochem. Soc. (USA)* vol.133 no.12 (1986) p.2656-8 ]

[33] D. Hofmann, G. Myller, N. Streckfuss [ *Appl. Phys. A (West Germany)* vol.48 no.4 (1989) p.315-20 ]

[34] H.K. Kim, J.S. Hwang, S.K. Noh, C.H. Chung [ *Jpn. J. Appl. Phys. 2 (Japan)* vol.25 no.11 (1986) p.L888-90 ]

[35] C.D. Brandt, A.M. Hennel, T. Bryskiewicz, K.Y. Lo, L.M. Pawlosicz, H.C. Gatos [ *J. Appl. Phys. (USA)* vol.65 no.9 (1989) p.3459-69 ]

[36] W.A. Gault, E.M. Monberg, J.E. Clemans [ *J. Cryst. Growth (Netherlands)* vol.74 no.3 (1986) p.491-506 ]

[37] E. Kubota, A. Katsui, Y. Ohmori, K. Sugii [ *J. Cryst. Growth (Netherlands)* vol.82 no.4 (1987) p.737-46 ]

[38] E. Kubota, A. Katsui [ *Jpn. J. Appl. Phys. 2 (Japan)* vol.25 no.3 (1986) p.L249-51 ]

[39] D.J. Dowling, R.A. Brunton, D.A.E. Crouch, A.J. Thompson, J.E. Wardill [ *J. Cryst. Growth (Netherlands)* vol.87 no.1 (1988) p.137-41 ]

[40] K.T. Ip, L.F. Eastman, V.L. Wrick [ *Electron. Lett. (UK)* vol.13 no.22 (1977) p.682-3 ]

[41] M.G. Astles, F.G.H. Smith, E.W. Williams [ *J. Electrochem. Soc. (USA)* vol.120 no.12 (1973) p.1750-7 ]

[42] G.G. Baumann, K.W. Benz, M.K. Pilkuhn [ *J. Electrochem. Soc. (USA)* vol.123 no.8 (1976) p.1232-5 ]

[43] S.H. Groves, M.C. Plonko [ *Inst. Phys. Conf. Ser. (UK)* no.45 (1979) p.71-7 ]

[44] W. Korber, J. Weber, A. Hangleiter, K.W. Benz, H. Ennen, H.D. Muller [ *J. Cryst. Growth (Netherlands)* vol.79 (1986) p.741-4 ]

[45] S. Kondo, T. Amano, H. Nagai [ *Jpn. J. Appl. Phys. 1 (Japan)* vol.26 no.12 (1987) p.1997-2001 ]

[46] H. Beneking, N. Emeis [ *IEEE Electron Device Lett. (USA)* vol.EDL-7 no.2 (1986) p.98-100 ]

[47] M.J. McCollum, M.H. Kim, S.S. Bose, B. Lee, G.E. Stillman [ *Appl. Phys. Lett. (USA)* vol.53 no.19 (1988) p.1868-70 ]

[48] P.J. Wang, B.W. Wessels [ *Appl. Phys. Lett. (USA)* vol.49 no.10 (1986) p.564-6 ]

[49] S. Kondo, T. Amano, H. Nagai [ *Jpn. J. Appl. Phys. 2 (Japan)* vol.25 no.9 (1986) p.L770-2 ]

[50] S.J.J. Teng, J.M. Ballingall, F.J. Rosenbaum [ *Appl. Phys. Lett. (USA)* vol.48 no.18 (1986) p.1217-9 ]

[51] N.D. Gerrard, D.J. Nicholas, J.O. Williams, A.C. Jones [ *Chemtronics (UK)* vol.3 no.1 (1988) p.17-30 ]

[52] M.J. Ludowise [ *J. Appl. Phys. (USA)* vol.58 no.8 (1985) p.R31-55 ]

[53] A.R. Clawson, D.I. Elder [ *J. Electron. Mater. (USA)* vol.15 no.2 (1985) p.111-6 ]

[54] C.H. Chen, C.A. Larsen, G.B. Stringfellow [ *Inst. Phys. Conf. Ser. (UK)* no.83 (1987) p.75-80 ]

[55] Y. Takeda, S. Araki, S. Noda, A. Sasaki [ *Jpn. J. Appl. Phys. 1 (Japan)* vol.29 no.1 (1990) p.11-8 ]

[56] M. Naitoh, T. Soga, T. Jimbo, M. Umeno [ *J. Cryst. Growth (Netherlands)* vol.93 (1988) p.52-5 ]

[57] A. Molassioti, M. Moser, A. Stapor, F. Scholz, M. Hostalek, L. Pohl [ *Appl. Phys. Lett. (USA)* vol.54 no.9 (1989) p.857-8 ]

[58] S.J. Bass, M.S. Skolnick, H. Chudzynska, L. Smith [ *J. Cryst. Growth (Netherlands)* vol.75 no.2 (1986) p.221-6 ]

[59] L.D. Zhu, K.T. Chan, J.M. Ballantyne [ *J. Cryst. Growth (Netherlands)* vol.73 no.1 (1985) p.83-95 ]

[60] C.H. Chen, M. Kitamura, R.M. Cohen, G.B. Stringfellow [ *Appl. Phys. Lett. (USA)* vol.49 no.15 (1986) p.963-5 ]

[61] M. Razeghi, Ph. Maurel, M. Defour, F. Omnes, G. Neu, A. Kozacki [ *Appl. Phys. Lett. (USA)* vol.52 no.2 (1988) p.117-9 ]

[62] E. Woelk, H. Beneking [ *J. Appl. Phys. (USA)* vol.63 no.8 (1988) p.2874-5 ]

[63] A.R. Clawson, T.T. Vu, D.I. Elder [ *J. Cryst. Growth (Netherlands)* vol.83 no.2 (1987) p.211-8 ]

[64] K.L. Hess, S.W. Zehr, W.H. Cheng, D. Perrachione [ *J. Electron. Mater. (USA)* vol.16 no.2 (1987) p.127-32 ]

[65] H. Horikawa, Y. Ogawa, Y. Kawai, M. Sakuta [ *Appl. Phys. Lett. (USA)* vol.53 no.5 (1988) p.397-9 ]

[66] M.K. Lee, D.S. Wuu, H.H. Tung, J.H. Chang, L.F. Lin [ *Appl. Phys. Lett. (USA)* vol.53 no.2 (1988) p.107-9 ]

# 3.4 Resistivity of bulk InP

G.W. Iseler, 1990; revised by R.K. Ahrenkiel, 1998

## A INTRODUCTION

The resistivities of InP samples are determined by their impurity content, which determines the carrier concentration. For bulk crystals, which are grown from the melt, the reported room temperature resistivities range from about $3 \times 10^{-4}$ ohm cm for the most heavily doped n-type samples to about $1 \times 10^{8}$ ohm cm for semi-insulating (SI) samples doped with Fe. This Datareview first summarises the resistivities that can be obtained for n- and p-type conducting samples and for SI samples. Because the incorporation of an impurity in melt-grown crystals is primarily controlled by its liquid-solid distribution coefficient, values of this coefficient are listed for the common dopants in InP. The uniformity of resistivity in bulk crystals is considered, and the Datareview concludes with a discussion of the thermal stability of SI samples, which can experience a reduction in room temperature resistivity as a result of being heated to the elevated temperatures used for epitaxial growth or ion implant activation.

## B RESISTIVITY OF n-TYPE InP

The concentration of valence-band holes is so low that the resistivity is given by the expression:

$$\rho = (q\, n\, \mu_n)^{-1} \qquad (1)$$

Therefore, these nominally undoped crystals generally have resistivities of 0.05 to 1.5 ohm cm. Currently, these are the highest resistivity values that can be obtained consistently for conducting n-type samples. Higher values would be achieved if the residual impurity content could be reduced. In principle, n-type samples with somewhat higher resistivities could also be obtained by doping with shallow-acceptor impurities to partially compensate the residual donors, but the residual concentration is not sufficiently reproducible for this procedure to be practical.

n-Type samples with lower resistivities are readily obtained by doping with impurities. The most commonly used dopants are Sn, which substitutes for In, and S, which substitutes for P. At room temperature, each shallow donor atom ionises to yield one electron, so that n is equal to the sum of the intentional and residual donor concentrations. For each donor species, n can be varied continuously up to the limit determined by the solid solubility of that donor in InP. As n increases, $\mu_n$ is reduced by ionised-impurity scattering, which increases with increasing donor concentration but varies little, if at all, with donor species.

The dependence of resistivity on electron concentration is shown in FIGURE 1. In the figure, resistivity is plotted with selected values of the free-electron concentration from $1 \times 10^{15}$ to $3 \times 10^{19}$ $cm^{-3}$. The latter is the highest reported value for InP, and was obtained by S-doping [1]. The free-electron concentration n is related to the donor density $N_D$ by the expression:

$$n = \eta N_D \qquad (2)$$

where $\eta$ is defined as the ionisation efficiency. For shallow donors and acceptors, $\eta \sim 1.0$ at room temperature. For deeper donors and acceptors, $\eta$ can be appreciably less than 1.0 at cryogenic temperatures as will be shown later.

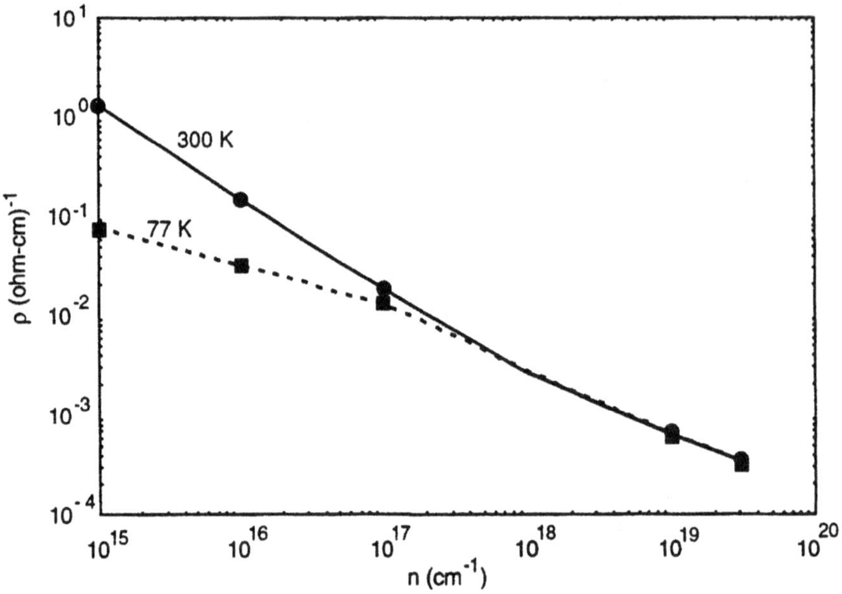


FIGURE 1 Values of the resistivity of n-type InP at 77 K and 300 K.

FIGURE 1 shows values of ρ that were calculated from the mobility values of Figure 1 in Datareview 3.3 and using EQN (1). The figure shows the values at 77 K and 300 K. For a given sample, n is the same at 77 K as at room temperature, because the ionisation energy of shallow donors is so low that deionisation is not significant down to 77 K [2]. Therefore, any changes in ρ that occur on cooling result from changes in $\mu_n$. Samples with the lowest values of n, for which the room temperature mobility is dominated by lattice-scattering, exhibit a strong increase in $\mu_n$. A corresponding decrease in ρ on cooling to 77 K occurs as the phonon contribution weakens and the mobility increases to about 300,000 $cm^2V^{-1}s^{-1}$ [3]. For even the largest values of $\mu_n$ observed at 77 K, the mobility is dominated by ionised-impurity scattering. For samples with the largest values of n, $\mu_n$ and ρ do not change between room temperature and 77 K, because the mobility is dominated by impurity scattering at both temperatures.

Data on the temperature dependence of resistivity above room temperature are limited. The intrinsic density, $n_i$, has been measured at about $5 \times 10^7$ $cm^{-3}$ at 300 K [4] by Hall measurements on InP:Fe semi-insulating crystals. The intrinsic conductivity is therefore an insignificant factor at 300 K. The value of n remains constant until the temperature becomes high enough for significant production of hole-electron pairs by thermal excitation, i.e. $n_i$ increases to a value comparable to doping levels. Therefore, the variation in ρ is initially determined by the change in $\mu_n$ with temperature. For values of n low enough for lattice-scattering to be dominant at room temperature, the decrease in lattice mobility with increasing temperature initially causes $\mu_n$ to decrease and ρ to increase correspondingly, while for heavily doped samples, $\mu_n$ and ρ are initially sensitive to temperature. Eventually, the increase in intrinsic carrier concentration $n_i$ causes ρ to decrease, although even at 300°C the value of $n_i$ is only of the order of magnitude of $1 \times 10^{13}$ $cm^{-3}$, much less than the shallow donor concentration in nominally undoped samples. When InP melts, there is a sharp discontinuity in resistivity, with the value of ρ for the liquid just above the melting point being about sixty-five times less than the value for the solid just below it [5].

## C RESISTIVITY OF p-TYPE InP

As before, the p-type resistivity values were calculated from EQN (1) and used the hole-mobility values of Figure 2 in Datareview 3.3. FIGURE 2 shows the room temperature resistivity of p-type

InP for selected values of p from $1 \times 10^{15}$ to $7 \times 10^{18}$ $cm^{-3}$. The values range from about 0.02 ohm cm [6] to 200 ohm cm [7]. To illustrate the dependence of ρ on hole concentration p and temperature, FIGURE 2 shows the ρ values at both 77 K and 300 K. Data for hole concentrations of $8 \times 10^{18}$ $cm^{-3}$ are shown, which are the highest reported for InP. These concentrations were obtained by Zn-doping [8]. The mobility values provided by Figure 2 of Datareview 3.3 were used to calculate the values of ρ with EQN (1). In studies of Zn-doping by diffusion to provide p-type conductivity, uniform resistivities as low as 0.02 ohm cm have been obtained in samples up to 100 μm thick [9,10].

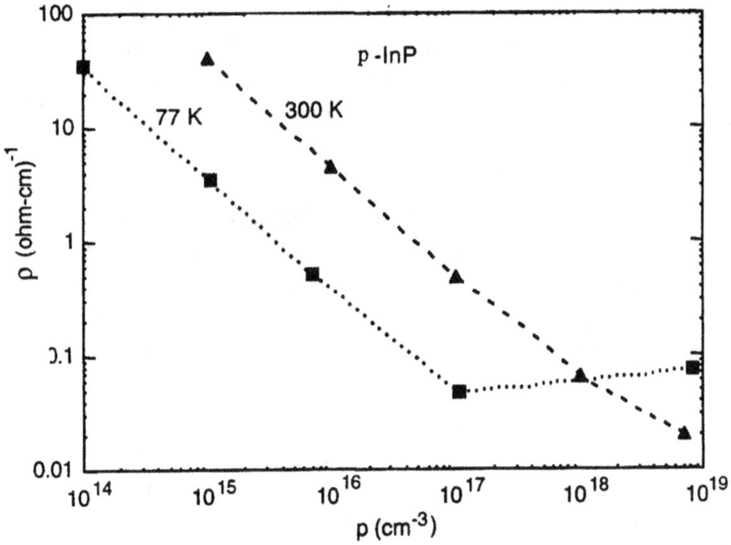


FIGURE 2 Values of the resistivity of p-type InP at 77 K and 300 K.

As noted earlier, the temperature dependence of the mobility and free-carrier concentration is markedly different for p-type InP than for n-type InP. In p-type InP, the energy levels of shallow acceptors are somewhat deeper than those of shallow donors and carrier 'freeze-out' or deionisation occurs. For samples with $p < 1 \times 10^{18}$ $cm^{-3}$ at 300 K, acceptor deionisation results in ionisation efficiency at 77 K of about 0.1 or less. As noted, the free-hole concentration decreased by factors of ten to fourteen at 77 K, for the sample set of FIGURE 2. For $p < 1 \times 10^{17}$ $cm^{-3}$, $\mu_p$ increases by a factor of twelve to fourteen in cooling from 300 to 77 K. However, the resistivity changes by less than 40% between the two temperatures, owing to deionisation. For the most highly doped samples, p is almost the same at 300 and 77 K because of impurity banding. Because $\mu_p$ decreases by about a factor of four, ρ is about four times greater at 77 K.

## D SEMI-INSULATING InP

Many electronic applications require semi-insulating (SI) InP with resistivities from $10^6$ to $10^8$ ohm cm. To achieve such resistivities in nominally undoped crystals would require the residual donor concentration to be reduced by a factor of at least $10^6$ from state-of-the-art values. Because there is no possibility of achieving this degree of additional purification, a method is required that employs acceptor doping to compensate the residual donors. In principle, any acceptor can compensate the donors. However, because bulk InP crystals commonly have a short-range variation of at least 5% in donor and acceptor concentration, the maximum resistivity that can be obtained by compensation with shallow acceptors, in the absence of p-n junction formation, is only about 15 ohm cm. Resistivities in the SI range are usually obtained by doping with the deep-acceptor Fe. The energy level of Fe is near the centre of the energy gap, or about 0.65 eV below the edge of the conduction band [9]. Therefore, when the Fe concentration is greater than the residual-donor concentration, the vast majority of electrons originating from the donors are localised on the Fe centres. In thermodynamic terms, the Fermi level 'pins' to the deep acceptor level and the electron concentration becomes nearly intrinsic.

For a typical SI crystal containing $10^{16}$ cm$^{-3}$ Fe atoms and $5 \times 10^{15}$ cm$^{-3}$ residual net donors, the concentration of conduction-band electrons (300 K) is only about $10^8$ cm$^{-3}$. This value is comparable to the intrinsic carrier density of $1 \times 10^7$ cm$^{-3}$. Because $\mu_n$ is generally between 2000 and 4000 cm$^2$V$^{-1}$s$^{-1}$ in SI InP [11-14], the resistivity of a crystal with $10^8$ cm$^{-3}$ carriers is between $1.6 \times 10^7$ and $3.1 \times 10^7$ ohm cm. Other deep acceptor levels in InP, including those of Cr at about 0.4 eV below the conduction band edge [15] and Mn and Co at ~0.3 eV above the valence band edge [16], are not deep enough to produce resistivities greater than ~$10^5$ ohm cm. Titanium has recently been identified as a deep donor lying ~0.6 eV below the conduction band edge [17]. Ti-doped crystals with resistivities greater than $10^6$ ohm cm have been grown [18-20]. Because the Ti donors cannot compensate the residual donors, these crystals are also doped with a shallow acceptor at a concentration exceeding the residual donor concentration. More recently, the gas source molecular beam epitaxy (GSMBE) technique has grown InP:Fe layers exceeding $5 \times 10^7$ ohm cm by Ti/Be co-doping [21].

The successful growth of SI GaAs without intentional impurity doping has motivated attempts to obtain SI behaviour in nominally undoped InP crystals [22,23]. In the case of SI GaAs, the native defect EL2, which is a deep donor, compensates the residual impurities, which are mainly acceptors. However, no comparable defect, which would need to be an acceptor to compensate the donor impurities, has been found in InP at concentrations sufficient to produce reproducible SI behaviour in as-grown undoped crystals.

One of the problems associated with Fe-doped substrates is the possible diffusion of Fe into the epitaxial layer with adverse effects on the devices. In 1989, SI behaviour was reported for nominally undoped InP crystals after annealing [4]. For samples with $n < \sim 3 \times 10^{15}$ cm$^{-3}$, the shallow donor concentration was decreased by annealing at about 900°C for 80 hours. An unidentified acceptor level located near the centre of the bandgap then compensated the remaining donors to produce a resistivity of $2 \times 10^7$ ohm cm. Recently, lightly Fe-doped wafers ($N_{Fe} = 5 - 8 \times 10^{15}$ cm$^{-3}$) were shown to be highly resistive after annealing at 900°C in a phosphorus ambient [24]. Many of these samples became highly resistive while the mobility also dramatically increased in some cases. In several cases, the resistivity increased by a factor of more than $1 \times 10^7$. The carrier concentration of some samples, as measured by the Hall effect, was in the range 2 to $6 \times 10^7$ cm$^{-3}$ which is very near the intrinsic concentration.

Because of the depth of the Fe acceptor and Ti donor levels below the conduction band edge, the temperature dependence of the resistivity of SI InP is very different from that of conducting samples. Near room temperature, an increase of 30°C reduces the resistivity by about an order of magnitude because of the thermal excitation of electrons from the deep levels to the conduction band.

## E DOPANT DISTRIBUTION COEFFICIENTS

To provide a specified carrier concentration and resistivity in a bulk InP crystal grown from the melt, the appropriate amount of dopant must be loaded with the InP charge into the growth vessel. The dopant concentration C that is incorporated into the growing crystal is usually different from the concentration Cl in the melt. The values of the effective distribution coefficient k = C/Cl reported for some common dopants in InP are listed in TABLE 1. Also listed are the maximum carrier concentrations that have been obtained with shallow impurities and the estimated solubility limits for deep impurities. Nearly all of the data were obtained from liquid-encapsulated Czochralski (LEC) grown wafers, so the values of k may be influenced by the $B_2O_3$ encapsulant. For example, Mg and Ca are removed from the melt by $B_2O_3$ [25], and Si and Ti are expected to be oxidised by $B_2O_3$.

TABLE 1 Dopants in InP.

| Dopants | Effective distribution coefficient | Maximum reported carrier concentration ($cm^{-3}$) | Solubility ($cm^{-3}$) |
|---|---|---|---|
| | | Donors | |
| Sn | 0.02 | $1 \times 10^{19}$ | - |
| Ge | 0.02 | $2 \times 10^{19}$ | - |
| Te | 0.04 | $2 \times 10^{19}$ | - |
| S | 0.47 | $3 \times 10^{19}$ | - |
| Si | 0.001 | - | - |
| | 0.53* | - | - |
| | | Acceptors | |
| Be | 0.16 | $1 \times 10^{19}$ | - |
| Zn | 1.0 | - | - |
| | 0.85 - 1.3 | $8 \times 10^{18}$ | - |
| Cd | 0.23 | - | - |
| Mn | 0.4 | $6 \times 10^{16}$ | - |
| Co | $5 \times 10^{-4}$ | - | - |
| | | Deep levels | |
| Cr | $3 \times 10^{-4}$ | - | - |
| Fe | $1.6 \times 10^{-3}$ | - | $2.5 \times 10^{17}$ |
| | $2.5 \times 10^{-4}$ | - | - |
| Ti | $5 \times 10^{-4}$ | - | $2 \times 10^{17}$ |

*Gradient-freeze method.

## F UNIFORMITY OF ELECTRICAL PROPERTIES

An as-grown InP bulk crystal is rarely, if ever, uniform in resistivity over its length. Because the value of k is less than one for all known dopants except Zn, in most crystals the dopant concentration C increases from the first-to-freeze end toward the tail. If k is constant, the dopant profile is given by:

$$C = C_0(1 - g)^{k-1} \tag{3}$$

where $C_0$ is the first-to-freeze concentration and g is the fraction of the melt frozen. For crystals doped with an element whose distribution coefficient is very small such as Sn (k = 0.02), the ratio $C/C_0$ is much greater than two and five for g = 0.5 and 0.8, respectively. This condition is true provided that the solubility limit has not been exceeded. In the case of S-doping (k = 0.5) the variation is slower, with $C/C_0 >> 2$ for g = 0.75. For conducting n- and p-type crystals, then, the resistivity decreases continuously along the length from seed to tail, while for crystals doped with Fe or Ti, which have very low values of k, the resistivity usually increases because the deep level concentration increases more rapidly than the residual-donor concentration.

Short-range dopant non-uniformity can also result when $k \neq 1$, particularly for LEC growth, because the temperature gradient required for this method induces buoyancy-driven turbulent convection in the melt. For $k \neq 1$, the melt turbulence creates growth striations along the crystal-melt interface that are easily observed in longitudinal cross-section by chemical etching [26], X-ray topography [27], or infrared absorption mapping [8,28,29]. In LEC crystals, the carrier concentration and, therefore, resistivity may vary by up to several hundred percent over distances of tens of microns [10]. However, if the seed or crucible is rapidly rotated during growth to promote forced melt mixing, the short-range variation in dopant concentration can be reduced to about 5% [10]. Alternatively, short-range doping uniformity can be improved by imposing a magnetic field on the LEC melt to damp convection [8,30]. With this technique, homogeneous doping with Fe along the growth axis has been achieved [31]. Recently, bulk crystals with excellent short-range doping uniformity have also been grown by the vertical gradient-freeze method [32].

A different kind of microscopic non-uniform resistivity has been observed in commercial SI InP wafers, which exhibit filamentary conduction that occurs through 'pipes' about 10 μm in diameter with a density of approximately 100 $cm^{-2}$ [33].

## G THERMAL STABILITY OF RESISTIVITY

Heating an Fe-doped SI InP wafer to 600° and 700°C for several hours, as in the epitaxial growth process, may produce an electrically conducting layer at or near the wafer surface. Similar effects occur by heating to between 800° and 900°C for 30 min, as in an ion implant activation anneal. As shown by secondary ion mass spectrometry, this layer is due to the motion of Fe from a depth of several microns toward the surface, leaving a depleted region in which the Fe concentration no longer exceeds the donor concentration [34-36]. Such thermal instability was the primary motivation for the development of Ti-doped SI InP. In high-temperature annealing experiments on SI InP co-doped with Ti and Hg [37] or with Ti and Zn [19], no redistribution of the dopants was observed, although Zn is normally a fast-diffusing impurity in InP. In studies on InP layers grown by metal-organic chemical vapour deposition, immobilisation has been observed for Zn acceptors paired with Si or S donors [38] and for complexes of Mg acceptors and Fe acceptors [39]. Perhaps the thermal stability of InP co-doped with Ti and Zn may also be due to pairing.

In InP doped with Zn, Cd, Be or Mg to provide p-type conductivity, the acceptors diffuse very rapidly at elevated temperatures. Therefore, the thermal stability of such p-type material is unsatisfactory for some high-temperature device-fabrication procedures. In a study of InP implanted with Hg, it was found that this acceptor diffuses much more slowly [40], so bulk InP crystals doped with Hg should exhibit superior thermal stability.

## H CONCLUSION

Bulk n- and p-type InP, with a wide range of resistivities, can be obtained by doping with shallow donors and shallow acceptors, respectively. For each conductivity type, the dependence of resistivity on carrier concentration at 300 and 77 K is well established. Semi-insulating InP with resistivities in the $10^6$ to $10^8$ ohm cm range can be obtained by doping with Fe or by co-doping with Ti and a shallow acceptor. Bulk crystals vary in resistivity along their length because the dopant distribution coefficients differ from 1. In addition, crystals grown by the conventional LEC method generally exhibit marked short-range non-uniformities in resistivity, which can be reduced by modifying the LEC technique or using the vertical gradient-freeze method. The resistivity of Fe-doped SI crystals can be significantly decreased near the surface by high-temperature processing, but Ti-doped crystals appear to be more stable.

## ACKNOWLEDGEMENT

Compilation of this review was sponsored by the US Department of the Air Force.

## REFERENCES

[1] A.G. Thompson, B. Ross [ *Final Technical Report, no.TR-6* (Bell and Howell, October 1970) ]

[2] J. Leloup, H. Djerassi, J.H. Albany, J.B. Mullin [ *J. Appl. Phys. (USA)* vol.49 no.6 (1978) p.3359-62 ]

[3] W. Walukiewicz et al [ *J. Appl. Phys. (USA)* vol.51 no.5 (1980) p.2659-68 ]

[4] J. Betko [ *Phys. Status Solidi (Germany)* vol.127 (1991) p.451 ]

[5] V.M. Glazov, A.Ya. Nashel'skii, F.R. Rzaev, M.M. Mamedov [ *Sov. Phys.-Semicond. (USA)* vol.10 no.5 (1976) p.498-501 ]

[6] M. Morioka, K. Tada, S. Akai [ *Annu. Rev. Mater. Sci. (USA)* vol.17 (1987) p.75 ]

[7] R.E. Nahory, A.A. Ballman, H. Brown, M.R. Wilson [ *Inst. Phys. Conf. Ser. (UK)* no.65 (1983) p.7-14 ]

[8] B. Tuck, M.D. Zahari [ *J. Phys. D (UK)* vol.10 no.18 (1977) p.2473-9 ]

[9] L. Eaves, A.W. Smith, P.J. Williams, B. Cockayne, W.R. MacEwan [ *J. Phys. C (UK)* vol.14 no.33 (1981) p.5063-8 ]

[10] R.S. Williams, P.A. Barnes, L.C. Feldman [ *Appl. Phys. Lett. (USA)* vol.36 no.9 (1980) p.760-2 ]

[11] S. Fung, R.J. Nicholas, R.A. Stradling [ *J. Phys. C (UK)* vol.12 no.23 (1979) p.5145-55 ]

[12] L.A. Demberel, A.S. Popov, D.B. Kushev, N.N. Zheleva [ *Phys. Status Solidi A (Germany)* vol.52 no.1 (1979) p.341-5 ]

[13] G.W. Iseler [ *Inst. Phys. Conf. Ser. (UK)* no.45 (1979) p.144-53 ]

[14] H.K. Kim, J.S. Hwang, S.K. Noh, C.H. Chung [ *Jpn. J. Appl. Phys. 2 (Japan)* vol.25 no.11 (1986) p.L888-90 ]

[15] D.N. Nichols, I. Odeh, R.J. Sladek [ *Solid State Commun. (USA)* vol.34 no.8 (1980) p.621-3 ]

[16] M.S. Skolnick, R.G. Humphreys, P.R. Tapster, B. Cockayne, W.R. MacEwan [ *J. Phys. C (UK)* vol.16 no.36 (1983) p.7003-18 ]

[17] C.D. Brandt et al [ *Appl. Phys. Lett. (USA)* vol.48 no.17 (1986) p.1162-4 ]

[18] G.W. Iseler, B.S. Ahern [ *Appl. Phys. Lett. (USA)* vol.48 no.24 (1986) p.1656-7 ]

[19] A. Katsui [ *J. Cryst. Growth (Netherlands)* vol.89 no.4 (1988) p.612-3 ]

[20] D.D. Nolte, D.H. Olsen, E.M. Monberg, P.M. Bridenbaugh, A.M. Glass [ *Opt. Lett. (USA)* vol.14 no.22 (1989) p.1278-80 ]

[21] S. Salaun et al [ *Electron. Lett. (UK)* vol.30 (1994) p.2076 ]

[22] P.B. Klein, R.L. Henry, T.A. Kennedy, N.D. Wilsey [ *Mater. Sci. Forum (Switzerland)* vol.10-12 pt.3 (1986) p.1259-64 ]

[23] D. Hoffmann, G. Muller, N. Streckfuss [ *Appl. Phys. A (Germany)* vol.48 no.4 (1989) p.315-20 ]

[24] R. Romari, A. Zapettini, E. Combia, R. Mosca, K. Cherkaoui, G. Marrakchi [ *J. Appl. Phys. (USA)* vol.81 (1997) p.7604 ]

[25] E. Kubota, A. Katsui [ *Jpn. J. Appl. Phys. 2 (Japan)* vol.24 no.5 (1985) p.L344-6 ]

[26] K. Naukkarinen, T. Tuomi, V.-M. Airaksinen, K.-M. Laakso, J.A. Lahtinen [ *J. Cryst. Growth (Netherlands)* vol.64 no.3 (1983) p.485-91 ]

[27] H. Miyairi, T. Inada, M. Eguchi, T. Fukuda [ *J. Cryst. Growth (Netherlands)* vol.79 (1986) p.291-5 ]

[28] R.L. Barns [ *J. Electron. Mater. (USA)* vol.18 no.6 (1989) p.703-10 ]

[29] G.W. Iseler [ *J. Electron. Mater. (USA)* vol.13 no.6 (1984) p.989-1011 ]

[30] S. Bachowski, D.F. Bliss, B. Ahern, R.M. Hilton, J. Adamski, D.J. Carlson [ *Proc. 2nd Int. Conf. on Indium Phosphide and Related Materials* (IEEE, New York, USA, 1990) p.30 ]

[31] S. Ozawa, T. Kimura, J. Kobayashi, T. Fukuda [ *Appl. Phys. Lett. (USA)* vol.50 no.6 (1987) p.329-31 ]

[32] E.M. Monberg, W.A. Gault, F. Dominguez, F. Simchock, S.N.G. Chu, C.M. Stiles [ *J. Electrochem. Soc. (USA)* vol.135 no.2 (1988) p.500-3 ]

[33] V. Montgomery, J.G. Swanson, P.J. Claxton [ *J. Cryst. Growth (Netherlands)* vol.94 no.3 (1989) p.721-5 ]

[34] H. Kamada, S. Shinoyama, A. Katsui [ *J. Appl. Phys. (USA)* vol.55 no.8 (1984) p.2881-4 ]

[35] B. Cockayne, G.T. Brown, W.R. MacEwan, G.W. Blackmore [ *J. Mater. Sci. Lett. (UK)* vol.2 no.7 (1983) p.309-13 ]

[36] M. Gauneau, R. Chaplain, A. Rupert, E.V.K. Rao, N. Duhamel [ *J. Appl. Phys. (USA)* vol.57 no.4 (1985) p.1029-35 ]

[37] G.W. Iseler, H.R. Clark Jr. [ *Proc. 2nd Int. Conf. on Indium Phosphide and Related Materials* (IEEE, New York, USA, 1990) p.25 ]

[38] C. Blaauw, F.R. Shepherd, D. Eger [ *J. Appl. Phys. (USA)* vol.66 no.2 (1989) p.605-10 ]

[39] C. Blaauw, R.A. Bruce, C.J. Miner, A.J. Howard, B. Emmerstorfer, A.J. SpringThorpe [ *J. Electron. Mater. (USA)* vol.18 no.5 (1989) p.567-72 ]

[40] P.N. Favennec, H. L'Haridon, J.M. Roquais, M. Salvi, X. Le Cleach, L. Gouskov [ *Appl. Phys. Lett. (USA)* vol.48 no.2 (1986) p.154-6 ]

# 3.5 Minority-carrier lifetime of InP

R.K. Ahrenkiel

June 1998

## A INTRODUCTION

The minority carrier lifetime ($\tau$) in direct-bandgap materials such as InP is controlled by a combination of radiative and non-radiative recombination processes. Radiative recombination occurs when an electron-hole pair annihilates with the production of a photon. Non-radiative recombination occurs through several mechanisms. Non-radiative or Shockley-Read-Hall (SRH) recombination occurs at forbidden-gap energy levels that arise from both chemical and mechanical defects. The free electron-hole pair is captured at a defect-produced energy level and the energy is transformed into heat. These defects are present in the volume and at the surface of the material. Recombination at the surface is described by a parameter called the surface-recombination velocity [1]. At very high carrier-concentrations, Auger recombination transforms the electron-hole pair energy into an energetic majority carrier. Measurable Auger recombination has not been reported for InP in the literature.

The intrinsic and often dominant recombination mechanism in direct-bandgap materials is radiative recombination. The radiative recombination rate, R, is given by theory as $R = Bnp$, where n and p are the electron and hole concentrations, respectively, and B is a coefficient of the particular semiconductor. For direct-bandgap semiconductors, the B-coefficient is orders of magnitude larger than for indirect semiconductors such as silicon. The radiative lifetime of any semiconductor is determined by the B-coefficient, and may be calculated from band theory or experimentally measured. The B-coefficient can be calculated by quantum mechanics and is derived from the dipole matrix elements connecting valence band and conduction band states. This coefficient was calculated by quantum mechanics [2] for GaAs and found to be about $2 \times 10^{-10}$ $cm^3/s$. Quantum-mechanical calculations of B have not been reported for InP. The van Roosbroeck-Shockley [3] equation uses a thermodynamic approach to equate the radiative lifetime of a semiconductor to an integral over the absorption spectrum. Using tabulated absorption data for InP, Bothra and co-workers [4] calculated B as:

$$B = 6.25 \times 10^{-10}\ (cm^3/s) \tag{1}$$

For low injection, when the injected electron-hole density, $\rho$, is much less than the doping density, the radiative lifetimes are:

$$\tau_R = \frac{1}{BN_d} \text{ or } \frac{1}{BN_a} \tag{2}$$

for n-type or p-type semiconductors, respectively. At high injection, the decay becomes non-linear and the instantaneous lifetime is:

$$\tau_R = \frac{1}{B\rho^2} \tag{3}$$

When radiative recombination is the dominant mechanism, the high-injection carrier decay is non-exponential until $\rho$ is less than N, the doping level.

As stated above, the radiative lifetime at low injected-carrier densities is 1/(BN), where N is the majority-carrier concentration. However, this formula neglects an effect called photon recycling [5] which is produced by the self-absorption of radiative recombination. In this process, higher energy photons ($h\nu > E_g$), which are emitted within the material, are partially or fully absorbed, generating new electron-hole pairs. These in turn recombine to produce another photon, and the cycle continues until the photon escapes or a non-radiative event occurs. The energy may cycle numerous times between the two states, electron-hole pairs, and photons. The recycling increases with increasing sample thickness and the factor $\phi$, discussed below, has been calculated for sample thicknesses up to 10 μm for GaAs. As the optical constants for $h\nu > E_g$ are similar for GaAs and InP, one expects similar increases in the effective radiative lifetime. Early evidence of photon recycling [6] in InP occurred while measuring diffusion lengths and using photoluminescence as a diagnostic. The diffusion lengths were considerably larger than expected. As the square root of the diffusion length is proportional to the lifetime, the larger measured values were attributed to photon recycling.

The effect of photon recycling is introduced by modifying the B-coefficient by a factor, $\phi$, that is a function of sample geometry. This factor may be calculated using elementary optics [7], although the calculation is somewhat tedious. The effective B-coefficient, B', is then:

$$B' = \frac{B}{\phi} \tag{4}$$

The low-injection radiative lifetime is expressed by:

$$\tau_R = \frac{\phi}{BN_d} \text{ or } \frac{\phi}{BN_a} \tag{5}$$

Experiments on samples that are dominated by radiative lifetime measure B' rather than B. B can be determined experimentally by measuring lifetimes over a range of sample thicknesses [8]. Photon recycling has been shown to increase the lifetime in GaAs epitaxial layers by as much as a factor of 7.5 in 8 μm thick GaAs epilayers [9,10]. For epitaxial layers with the substrate removed, the recycling effect is increased as the critical angle for total internal reflection increases at the film/air interface [11].

The total carrier lifetime in InP is controlled by a combination of bulk and surface recombination mechanisms. The surface effects for a planar sample that is very long in two dimensions (relative to the thickness) can be accounted for by the lifetime expression:

$$\frac{1}{\tau} = \frac{1}{\tau_B} + \frac{2S}{d} \tag{6}$$

where $\tau_B$ is the bulk lifetime, S is the surface or interface recombination velocity and d is the sample thickness.

The database on minority-carrier lifetime in both n- and p-type InP is limited compared to that for GaAs, but has expanded since 1990. Data have been published for both single-crystal wafers and epitaxial material. Unlike the case of GaAs [12], the reported lifetime values in wafer material have often been larger than in comparably doped epitaxial material.

Surface recombination often dominates the minority-carrier lifetime in III-V materials. Surface passivation for many semiconducting materials has been achieved by either epitaxial window layers or passivating glasses. For example, the ternary semiconductors $Al_xGa_{1-x}As$ and $Ga_{0.5}In_{0.5}P$, which are lattice-matched to GaAs, successfully passivate the surface and confine carriers to the GaAs. No comparable epitaxial material has been reported for InP, so the surface-recombination effects influence most measurements. Some reports [13] describe chemical passivation of InP surfaces by the electrolytic deposition of sulphur compounds, but these methods are not widely adopted in the technology.

Early measurements have used a variety of non-optical techniques to deduce the carrier lifetime of InP, and the scatter in the data is very severe. Yamaguchi [14] and co-workers grew n-type InP wafers by the liquid-encapsulated Czochralski (LEC) method, producing some dislocation-free crystals. At $|N_d - N_a| = 10^{16}\ cm^{-3}$, they reported hole lifetimes of 3 μs by the junction-recovery technique. Beneking and Emeis [15] grew liquid phase epitaxial (LPE) n-type epilayers on InP substrates in the doping range $N_d = 0.4$ to $2 \times 10^{16}\ cm^{-3}$. Hole lifetimes were extracted from the analysis of diffusion capacitance of junctions and ranged from 1.4 to 4.3 ns. By adding isoelectronic As to the epilayer in concentrations greater than $1 \times 10^{19}\ cm^{-3}$, the dislocation density was reduced and the lifetime increased to 5.7 to 8.6 ns. These data indicate a lattice mismatch between the wafer material and the LPE layers.

An improvement in the diagnostic technology occurred with the increasing popularity of the time-resolved photoluminescence (TRPL) [12] technique to measure carrier lifetimes. There are several early reported measurements of TRPL of InP, including the work by Garbuzov [16] and co-workers. Garbuzov et al employed a krypton laser beam, that was modulated by a mechanical chopper, producing 10 ms pulses. Time-resolution problems were limiting, but they measured the PL lifetime of bulk p-InP (Sn-doped to $2 \times 10^{18}\ cm^{-3}$) as about 100 ns. These lifetimes were longer than expected and associated with photon-recycling effects by the authors.

The advent of short-pulse lasers greatly aided the time resolution of this technique and improved the measurements. TRPL experiments [17] used a pulsed argon-ion laser on thick (30 μm, $N_d \sim 1.5 \times 10^{16}\ cm^{-3}$) LPE-grown layers. The PL decay was measured under conditions of high injection and therefore the measured lifetimes (about 200 ns) may be shorter than the low-injection lifetime.

The first TRPL measurements of electron lifetimes in bulk wafers of p-InP, ($\sim 5 \times 10^{16}\ cm^{-3}$) were measured by the author and co-workers in 1988 [18]. The measured PL lifetimes were about 200 ns, and fitting the data for surface recombination, a bulk lifetime of 300 ns was calculated.

Keyes and co-workers [19,20] used the TRPL diagnostic technique to measure the carrier lifetime of n-type InP wafers over the temperature range from 80 K to 290 K as shown in FIGURE 1. The doping levels of the two samples are A: $5 \times 10^{15}\ cm^{-3}$ and B: $1 \times 10^{17}\ cm^{-3}$, respectively. The solid lines in the curves represent the theoretical expression for radiative recombination that states that B varies as:

$$B(T) = A\, E_g^2(T) / T^{1.5} \qquad (7)$$

where A is a constant, $E_g(T)$ is the bandgap at temperature T, and T is the absolute temperature. Therefore, the radiative recombination lifetime decreases strongly with temperature.

The data show that the lifetime varied with temperature as predicted by dominant radiative recombination. By fitting the lifetime-temperature data, they derived B-coefficients for InP as:

$$\begin{aligned} B/\phi &= 7.0 \times 10^{-11}\ cm^3/s,\ N_d = 5 \times 10^{15}\ cm^{-3} \\ B/\phi &= 3.4 \times 10^{-11}\ cm^3/s,\ N_d = 1 \times 10^{17}\ cm^{-3} \end{aligned} \qquad (8)$$

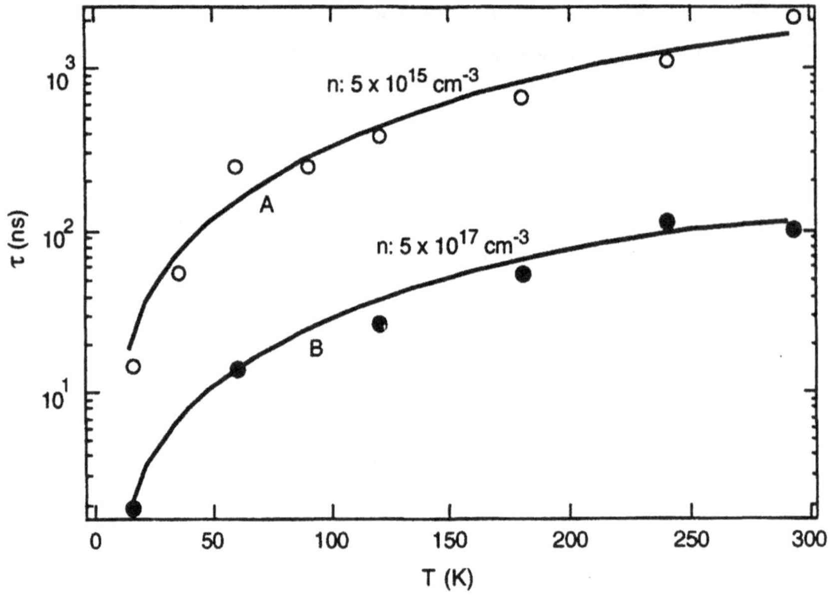


FIGURE 1 The minority-carrier lifetime of two n-type InP wafers measured by time-resolved photoluminescence [20].

These B/ϕ values are 8.9 and 18.4 times smaller than the B-value calculated by Bothra et al [4]. The photon-recycling factor has not been calculated for thick (350 μm) samples used in these studies. However, values of ϕ of about ten have been calculated [5] for 10 μm films of GaAs. Also, these calculations show that ϕ decreases with increasing doping density for a given sample thickness. In summary, the radiative lifetime is the intrinsic, upper limit for the InP lifetime when corrected for the photon-recycling effect.

## B n-TYPE InP

A number of measurements have been made in recent years on n-type InP using the TRPL technique. A composite of these data for n-InP is shown in FIGURE 2. Extensive lifetime measurements for both n- and p-type InP wafers, over a wide range of doping densities, were reported by Jenkins and co-workers [21]. These data are shown by the ▲ symbols in FIGURE 2. The room-temperature measurements of Keyes and co-workers, from Figure 1 in Datareview 3.3, are also shown in FIGURE 2. Rosenwaks and co-workers [22] used TRPL to measure the carrier lifetime and surface recombination velocity (S) on several InP wafers. The S-data will be described in Datareview 3.7. The bulk lifetime in n-type wafers doped at 4 - 5 × $10^{16}$ $cm^{-3}$ is shown in FIGURE 2. Curve A is a calculated value of the radiative lifetime over this doping range using the B/ϕ value of Keyes et al. Curve B is the radiative lifetime using the B-value of Bothra and co-workers and ignores photon recycling. We see that data from various sources are in fair agreement with the lifetimes calculated using this B/ϕ value of Keyes et al. The measured lifetime values of Jenkins and co-workers is larger than the calculated values at doping concentrations greater than 1 × $10^{17}$ $cm^{-3}$. On the other hand, the measured values of Jenkins et al are lower than the radiative lifetime at the two lowest doping levels. In this range, the lifetime may be limited by SRH-recombination rather than by radiative recombination.

As for epitaxial materials, Bothra and co-workers [4] measured the carrier lifetime in thin epitaxial layers (~3 × $10^{15}$ $cm^{-3}$) of both n- and p-type and found lifetimes in the 20 to 40 ns range using microwave reflection. These lower values may indicate either that epitaxial quality is not comparable to bulk, or that the surface-recombination effect dominates thin-film measurements. The interface between epitaxial layers and the substrate could also contribute to recombination observed in thin films.

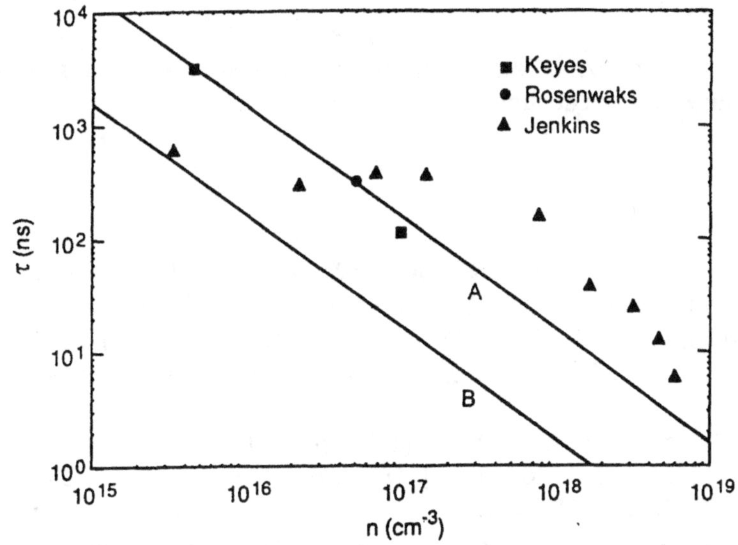


FIGURE 2 Data on the minority-carrier lifetime of n-type InP. Curve A is the calculated lifetime that included the photon-recycling effect. Curve B is the radiative lifetime using the calculated B-coefficient.

## C p-TYPE InP

Work by the author and co-workers [18] used TRPL to measure the lifetime of some p-type wafers of InP with doping levels of about $5 \times 10^{16}$ cm$^{-3}$, and found lifetimes of 200 to 300 ns. Rosenwaks and co-workers [22] found much smaller values of carrier lifetime in their p-type wafers at doping levels ranging from $6 \times 10^{15}$ cm$^{-3}$ to $4 - 5 \times 10^{17}$ cm$^{-3}$. Jenkins and co-workers also measured carrier lifetime in p-InP wafers over a range of doping levels (FIGURE 3). Data from curves A and B are also shown on this figure. We see that most of the measured values are smaller than the calculated radiative limit, either including or ignoring photon recycling. The origin of these low lifetime values is not known and has been the subject of a number of theories. Early work [23] showed that the recombination is larger in p-type than in n-type InP and surface recombination has been suggested. The work of Rosenwaks and co-workers, however, finds comparable S-values in p- and n-type InP. They suggest that bulk defects are much more numerous or effective in p-type InP.

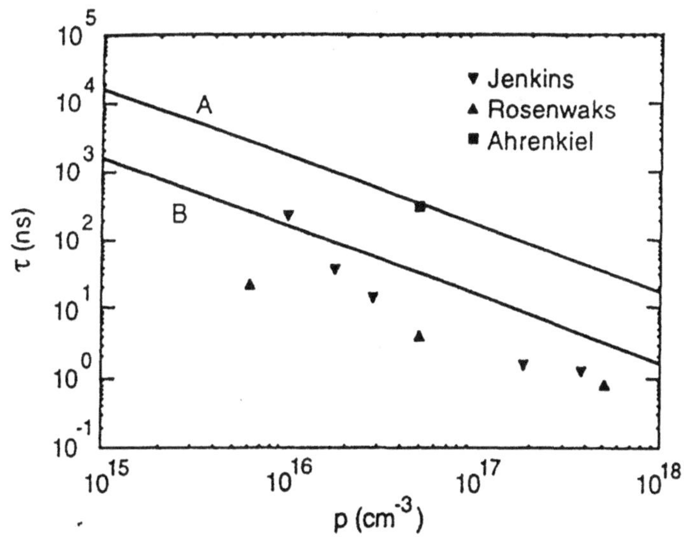


FIGURE 3 Data showing the minority-carrier lifetime of p-InP. Curve A is the calculated lifetime that included the photon recycling effect. Curve B is the radiative lifetime using the calculated B-coefficient.

**ACKNOWLEDGEMENT**

The author would like to thank Dr. B. Keyes for the unpublished data from his thesis and also for his helpful comments concerning this document.

**REFERENCES**

[1] R.K. Ahrenkiel [ Datareview in this book: *3.7 Surface recombination velocity in InP* ]
[2] H.C. Casey Jr., F. Stern [ *J. Appl. Phys. (USA)* vol.47 (1976) p.631 ]
[3] W. van Roosbroeck, W. Shockley [ *Phys. Rev. (USA)* vol.94 (1954) p.1558 ]
[4] S. Bothra, S. Tyagi, S.K. Ghandhi, J.M. Borrego [ *Solid-State Electron. (UK)* vol.34 (1991) p.47 ]
[5] P. Asbeck [ *J. Appl. Phys. (USA)* vol.48 (1977) p.820 ]
[6] S.D. Lester, T.S. Kim, B.G. Streetman [ *Appl. Phys. Lett. (USA)* vol.52 (1988) p.474 ]
[7] T. Kuriyama, T. Kamiya, H. Yanai [ *Jpn. J. Appl. Phys. (Japan)* vol.16 (1977) p.465 ]
[8] R.K. Ahrenkiel, B.M. Keyes, G.B. Lush, M.R. Melloch, M.S. Lundstrom, H.F. MacMillan [ *J. Vac. Sci. Technol. A (USA)* vol.10 (1992) p.990 ]
[9] R.K. Ahrenkiel et al [ *Appl. Phys. Lett. (USA)* vol.55 (1989) p.1088 ]
[10] R.K. Ahrenkiel, D.J. Dunlavy, B.M. Keyes, S.M. Vernon, S.P. Tobin, T.M. Dixon [ *Proc. 21st Photovoltaic Specialist Conference* (IEEE, New York, USA, 1990) ]
[11] G.B. Lush, M.R. Melloch, M.S. Lundstrom, D.H. Levi, R.K. Ahrenkiel, H.F. MacMillan [ *Appl. Phys. Lett. (USA)* vol.61 (1992) p.2440 ]
[12] R.K. Ahrenkiel [ *Semicond. Semimet. (USA)* vol.39 (1993) ch.2 ]
[13] R. Iyer, R.R. Chang, D.L. Lile [ *Appl. Phys. Lett. (USA)* vol.53 (1988) p.134 ]
[14] M. Yamaguchi, S. Shinoyama, C. Uemura [ *J. Appl. Phys. (USA)* vol.52 (1981) p.6429 ]
[15] H. Beneking, N. Emeis [ *IEEE Electron Device Lett. (USA)* vol.EDL-7 (1986) p.98 ]
[16] D.Z. Garbuzov, V.V. Agaev, A.T. Gorelenok [ *Sov. Phys.-Semicond. (USA)* vol.16 (1982) p.986 ]
[17] V. Diadiuk, S.H. Groves, C.A. Armiento, C.E. Hurwith [ *Appl. Phys. Lett. (USA)* vol.42 (1983) p.892 ]
[18] R.K. Ahrenkiel, D.J. Dunlavy, T. Hanak [ *J. Appl. Phys. (USA)* vol.64 (1988) p.1916 ]
[19] B.M. Keyes et al [ *J. Appl. Phys. (USA)* vol.75 (1994) p.4249 ]
[20] B.M. Keyes [ PhD Thesis, University of Colorado, Boulder, 1997, unpublished ]
[21] P. Jenkins, G.A. Landis, I. Weinberg, K. Kneisel [ *Proc. 22nd Photovoltaic Specialist Conference* (IEEE, New York, USA, 1991) p.177 ]
[22] Y. Rosenwaks, Y. Shapira, D. Huppert [ *Phys. Rev. B (USA)* vol.44 (1991) p.13097 ]
[23] H.C. Casey Jr., E. Buehler [ *Appl. Phys. Lett. (USA)* vol.30 (1977) p.247 ]

# 3.6 Diffusion lengths in InP

R.K. Ahrenkiel

June 1998

## A INTRODUCTION

The diffusion length may be described as the average distance that a minority carrier diffuses in a medium prior to recombining with a majority carrier. For example, analysis of the spatial distribution of a sheet of excess minority carriers injected at a plane at x = 0 in a bulk material shows that the profile decays as exp(-x/L), where L is the minority-carrier diffusion length. The minority-carrier diffusivity D is related to the mobility through the Einstein relationship. The latter shows that $D = kT/q\mu$, where k is Boltzmann's constant, T is the absolute temperature, q is the electron charge and $\mu$ is the carrier mobility in units of $cm^2/V$ s. As the minority-carrier mobility is not routinely measured, workers often assume that it is equal to the majority-carrier mobility at the same doping concentration that may be measured by a Hall-effect experiment. Theory and recent experiments, however, indicate that the minority carrier mobility may be considerably smaller than the majority carrier mobility. The minority carrier diffusion length, diffusivity, and lifetime are related by the relationship $L^2 = D\tau$. As L and $\tau$ are more directly measured, D is often determined from this relationship.

## B ELECTRON DIFFUSION LENGTH IN p-InP

Most of the current diffusion length measurements have been obtained by the surface photovoltage technique (SPV). A shortcoming of the standard analysis of SPV data [1] is that it ignores recombination in the space-charge region that would influence the data fit. Gouskov and co-workers [2] studied the electron-diffusion length in indium tin oxide (ITO)/p-InP photodiodes by electron beam induced currents (EBIC) and by the SPV technique [1,3]. Hanak and Ahrenkiel [4] used a surface photocurrent technique (SPC) and ITO/p-InP heterojunctions to measure electron-diffusion lengths. For deeply penetrating light, they found that $L_n$ increases significantly. Near the ITO, they found a 'dead layer' and much shorter diffusion lengths. Chiang and Wagner did extensive work on p-InP using Mn as the dopant [5]. Their work was done over a large range of doping levels using the SPV technique.

TABLE 1 Electron diffusion length in p-InP (T = 300 K).

| $N_a$ ($cm^{-3}$) | $L_n$ (μm) | Material | Technique | Ref |
|---|---|---|---|---|
| $1.0 \times 10^{16}$ | 24 - 28 | wafer-Zn | SPV | [2] |
| $5.0 \times 10^{16}$ | 9.4 - 11 | wafer-Zn | SPV | [2] |
| $5.0 \times 10^{17}$ | 2 | wafer-Zn | SPV | [2] |
| $5.0 \times 10^{16}$ | 7.5 | wafer-Zn | SPC | [5] |
| $4.1 \times 10^{14}$ | 11 | wafer-Mn | SPV | [6] |
| $1.3 \times 10^{15}$ | 7 | wafer-Mn | SPV | [6] |
| $1.7 \times 10^{17}$ | 1.9 | wafer-Mn | SPV | [6] |
| $2.8 \times 10^{16}$ | 2.65 | wafer-Mn | SPV | [6] |
| $2.65 \times 10^{16}$ | 1.27 | wafer-Mn | SPV | [6] |
| $5.3 \times 10^{16}$ | 0.75 | wafer-Mn | SPV | [6] |
| $5.3 \times 10^{16}$ | 1.6 | wafer-Mn | SPV | [6] |

## C HOLE DIFFUSION LENGTH IN n-InP

The concepts involving diffusion length have been discussed above. Here the square of the hole diffusion length equals the hole diffusivity times the hole lifetime. The diffusion length was described as the average distance that a minority-carrier diffuses in a medium prior to recombining with a majority carrier. The minority carrier diffusivity D is related to the mobility through the Einstein relationship. The latter shows that D = kT/qμ, where k is Boltzmann's constant, T is the absolute temperature, q is the electron charge and μ is the carrier mobility in units of $cm^2/V$ s.

TABLE 2 Hole diffusion length in n-InP (T = 300 K).

| $N_d$ ($cm^{-3}$) | Material | L (μm) | Technique | Ref |
|---|---|---|---|---|
| undoped | wafer | 1.8 | SPV | [6] |
| $5 \times 10^{15}$ | wafer | 3.7 | EBIC | [7] |
| $5 \times 10^{15}$ | wafer | >20 | PL | [8] |
| $1.5 \times 10^{16}$ | LPE | 12.0 | PC | [9] |
| $3 \times 10^{17}$ | epi | 0.24 | SPC | [10] |
| $1 \times 10^{18}$ | wafer | 2.0 | PL | [11] |

Most of the data on the diffusion length (L) in n-InP have been acquired from the same two measurement techniques described above. The SPV technique [12] is the most commonly reported. A temporary, liquid (electrolytic) [1] Schottky junction is often used. As noted earlier, the Goodman SPV analysis ignores junction recombination and may need to be modified [1] to produce correct values of L. EBIC is another technique that has been described. If the surface-recombination velocity at the e-beam injection point is appreciable, the surface effects must be corrected for in the data analysis model. Other techniques used for diffusion length determination include surface photocurrent and photoluminescence. The photoconductive quantum efficiency was also used to calculate L in deep junctions.

## D ELECTRON DIFFUSION COEFFICIENT IN p-InP

The minority-carrier diffusivity is much more difficult to measure than the majority-carrier mobility. The former requires a special measurement technique whereas the latter is routinely done by means of the Hall effect. Workers often extrapolate majority-carrier mobility values to minority-carrier values at the same doping level. For example, to estimate the minority-electron mobility at doping level $N_a$, one uses the electron mobility in n-type material at $N_d = N_a$. Theory [12] and experiment [13] applied to GaAs have shown that the minority-carrier mobility (and diffusivity) may be quite different. In this work, heavy hole scattering lowers the electron mobility much below that of n-type GaAs at the same doping level.

Hoffman [14] and co-workers used a picosecond Nd:glass laser (λ = 1.06 μm) to excite and monitor the surface minority-carrier plasma by a diffraction effect. By fitting the plasma decay to a model, they derived D.

TABLE 3 Electron diffusion coefficient in p-InP (T = 300 K).

| $N_a$ ($cm^{-3}$) | $D_n$ ($cm^2/s$) | Technique | Ref |
|---|---|---|---|
| $7.0 \times 10^{17}$ | 4.1 | Plasma decay | [14] |

## REFERENCES

[1] C. Chiang, S. Wagner [ *IEEE Trans. Electron Devices (USA)* vol.ED-32 (1985) p.1722 ]

[2] L. Gouskov, H. Luquet, L. Soonckindt, A. Oemry, M. Boustani [ *J. Appl. Phys. (USA)* vol.53 (1982) p.7014 ]

[3] A.M. Goodman [ *J. Appl. Phys. (USA)* vol.32 (1961) p.2550 ]

[4] T. Hanak, R.K. Ahrenkiel [ *J. Appl. Phys. (USA)* vol.64 (1988) p.3528 ]

[5] C.L. Chiang, S. Wagner [ *Mater. Lett. (Netherlands)* vol.1 (1983) p.145 ]

[6] S.S. Li [ *Appl. Phys. Lett. (USA)* vol.29 (1976) p.126 ]

[7] K. Ando, M. Yamaguchi, C. Uemura [ *Jpn. J. Appl. Phys. (Japan)* vol.23 (1984) p.1406 ]

[8] S.D. Lester, T.S. Kim, B.G. Streetman [ *Appl. Phys. Lett. (USA)* vol.52 (1988) p.474 ]

[9] V. Diadiuk, S.H. Groves, C.A. Armiento, C.E. Hurwitz [ *Appl. Phys. Lett. (USA)* vol.42 (1983) p.892 ]

[10] A. Etcheberry, M. Etman, B. Fotouhi, J. Gautron, J. Scolfort, P. Lemasson [ *J. Appl. Phys. (USA)* vol.53 (1982) p.8867 ]

[11] A. Etcheberry, J. Vigeron, J.L. Sculfort, J. Gautron [ *Appl. Phys. Lett. (USA)* vol.55 (1989) p.145 ]

[12] W. Walukiewicz, J. Lagowski, L. Jastrzebski, M. Lichtensteiger, H. Gato [ *J. Appl. Phys. (USA)* vol.50 (1979) p.899 ]

[13] R.K. Ahrenkiel, D.J. Dunlavy, D. Greenberg, J. Schlupmann, H.C. Hamaker, H.F. MacMillan [ *Appl. Phys. Lett. (USA)* vol.51 (1987) p.776 ]

[14] C.A. Hoffman, K. Jarasiusnas, H.J. Gerritsen, A.V. Nurmikko [ *Appl. Phys. Lett. (USA)* vol.33 (1978) p.536 ]

# 3.7 Surface recombination velocity in InP

R.K. Ahrenkiel

June 1998

The recombination of electron-hole pairs through near midgap levels at a surface is called surface recombination. Surface recombination often is the dominant recombination mechanism in semiconductor device structures. The physics of surface recombination is described by the Shockley-Read-Hall (SRH) mechanism, which is applied to the two-dimensional surface. The rate at which minority carriers recombine at a surface can be expressed as the product of excess minority-carrier density at the surface, and the surface recombination velocity S that is characteristic of the material. Application of SRH theory to a surface with N midgap states ($cm^{-2}$) with capture cross-section $\sigma$ and minority-carrier thermal velocity V shows that $S = N \sigma V$ (cm/s).

The analysis of PL [1] intensity indicated that the free surface of n-InP has a recombination velocity (S) of less than $10^3$ cm/s by comparing the intensity with n-type GaAs. The cited work by Casey and Buehler also shows that the n-type photoluminescence intensity is about one hundred times greater than that of p-type InP. The authors do not identify the source of this disparity, i.e. whether this is a bulk or surface effect. Subsequent reports indicate that S may be smaller in n-type than in p-type InP and that S is concentration-dependent. The concentration variation of S is attributed [2] to the combination of band bending and Fermi level 'pinning'. Data [3] indicate that the Fermi level of the free InP surface is 'pinned' near the conduction band. The 'pinning' occurs near the maximum density of dangling bond states [4] that lie below $E_C$. Data [5] on electrolytic/p-InP interfaces used surface photovoltage spectroscopy to locate surface states at $E_v$ + 1.1 eV and $E_v$ + 1.2 eV. Diadiuk and co-workers [6] grew thick (~30 μm) n-type mesa structures by liquid-phase epitaxy (LPE) and deduced S from photoconductivity measurements. Their data indicated that S is approximately 100 cm/s. Hoffman [7] and co-workers used a picosecond Nd:glass laser ($\lambda$ = 1.06 μm) to excite and monitor the surface minority-carrier plasma by a diffraction effect. By fitting the decay data to a model, they calculated values of S on n-type and p-type InP. Garbuzov [8] and co-workers analysed the PL efficiency in thin n-InP films to calculate S. Transient photoconductivity measurements [9] used microwave reflection to calculate S on MOCVD-grown layers ($N_d = 3 \times 10^{15}$ $cm^{-3}$). New time-resolved PL measurements [10] on both n-type and p-type InP wafers conclude that the surface recombination is very small on both types. These crystals were etched with aqua regia etch prior to measurement. This work analyses the shape of the time-resolved photoluminescence decay curves that are a function of S. Analysis of the experiments indicated that S is about 200 cm/s for n-type and 500 cm/s for p-type InP. TABLE 1 gives literature values of S for both n- and p-type InP. Also listed is the experimental technique used to ascertain S and the reference used. The table shows a general trend of lower S for n-type than for p-type InP.

TABLE 1 Surface recombination velocity (T = 300 K).

| N ($cm^{-3}$) | S (cm/s) | Technique | Ref |
|---|---|---|---|
| n: $5 \times 10^{17}$ | $<10^3$ | PL intensity | [1] |
| n: $1.5 \times 10^{16}$ | $10^2$ | Quantum efficiency | [6] |
| p: $5 \times 10^{16}$ | 500 | TRPL | [10] |
| n: $5 \times 10^{16}$ | 200 | TRPL | [10] |
| n: $1 \times 10^{17}$ | $2 \times 10^4$ | Diffraction effect | [7] |
| p: $7 \times 10^{17}$ | $1.5 \times 10^5$ | Diffraction effect | [7] |
| n: $2 \times 10^{18}$ | $4.6 \times 10^3$ | Quantum efficiency | [8] |
| n: $3 \times 10^{15}$ | $3 - 5 \times 10^3$ | Transient microwave | [9] |

A remarkable feature of these data is that n-type InP has a free surface recombination velocity that is very small compared to that of GaAs and other III-V semiconductors.

Passivation is the reduction of S by growing a glass, native oxides (such as $SiO_2$ for Si) or epitaxial layers (AlGaAs on GaAs) to reduce S to lower values. Surface state passivation has greatly increased the quality of devices in other semiconductor technologies. In recent years, sulphide surface treatments have been used [11-13] to reduce the surface-state density on InP to less than $10^{11}$ $cm^{-2}$ $eV^{-1}$. A recent result describes using the sulphide-passivation treatment in the fabrication of metal-insulator-semiconductor devices built on InP to improve the performance [14]. Surface-state studies on these structures indicate midgap densities of about $4 \times 10^{10}$ $cm^{-2}$ $eV^{-1}$.

## REFERENCES


[1] H.C. Casey Jr., E. Buehler [ *Appl. Phys. Lett. (USA)* vol.30 (1977) p.247 ]

[2] D.D. Nolte [ *Solid-State Electron. (UK)* vol.33 (1990) p.295 ]

[3] J.M. Moison, M. Van Rompay, M. Bensoussan [ *Appl. Phys. Lett. (USA)* vol.48 (1986) p.1362 ]

[4] L.J. Brillson, Y. Shapira, A. Heller [ *Appl. Phys. Lett. (USA)* vol.43 (1983) p.174 ]

[5] Y. Byun, B.W. Wessels [ *Appl. Phys. Lett. (USA)* vol.52 (1988) p.1352 ]

[6] V. Diadiuk, S.H. Groves, C.A. Armiento, C.E. Hurwith [ *Appl. Phys. Lett. (USA)* vol.42 (1983) p.892 ]

[7] C.A. Hoffman, K. Jarasiusnas, H.J. Gerritsen, A.V. Nurmikko [ *Appl. Phys. Lett. (USA)* vol.33 (1978) p.536 ]

[8] D.Z. Garbuzov, V.V. Agaev, A.T. Gorelenok [ *Sov. Phys.-Semicond. (USA)* vol.16 (1983) p.986 ]

[9] S. Bothra, S.D. Tyagi, S.K. Ghandhi, J.M. Borrego [ *Proc. 21st Photovoltaic Specialist Conference* (IEEE, New York, USA, 1990) ]

[10] Y. Rosenwak, Y. Shapira, D. Huppert [ *Phys. Rev. B (USA)* vol.44 (1991) p.13097 ]

[11] R. Iyer, R.R. Chang, D.L. Lile [ *Appl. Phys. Lett. (USA)* vol.53 (1988) p.134 ]

[12] D. Gallet, G. Hollinger [ *Appl. Phys. Lett. (USA)* vol.62 (1993) p.982 ]

[13] Y.H. Jeong, G.T. Kim, S.T. Kim [ *J. Appl. Phys. (USA)* vol.69 (1991) p.6699 ]

[14] Y.H. Jeong, B.H. Lee, S.K. Jo, M.Y. Jeong, T. Sugano [ *Jpn. J. Appl. Phys. (Japan)* vol.34 (1995) p.1329 ]

CHAPTER 4

# BAND STRUCTURE

# 4.1 Band structure of InP: overview

E.P. O'Reilly

This Datareview appeared in *Properties of Indium Phosphide* (INSPEC, IEE, London, UK, 1991) and is reproduced here for the reader's convenience.

## A INTRODUCTION

We provide an overview of the main features of the band structure of InP here, with the next four Datareviews [1-4] focusing on particular aspects of the band structure. InP has a direct bandgap, with both the valence band maximum and the conduction band minimum at the centre of the Brillouin zone ($\Gamma$ point). The conduction band minimum is doubly degenerate ($\Gamma_6$ symmetry), and is separated by an energy $E_o$ from the valence band maximum which is four-fold degenerate ($\Gamma_8$ symmetry). Away from the zone centre, the highest valence bands separate into two sets of bands, referred to as the heavy-hole (HH) and light-hole (LH) bands, because of their large and small effective masses respectively. The next set of valence bands, referred to as the spin-split-off (SO) bands, also have their maximum at $\Gamma$, where they are doubly degenerate (of $\Gamma_7$ symmetry), separated by an energy $\Delta_o$ from the $\Gamma_8$ states.

The most important directions in the Brillouin zone lie along the [111] ($\Lambda$) direction, the [001] ($\Delta$) direction and also the [110] ($\Sigma$) direction, which reach the edge of the first Brillouin zone at the L, X and K points respectively. While the conduction band minimum is situated at $\Gamma$, higher sets of minima are found at L and near X (about 10% away from the zone boundary along $\Delta$) which are important for optical and electronic properties.

We present information in Section B on the energies of states at high symmetry points in the Brillouin zone ($\Gamma$, X and L), focusing in particular on the lowest conduction band and the three highest sets of valence bands but including also details of higher-lying conduction states. Section C presents details of carrier effective masses at the $\Gamma$ conduction band minimum and for the heavy- and light-hole valence bands.

## B SYMMETRY POINT ENERGIES

The energies of interband transitions between valence and conduction states in InP have been measured by a wide variety of optical techniques, including photoluminescence (for the $E_o$ gap), absorption, reflectance, and several reflectance-modulation techniques, including thermoreflectance, electroreflectance and wavelength-modulated reflectance. A review of the $E_o$ data and its temperature dependence is presented by Adams [1], while the optical spectra as a function of photon energy are discussed in [5]. An extensive list of experimental references is included in e.g. Lautenschlager et al [6], who used spectroscopic ellipsometry to determine the temperature dependence of transition energies from 30 K to 750 K. The various optical techniques provide information on the separation of valence and conduction states which are associated with the same wavevector (often $\Gamma$, X or L) [7]. The relative energies of states at different wavevectors must then be found by other techniques.

The energies at high symmetry points in the valence band have been determined by high resolution X-ray photoelectron spectroscopy [8], with the dispersion along $\Gamma$-K-X determined also by angle-resolved photoemission [9,10]. The gap between the valence band maximum and the lowest conduction L state has been estimated among other ways by an analysis of the energy distribution of photoelectrons [11], while the gap to the lowest conduction state at X can be inferred from high pressure photoluminescence measurements [12]. The energies at some high symmetry points in the conduction band can also be determined by combining optical transition [6] and photoemission data [8].

The band structure of InP has been calculated empirically by a number of theoretical techniques, with the band energies at various high symmetry points fitted to the experimental data. The non-local empirical pseudopotential method [13] gives symmetry point energies in the highest valence and lowest conduction bands to within about 0.2 eV, while good agreement is also obtained using local empirical pseudopotentials, e.g. [14]. Tight-binding parameters have also been developed which describe well the valence band structure and the lowest conduction band, e.g. [15].

We present in TABLE 1 the experimentally determined symmetry point energies of InP, as well as the energies calculated using the non-local empirical pseudopotential method (NEPM) [13]. The valence band maximum is chosen as the zero of energy. The direct and indirect bandgap energies are taken from low temperature measurements. The valence band energies were measured at room temperature. As it is expected that the relative valence band energies do not vary significantly with temperature [21], the values listed in TABLE 1 can be regarded as the low temperature band structure. The temperature dependence of the direct gap is discussed by Adams and Lancefield in [1], while the temperature dependence of the interband critical-point energies between 1.3 and 5.5 eV is analysed by Lautenschlager et al [6].

TABLE 1 Energies of symmetry points in GaAs band structure (in eV).

| Symmetry point | Experimental energy | NEPM [13] |
|---|---|---|
| $\Gamma_6^v$ | -11.0 [8] | -11.42 |
| $\Gamma_7^v$ | -0.108 [16] | -0.21 |
| $\Gamma_8^v$ | 0.0 | 0.0 |
| $\Gamma_6^c$ | 1.432 [1] | 1.50 |
| $\Gamma_7^c$ | 4.78 [6]* | 4.64 |
| $\Gamma_8^c$ | 4.85 [17] | 4.92 |
| $X_6^v$ | -8.9 [8] | -8.91 |
| $X_6^v$ | -5.9 [8] | -6.01 |
| $X_6^v$ | -2.0 [8]* | -2.09 |
| $X_7^v$ | -2.0 [8]* | -2.06 |
| $X_6^c$ | 2.4 [12,18] | 2.44 |
| $X_7^c$ | | 2.97 |
| $L_6^v$ | -10.0 [8] | -9.67 |
| $L_6^v$ | -5.9 [8] | -5.84 |
| $L_6^v$ | -1.0 [8]* | -1.09 |
| $L_{4,5}^v$ | -1.0 [8]* | -0.94 |
| $L_6^c$ | 2.04 [11] | 2.19 |
| $L_6^c$ | | 5.58 |
| $L_{4,5}^c$ | | 5.70 |

*The spin-orbit splittings between the highest valence bands were not determined by Ley et al [8]. The $L_{4,5}^v$ - $L_6^v$ splitting has been measured as 0.13 eV by thermoreflectance [19] and as 0.15 eV by electroreflectance [20].

## C BAND EDGE EFFECTIVE MASSES

### C1 Electron Mass

The conduction band dispersion very close to the band minimum at Γ is parabolic and isotropic. The electron effective mass m* is almost directly proportional to the direct energy gap, $E_o$, and so is expected to decrease with temperature, following the decrease in $E_o$ with temperature. Values of the effective mass measured by a variety of techniques and at different temperatures are listed in TABLE 2.

TABLE 2 Electron effective mass (in units of $m_o$).

| Temperature (K) | m* | Technique |
|---|---|---|
| 4.2 | 0.079 | Piezomodulated photovoltaic effect [16] |
| 9 | 0.0803 | Magnetospectroscopy of donor transitions [22] |
| 10 | 0.0810 | Cyclotron resonance [23] |
| 50 | 0.0817 | Cyclotron resonance [23] |
| 77 | 0.0820 | Magnetophonon data [24] |
| 80 | 0.0813 | Magnetophonon data [25] |
| 282 | 0.0763 | Magnetophonon data [25] |
| 300 | 0.078 | Magnetophonon data [24] |

The average of the measurements up to 80 K in TABLE 2 gives a low temperature effective mass of $m^* = 0.081 \pm 0.001$, with m* decreasing to $0.077 \pm 0.001$ at room temperature. A more detailed discussion of the temperature dependence of the effective mass can be found in [25].

## C2 Heavy- and Light-Hole Masses

The heavy- and light-hole valence bands are anisotropic in III-V semiconductors at the valence band maximum. This effect is most pronounced for the HH band which has a strongly directionally dependent effective mass, with a larger mass along the (111) direction compared to the (001) direction, $m_H^*(111) > m_H^*(001)$. The bulk valence band effective masses are often defined in terms of the Luttinger $\gamma$ parameters [26], $\gamma_1$, $\gamma_2$, $\gamma_3$, with

$$
\begin{aligned}
m_H^*(001) &= (\gamma_1 - 2\gamma_2)^{-1} \\
m_H^*(111) &= (\gamma_1 - 2\gamma_3)^{-1} \\
m_L^*(001) &= (\gamma_1 + 2\gamma_3)^{-1} \\
m_L^*(111) &= (\gamma_1 + 2\gamma_3)^{-1}
\end{aligned} \tag{1}
$$

The gamma values have been determined by cyclotron resonance and magnetoreflectance [27,28] in which the cyclotron hole mass or exciton mass are respectively measured as a function of magnetic field. The $\gamma_1$ and $\gamma_2$ values measured by the two techniques are in good agreement with each other, with poorer agreement between the $\gamma_3$ values. Leotin et al [27] deduce $\gamma_1 = 5.04$, $\gamma_2 = 1.56$ and $\gamma_3 = 1.73$, while Bimberg et al [28] find $\gamma_1 = 4.95 \pm 0.20$, $\gamma_2 = 1.65 \pm 0.20$ and $\gamma_3 = 2.35 \pm 0.20$, the average values then being $\gamma_1 = 5.0$, $\gamma_2 = 1.6$ and $\gamma_3 = 2.04$.

## REFERENCES

[1] A.R. Adams, D. Lancefield [ Datareview in this book: *4.2 Direct bandgap of InP, temperature dependence* ]

[2] A.R. Adams, D. Lancefield [ Datareview in this book: *4.3 Direct bandgap of InP, pressure dependence* ]

[3] A.R. Adams, D. Lancefield [ Datareview in this book: *4.4 Electron effective mass in InP, pressure dependence* ]

[4] A.R. Adams, D. Lancefield [ Datareview in this book: *4.5 Hole effective mass in InP, pressure dependence* ]

[5] A.R. Forouhi, I. Bloomer [ *Properties of Indium Phosphide* (INSPEC, IEE, London, UK, 1991) p.103-37 ]
[6] P. Lautenschlager, M. Garriga, M. Cardona [ *Phys. Rev. B (USA)* vol.36 no.8 (1987) p.4813-20 ]
[7] M. Alouani, L. Brey, N.E. Christensen [ *Phys. Rev. B (USA)* vol.37 no.3 (1988) p.1167-79 ]
[8] L. Ley, R.A. Pollak, F.R. McFeely, S.P. Kowalczyk, D.A. Shirley [ *Phys. Rev. B (USA)* vol.9 no.2 (1974) p.600-21 ]
[9] L. Sorba, V. Hinkel, H.U. Middelmann, K. Horn [ *Phys. Rev. B (USA)* vol.36 no.15 (1987) p.8075-81 ]
[10] G.P. Williams, F. Cerrina, G.J. Lapeyre, J.R. Anderson, R.J. Smith, J. Hermanson [ *Phys. Rev. B (USA)* vol.34 no.8 (1986) p.5548-57 ]
[11] L.W. James, J.P. van Dyke, F. Herman, D.M. Chang [ *Phys. Rev. B (USA)* vol.1 no.10 (1970) p.3998-4004 ]
[12] S.W. Tozer, D.J. Wolford, J.A. Bradley, D. Bour, G.B. Stringfellow [ *Proc. 19th Int. Conf. on The Physics of Semiconductors* Warsaw, Poland, Ed. W. Zawadski (Inst. Phys., Polish Acad. Sci., 1988) p.881 ]
[13] J.R. Chelikowsky, M.L. Cohen [ *Phys. Rev. B (USA)* vol.14 no.2 (1976) p.556-82 ]
[14] I. Singh, G.P. Srivastava [ *Phys. Status Solidi B (Germany)* vol.108 no.2 (1981) p.467-73 ]
[15] P. Vogl, H.J. Hjalmarson, J.D. Dow [ *J. Phys. Chem. Solids (UK)* vol.44 no.5 (1983) p.365-78 ]
[16] P. Rochon, E. Fortin [ *Phys. Rev. B (USA)* vol.12 no.12 (1975) p.5803-10 ]
[17] K.L. Shaklee, M. Cardona, F.H. Pollak [ *Phys. Rev. Lett. (USA)* vol.16 (1966) p.48 ]
[18] A. Onton, R.J. Chicotka, Y. Yacoby [ *Proc. 11th Int. Conf. on The Physics of Semiconductors* Warsaw, Poland (Polish Scientific Publishers, Warsaw) p.1023 ]
[19] E. Matataguai, A.R. Thompson, M. Cardona [ *Phys. Rev. (USA)* vol.176 (1968) p.950 ]
[20] M. Cardona, K.L. Shaklee, F.H. Pollak [ *Phys. Rev. (USA)* vol.154 (1967) p.696 ]
[21] D. Auvergne, J. Camassel, H. Mathieu, M. Cardona [ *Phys. Rev. B (USA)* vol.9 no.12 (1974) p.5168-77 ]
[22] J.M. Chamberlain, P.E. Simmonds, R.A. Stradling, C.C. Bradley [ *Proc. 11th Int. Conf. on The Physics of Semiconductors* Warsaw, Poland (Polish Scientific Publishers, Warsaw) p.1016 ]
[23] J.M. Chamberlain et al [ *J. Phys. C (UK)* vol.4 no.2 (1971) p.L38-41 ]
[24] L. Eaves et al [ *J. Phys. C (UK)* vol.4 no.2 (1971) p.L42-7 ]
[25] T. Nakashima, C. Hamaguchi, J. Komeno, M. Ozeki [ *J. Phys. Soc. Jpn. (Japan)* vol.54 no.2 (1985) p.725-35 ]
[26] J.M. Luttinger [ *Phys. Rev. B (USA)* vol.102 (1956) p.1030 ]
[27] J. Leotin, R. Barbaste, S. Askenazy, M.S. Skolnick, R.A. Stradling, J. Tuchender [ *Solid State Commun. (USA)* vol.15 no.4 (1974) p.693-7 ]
[28] D. Bimberg, K. Hess, N.O. Lipari, J.U. Fischbach, M. Altarelli [ *Physica B&C (Netherlands)* vol.89 (1977) p.139-42 ]

# 4.2 Direct bandgap of InP, temperature dependence

A.R. Adams and D. Lancefield

September 1998

The direct bandgap of InP at the Γ point, $E_0$, has been measured as a function of temperature using various optical techniques. Turner et al [1] used absorption up to 300 K, Lautenschlager et al [2] employed spectroscopic ellipsometry to follow the $E_0$ transition up to 420 K, while more recently Hang et al [3] have managed to extend the measurement to 870 K using photoreflectance while Beaudoin et al [4] have used optical absorption. Recently the temperature dependence of a number of III-V semiconductors has been modelled by Zollner et al [5].

$E_0$ decreases with increasing temperature partly due to thermal expansion but primarily due to electron-phonon interactions. The variation is approximately quadratic at low temperatures and linear at higher temperatures. It can therefore be conveniently described by the empirical equation proposed by Varshni [6]

$$E(T) = E(T = 0\,K) - \frac{aT^2}{T + b} \tag{1}$$

where T is the absolute temperature.

The best fits to this equation for $E_0$ obtained by Hang et al [3] agree approximately with earlier workers and are given in TABLE 1.

TABLE 1 Best fit parameters to EQN (1).

| Best fit parameters | $E_0$ gap | $E_0 + \Delta_0$ gap |
|---|---|---|
| E(T = 0 K) (meV) | 1432 ± 7 | 1544 ± 10 |
| a (meV/K) | 0.41 ± 0.03 | 0.40 ± 0.06 |
| b (K) | 136 ± 60 | 120 ± 80 |

For the energy gap from the spin-split-off valence band maximum to the conduction band, $E_0 + \Delta_0$, Hang et al [3] get the parameters shown in TABLE 1.

Pavesi et al [7] have used an equation of the form

$$E(T) = E_B(T = 0\,K) - \alpha\left[1 + \frac{2}{\left\{\exp\left(\frac{\theta}{T}\right) - 1\right\}}\right] \tag{2}$$

containing the Bose-Einstein occupation factor for phonons where θ is related to the average phonon frequency and α describes the strength of the interaction. Their best fit parameters for $E_0$ are given in TABLE 2. With spectroscopic ellipsometry Lautenschlager et al [2] were able to follow a number of points in the Brillouin zone to about 700 K. Their best fit values for both $E_0$ and $E_1$ (at the L point) are also given in TABLE 2.

TABLE 2 Best fit parameters to EQN (2).

| Best fit parameters | $E_0$ gap [7] | $E_0$ gap [2] | $E_1$ gap [2] |
|---|---|---|---|
| $E_B$(T = 0 K) (meV) | 1453.9 | 1629 ± 111 | 3348 ± 7 |
| α (meV) | 35.9 | 217 ± 113 | 68 ± 10 |
| θ (K) | 209 | 697 ± 177 | 224 ± 30 |

They obtained a value of 0.136 eV for the spin-orbit splitting at the L point and found it to be independent of temperature.

Readers are referred to Lautenschlager et al [2] for more detailed discussion and bibliography.

## REFERENCES


[1] W.J. Turner, W.E. Reese, G.D. Pettit [ *Phys. Rev. (USA)* vol.136 (1964) p.A1467 ]

[2] P. Lautenschlager, M. Garriga, M. Cardona [ *Phys. Rev. B (USA)* vol.36 no.9 (1987) p.4813-20 ]

[3] Z. Hang, H. Shen, F.H. Pollak [ *Solid State Commun. (USA)* vol.73 no.1 (1990) p.15-8 ]

[4] M. Beaudoin, A.J.G. DeVries, S.R. Johnson, H. Laman, T. Tiedje [ *Appl. Phys. Lett. (USA)* vol.70 no.26 (1997) p.3540-2 ]

[5] S. Zollner, S. Gopalan, M. Cardona [ *Solid State Commun. (USA)* vol.77 no.7 (1991) p.485-8 ]

[6] Y.P. Varshni [ *Physica (Netherlands)* vol.34 (1967) p.149 ]

[7] L. Pavesi, F. Piazza, A. Rudra, J.F. Carlin, M. Ilegems [ *Phys. Rev. B (USA)* vol.44 no.16 (1991) p.9052-5 ]

# 4.3 Direct bandgap of InP, pressure dependence

A.R. Adams and D. Lancefield

September 1998

As with all tetrahedrally bonded semiconductors, the direct bandgap at the centre of the Brillouin zone, $E_0$, increases with pressure in InP. The variation can be measured optically [1-6] using high pressure diamond-anvil-cell techniques up to about 10.6 GPa [5] at which point it undergoes a structural phase transition.

The increase in $E_0$ with pressure P is found to be slightly sublinear and is normally fitted to a quadratic equation of the form

$$E_0(P) = E_0 + aP + bP^2 \quad (1)$$

Using both absorption and photoluminescence Müller et al [1] obtained values for a and b at room temperature while Tozer et al [5], using photoluminescence, fitted 10 K data. The results are given in TABLE 1.

TABLE 1 Best fit parameters to EQN (1) at room temperature and 10 K.

| Fitting parameter | Room temperature data [1] | 10 K data [5] |
|---|---|---|
| a ($meV.GPa^{-1}$) | $84 \pm 2$ | 83.2 |
| b ($meV.GPa^{-2}$) | $-18 \pm 3$ | -14.1 |

From these results it can be concluded that, within experimental error, the dependence of $E_0$ on pressure is independent of temperature. (Note that values for $E_0$ at different temperatures are given in Datareview 4.2 of this book.)

## REFERENCES

[1] H. Müller, R. Trommer, M. Cardona [ *Phys. Rev. B (USA)* vol.21 no.10 (1980) p.4879-83 ]

[2] T. Kobayashi, T. Tei, K. Aoki, K. Yamamoto, K. Abe [ *Phys. of Solids Under High Pressure* vol.141, Eds. J.S. Schilling, R.N. Shelton (Amsterdam, 1981) ]

[3] C.S. Menoni, H.D. Hochheimer, I.L. Spain [ *Phys. Rev. B (USA)* vol.33 no.8 (1986) p.5896-8 ]

[4] P. Lautenschlager, M. Garriga, M. Cardona [ *Phys. Rev. B (USA)* vol.36 no.9 (1987) p.4813-20 ]

[5] S.W. Tozer, D.J. Wolford, J.A. Bradley, D. Bour, G.B. Stringfellow [ *19th Int. Conf. Physics of Semiconductors* Ed. W. Zawadzki (Inst. Phys., Polish Acad. Sci., 1988) p.881 ]

[6] M. Leroux [ *Semicond. Sci. Technol. (UK)* vol.4 no.4 (1989) p.231-2 ]

# 4.4 Electron effective mass in InP, pressure dependence

A.R. Adams and D. Lancefield

September 1998

The electron effective mass in InP is almost directly proportional to the direct energy gap, $E_0$. This increases with pressure as described in Datareview 4.3 [1] leading to an increase in effective mass of $(1/m_e^*)(dm_e^*/dP) = 0.053\ GPa^{-1}$.

Shantharama et al [2] studied high purity InP grown by LPE on semi-insulating substrates. They determined the effective mass from magnetophonon oscillations in magnetic fields up to 9 T while applying hydrostatic pressure up to 2 GPa. By also measuring the photoconductive response they were able to determine $E_0$ and $m_e^*$ simultaneously as pressure was applied. Thus they were able to compare their results directly with **k.p** theory. After correcting for non-parabolicity and polaron enhancement effects, best fit was obtained with

$$\frac{dm_e^*}{dE_0} = 0.049 \pm 0.002\, m_0\, eV^{-1}$$

Taking the pressure coefficient of the direct bandgap from [1]

$$\frac{dE_0}{dP} = 83 \pm 2\ meV\ GPa^{-1}$$

gives

$$\frac{dm_e^*}{dP} = 41 \pm 3 \times 10^{-4}\, m_0\, GPa^{-1}$$

Taking the atmospheric pressure value of the effective mass $m_e^* = 0.0789\, m_0$ gives

$$\frac{1}{m_e^*}\frac{dm_e^*}{dP} = 0.053 \pm 0.004\, GPa^{-1}$$

## REFERENCES


[1] A.R. Adams, D. Lancefield [ Datareview in this book: *4.3 Direct bandgap of InP, pressure dependence* ]

[2] L.G. Shantharama, A.R. Adams, C.N. Ahmad, R.J. Nicholas [ *J. Phys. C (UK)* vol.17 no.25 (1984) p.4429-42 ]

# 4.5 Hole effective mass in InP, pressure dependence

A.R. Adams and D. Lancefield

September 1998

No direct measurement has yet been made of the hydrostatic pressure dependence of the effective mass of holes in InP. The heavy hole effective mass is highly anisotropic and its variation with pressure will depend on the direction of motion being considered. For hydrostatic pressures the valence band degeneracy is unaltered while for biaxial strains the hole masses can be calculated from the Kohn-Luttinger parameters for the valence band following the method described by Pollak [1].

Adams and Shantharama [2] managed to obtain a measure of the pressure dependence of an average effective mass by studying the pressure dependence of the hole mobility in relatively lightly doped samples of p-type material ($5 \times 10^{15}$ to $1 \times 10^{17}$ $cm^{-3}$) grown by LPE on semi-insulating substrates. Considering scattering by polar, non-polar and acoustic phonons they concluded that the effective mass $m_h$* varied according to

$$\frac{1}{m_h^*}\frac{dm_h^*}{dP} = +8.0 \times 10^{-3}\ GPa^{-1}$$

This compares reasonably well with a theoretical value of $+5 \times 10^{-3}$ $GPa^{-1}$ they obtained using the valence band parameters of Lawaetz [3].

According to **k.p** theory the pressure dependence of the light hole effective mass should approximately reflect that of the electron effective mass given in [4].

## REFERENCES


[1] F.H. Pollak [ *Semicond. Semimet. (USA)* vol.32 (1990) p.17-53 ]

[2] A.R. Adams, L.G. Shantharama [ *Physica B & C (Netherlands)* vol.139/140 (1986) p.419-22 ]

[3] P. Lawaetz [ *Phys. Rev. B (USA)* vol.4 no.10 (1971) p.3460-7 ]

[4] A.R. Adams, D. Lancefield [ Datareview in this book: *4.4 Electron effective mass in InP, pressure dependence* ]

# 4.6 Conduction band and valence band offsets at various InP/semiconductor interfaces

J.D. Lambkin, A.F. Phillips and D. Lancefield

August 1999

## A INTRODUCTION

The band edge discontinuities at semiconductor heterojunctions have an important effect on the electronic and optical properties of heterojunction based devices. Consequently considerable experimental and theoretical studies have been undertaken but the basic physics remains controversial [1]. Are band offsets determined by the bulk properties of the constituent materials or are they influenced by dipoles, segregation or strain at the interface? If they are an intrinsic property of the constituent materials they should be independent of the growth technique and the rules of band offset transitivity and commutativity would apply. Transitivity implies that the band offset between materials A and C (A-C) can be obtained by adding the offsets between materials A-B and the offsets between materials B-C. If true this provides a useful method of calculating band offsets. The commutativity rule implies that the interface A-B is equivalent to the interface B-A. If this is not the case then in multiple quantum wells and superlattices the structure must develop a built-in electric field by accumulating charge at the boundaries of these regions. This may cause energy levels to be Stark shifted and the appearance of parity-forbidden transitions as a result of the broken symmetry of the structure. In the sections that follow we have tried to distinguish between data collected from superlattice (SL), quantum well (QW) and heterojunction (HJ) structures. We have also tried to distinguish between 'direct' and 'inverse' interfaces. In the absence of any experimental data for either strained or unstrained interfaces, it is suggested that a theoretical estimate of $\Delta E_c$ and $\Delta E_v$ may be easily found using the model solid theory of Van de Walle [2].

## B $In_xGa_{1-x}As/InP$

Several groups have investigated the $In_xGa_{1-x}As/InP$ interface band offsets, by a variety of techniques. Most measurements have been reported for the lattice matched condition when x = 0.53. In this case a spread of values is reported ranging from around 160 meV to 290 meV for the conduction band offset $\Delta E_c$, with a corresponding spread in the valence band offset, $\Delta E_v$. Most authors quote either $\Delta E_c$ and $\Delta E_v$, or the fractional valence-band and conduction-band offsets, $Q_v$ and $Q_c$ for type I heterojunctions, where $Q_v = \Delta E_v/\Delta E_g$ and $Q_c = \Delta E_c/\Delta E_g$. We have used the total band offset $\Delta E_g = \Delta E_c + \Delta E_v = 600$ meV, at room temperature, to evaluate the other, in order to facilitate easy comparison between authors. The results obtained for various structures are summarised in TABLE 1. Further details are given below.

Cavicchi et al [3] used admission spectroscopy on a multi-QW (MQW) structure grown by gas source MBE with x = 0.37, x = 0.53 and x = 0.69 and found values for $\Delta E_c$ of 175 ± 25 meV, 210 ± 20 meV (implying $\Delta E_v$ = 390 meV) and 315 ± 25 meV, respectively. They point out that knowing the well thickness and strain is crucial to the analysis of their data and therefore they carried out a high resolution X-ray diffraction scan whose data were computer simulated using a kinematic step model. These results were backed up by high-resolution transmission electron microscopy. Photoconductivity measurements were conducted over a range of alloy compositions grown by gas source MBE by Gershoni et al [4] who observe, as expected, a cross-over from a heavy-hole valence band to a light-hole valence band as the material changes from biaxial compression to biaxial tension. Their data fit a model by Hrivnak [5] where $\Delta E_v$ has been linearly interpolated from the relative valence band energies of the unstrained binary compounds. This model predicts that at low values of x ($x \leq 0.2$) the quantum well is type II with the electrons being confined in the InP barriers.

TABLE 1 Summary of conduction band offsets for the various lattice matched InGaAs/InP structures indicated.

| Conduction band offset (meV) | | | Ref |
|---|---|---|---|
| Quantum well structure | 'Direct' heterojunction | 'Inverse' heterojunction | |
| 210 ± 20 | | | [3] |
| | 250 ± 10 | | [6] |
| | 160 ± 50 | 340 ± 60 | [7] |
| | 170, 186 | 272, 260 | [8] |
| | | 260 ± 10 | [9] |
| | 290 ± 40 | | [10] |
| | 203 ± 15 | 1600 ± 20 | [11] |

Lang and co-workers [6] also used admittance spectroscopy to determine the band offsets in lattice matched MOVPE-grown and MOMBE-grown structures containing a superlattice of HJs. They obtained values of 250 ± 10 meV and 346 ± 10 meV for the conduction and valence band offsets respectively. Using ultraviolet photoemission Landesman et al [7] measured $\Delta E_v$ = 440 ± 50 meV for lattice matched 'direct' InGaAs grown on InP by MOMBE, but $\Delta E_v$ = 260 ± 60 meV for 'inverse' InP grown on InGaAs. They conducted a detailed analysis of the core-level spectra and concluded that this asymmetry is caused by a difference in the chemical nature of the two interfaces. Seidel et al [8] also measured conduction band offsets in MBE grown InGaAs on InP and InP on InGaAs by conduction techniques and UV photoemission. They found conduction band offsets of 186 meV and 170 meV for InGaAs on InP by transport and photoemission respectively, and for the inverted interfaces conduction band offsets of 272 meV and 260 meV. However, Waldrop et al [9] measured $\Delta E_v$ to be 340 meV for InP grown on lattice matched InGaAs grown by MBE. They used X-ray photoemission spectroscopy where photoelectrons are only detected that originate within ~20 Å of the sample surface. Almeida et al [10] used internal photoemission (IPE) to measure the conduction band offset on CBE-grown samples. They scanned from low to high energy and observed a transition from the valence band of the lattice matched InGaAs on InP to the conduction band of the InGaAs, followed by a second transition corresponding to the InGaAs valence band to the InP conduction band. The difference between the two thus gave $\Delta E_c$, with a value of 290 ± 40 meV (corresponding to $\Delta E_v$ = 310 meV), a larger value than found by other groups. Almeida et al [10] also introduced a δ-doped Si layer 10 Å from the interface and found that the conduction band offset decreased by 110 ± 40 meV at 77 K. Also using IPE, Haase et al [11] measured $\Delta E_c$ = 203 ± 15 meV corresponding to a $\Delta E_v$ of 397 meV. They conducted their measurements between 135 K and 297 K and deduced a temperature dependence for $\Delta E_c$ of -0.2 ± 0.1 meV/K. Their samples were grown by liquid phase epitaxy (LPE). Higgs et al [12] used conductance measurements on mesas of different diameters fabricated on InGaAs lattice matched to InP grown by atmospheric pressure MOVPE and found values of $\Delta E_c$ in the range 140 meV - 190 meV, where the highest values were generally associated with the smallest mesas (50 μm) and the lowest values with the largest mesas (400 μm). Ogawa et al [13] used C-V profiling, backed up by computer simulation, and obtained a value for $\Delta E_c$ of 160 ± 20 meV for a VPE grown inverse HJ structure. Leibovitch et al [14] monitored the evolution of the surface photovoltage with overlayer thickness in their MOVPE-grown samples in conjunction with simulation, and determined values for $\Delta E_c$ and $\Delta E_v$ of 275 ± 30 meV and 325 ± 30 meV respectively. Theoretical studies have been carried out by Nag et al [15], who find that the best fit to experiment is for $\Delta E_c/\Delta E_g$ = 0.4, i.e. conduction and valence band offsets of 240 meV and 360 meV, respectively. Hybertsen [16] performed first principles calculations and found $\Delta E_v$ = 350 - 410 meV, giving $\Delta E_c$ between 250 and 190 meV.

In conclusion, a range of values for $\Delta E_c$ between 160 meV and 290 meV has been measured. Taking an overall view of these papers and references contained therein, the general consensus is that $\Delta E_c$ has a value of about 260 ± 20 meV, with a corresponding value for $\Delta E_v$ of 340 ± 20 meV.

## C UNSTRAINED $In_{1-x}Ga_xAs_yP_{1-y}$/InP ($y \approx 2.2x$)

Very little has been reported on this materials system in recent years. Forrest et al [17] used C-V profiling on uniformly n-doped heterojunctions grown by LPE across the whole spectrum of compositions where InGaAsP remains lattice matched to InP. They find that $\Delta E_c = (0.39 \pm 0.01)\Delta E_g$ consistently across the entire range. Ogura et al [18], however, report that $\Delta E_c$ varies between $0.50\Delta E_g$ and $0.70\Delta E_g$ using samples of InP/InGaAs also grown by LPE. Forrest et al [19] point out that the largest errors associated with this technique are in measuring the background concentration on both sides of the heterojunction and ensuring that the doping is uniform across the interface. Both Forrest et al [20] and Ogura et al [18] report that $\Delta E_c$ remains constant at high temperatures but tends towards zero below 150 K, a surprising result explained by Lang et al [21], who suggest that this is a measurement problem in the C-V technique itself due to the strongly temperature dependent series resistance of the InP depletion region. They used admittance spectroscopy to investigate superlattices grown by MOCVD to obtain $\Delta E_c = 250 \pm 10$ meV and $\Delta E_v = 346 \pm 10$ meV, in excellent agreement with the results of Forrest et al.

Most successful optical methods require observation of a free hole transition from the InP valence band edge to the first confined electron state in the well, and/or a transition from the first heavy hole level in the well to free electrons at the conduction band edge of the InP. Montie et al [22] used PLE to deduce $Q_c = 0.34 \pm 0.01$ in a single quantum well of InP/$In_{0.57}Ga_{0.43}As_{0.92}P_{0.08}$. Using low temperature photoconductivity Zachau et al investigated 100 Å InP/$In_{1-x}Ga_xAs_yP_{1-y}$ single quantum wells with y = 1.0, 0.835 and 0.755 [23]. They found $\Delta E_c$ to be 219, 183 and 178 meV respectively with $\Delta E_v$ = 390, 342 and 306 meV, supporting the result of Forrest et al [19]. In stark disagreement with the results previously mentioned, Brunemeier et al [24] found that $\Delta E_c = 0.65\Delta E_g$ at 77 K for all compositions of InGaAsP lattice matched to InP. However, it should be noted that they studied quantum wells grown by LPE. LPE quantum wells suffer from severe grading at the interfaces that could lead to incorrect interpretation of the luminescence spectrum.

## D AlInAs/InP AND AlInAs/InGaAs

We now turn our attention to the AlInAs/InP and the AlInAs/InGaAs material systems, of importance for high-power electronic device applications due to the high electron saturation velocity and mobility in the InP channel. $Al_xIn_{1-x}As$ is lattice matched to InP with x = 0.48. In spite of the type II nature of the band alignment for lattice matched AlInAs/InP, where the electrons are confined in the InP while the holes are confined in the AlInAs, it exhibits good optical properties. It is thus possible to use optical techniques such as PL, PLE or photocurrent measurements to measure the bandgap of this type II system which has a linear dependence on the band offsets, in contrast to the type I systems where equivalent measurements are rather insensitive to the offsets.

Hybertsen [16] performed first principle calculations and achieved reasonable agreement with experiment. He quotes a value of 250 meV for the valence band offset from his theory. Seidel et al [8] used a combination of internal photoemission, temperature dependent current voltage and photoluminescence excitation experiments on MBE-grown samples. They found a value for the conduction band offset of 375 meV for the 'direct' AlInAs grown on InP and 322 meV for the 'inverse' InP grown on AlInAs, showing evidence of non-commutativity. Photocurrent measurements conducted by Lugagne-Delpon et al [25] studied strained $Al_xIn_{1-x}As$, x = 0.53, samples grown by MOVPE. They suggest values for $\Delta E_c$ of 460 meV for the 'direct' HJ structure and 360 meV for the 'inverse' structure. They claim that this 100 meV difference is quite reproducible. The microscopic reason for the strong asymmetry is still unknown but several groups [26,27] discuss the role of the intrinsic asymmetry in the chemical bonds at the two interfaces because the materials share no common atom. The results may also depend on the growth process. Exchange mechanisms resulting in segregation have also been considered. Further optical measurements on samples of AlInAs grown on InP with different alloy compositions give a formula for the average conduction offset in superlattice structures as $\langle \Delta E_c \rangle = 410 \pm 20 + 1500\delta x$ (meV), where $\delta x$ is the deviation from the lattice matched composition of x = 0.48. Böhrer et al [28] report a smaller value for the conduction band offset in lattice matched material, grown by CBE, of 252 meV from photoluminescence measurements, Shubnikov-de-Haas measurements and self-consistent band structure measurements. Further, they report that the three conduction band discontinuities $In_xGa_{1-x}As/InP$, $InP/In_{1-y}Al_yAs$ and $In_{1-y}Al_yAs/In_xGa_{1-x}As$ are transitive. Abraham et al [29] report on the temperature dependence of $Al_{0.46}In_{0.54}As/InP$ photoluminescence transition energies using samples grown by MOVPE. They determine the band offsets from solving the Schrödinger and Poisson equations self-consistently. They find $\Delta E_c = 384$ meV and $\Delta E_v = 295$ meV at 4.5 K. They further investigate the temperature dependence of the band offsets, finding 349 meV and 272 meV for the conduction and valence band offsets at 300 K, respectively. They correct for the lattice mismatch quoting $\Delta E_c = 410$ meV at 4.5 K for $Al_{0.48}In_{0.52}As$. Further, they conclude that the large spread of band offset values reported in the literature is due to both the strong dependence of the PL energy on the growth conditions of the AlInAs/InP interface and the fact that the PL from AlInAs is dominated by impurities.

For the AlInAs/GaInAs interface there is a consensus for $\langle \Delta E_c \rangle$ of 500 ± 50 meV [30] although Waldrop et al [9] obtain a value of 470 meV and state that transitivity is satisfied.

## E AlInP/InGaAs AND AlInP/InP

Little has been published on the nature of the AlInP/InGaAs interface: indeed Zhu et al [31] claim to be the first to investigate the conduction band offsets as a function of Al composition in $Al_xIn_{1-x}P/Ga_{0.47}In_{0.53}As$. Their material was grown by MOVPE. AlInP is becoming important for use as a barrier material for InGaAs for use in devices such as metal-semiconductor-metal photodetectors (MSM-PDs), HEMTs and resonant tunnelling barrier diodes where there is a requirement for a large Schottky barrier height and/or a large conduction band offset. Zhu et al report that the conduction band offset is linear with Al composition and quote a value for $\Delta E_c$ of 620 meV for an Al composition of 28%. Further they report that the Schottky barrier height ($\Phi_B$) has the same Al composition dependence and for the same composition find a value of 0.90 eV. They determined these values from current-voltage and capacitance-voltage measurements on Schottky barrier diodes.

Lammasniemi et al [32] report measurements of the band offsets in AlInP/InP, also to determine its suitability as a barrier material. Absorption measurements were made on superlattice structures with 15% Al and multiple quantum well structures with 20% Al. All the samples were grown by gas source MBE. The superlattices had a conduction band offset of 260 meV while the value for the MQW structures was 337 meV. In the superlattice structure the heavy hole valence band offset was found to be 49 meV and the light hole offset -35 meV, i.e. the light hole valence band forms a staggered type II junction with InP. Similarly in the MQW structure the heavy hole offset was 85 meV and the light hole offset -29 meV. No estimate of errors is made. It should also be noted that in both cases the InP substrate was removed by selective etching.

## F $InAs_xP_{1-x}/InP$

In the last decade the $InAs_xP_{1-x}/InP$ material system has been studied for ultrafast optical modulators, lasers and solar cells operating between 0.9 μm and 1.6 μm wavelength. Beaudoin et al [33] measured the absorption spectra for MQW samples over the range of compositions $0.11 \leq x \leq 0.35$

grown by LP-MOVPE. They then fitted to the excitonic peak energy positions. The transition energies were determined from a solution to the Schrödinger equation in the envelope function formulation including strain effects. The band offsets were then deduced self-consistently. They obtained a conduction band offset that was a linear function of the As composition varying from about 75 meV for x = 0.12 to about 260 meV for x = 0.35. They observed no difference in the offset between 8 K and 300 K. Monier et al [34] fitted photocurrent and photoluminescence measurements for $In_{0.58}As_{0.42}P$/InP CBE-grown samples with well thicknesses of 24, 27 and 55 Å. They then applied the same fitting techniques to find $\Delta E_c$ over the As composition range $0.26 < x < 0.65$. The best agreement between experiment and theory was obtained for $\Delta E_c = (0.70 \pm 0.02)\Delta E_g$ for the entire composition range with no temperature dependence. In contrast to these findings, Anan et al [35] fitted photocurrent and absorption measurements and obtained $\Delta E_c$ = 135 meV for x = 0.35, $\Delta E_c$ = 163 meV for x = 0.45 and $\Delta E_c$ = 160 meV for x = 0.52 using samples grown by gas source MBE. They acknowledge that these results are somewhat smaller than previously reported values, but argue that their method does not necessitate knowledge of the exact values for the deformation potentials.

## G OTHER INTERFACES

Hu et al [36] have reported conduction band offsets of $GaAs_{0.5}SbO_{0.5}/In_{0.52}Ga_{0.48}As$ and GaAsSb lattice matched to InP of 0.36 eV and 0.18 eV, respectively. The samples were grown by MOVPE and the offsets obtained from an analysis of the type II PL obtained at 4.2 K.

Chang et al [37] have reported temperature dependent PL on MOVPE grown $Ga_{0.64}In_{0.36}As_{0.84}SbO_{0.16}$/InP heterostructures. They estimate $\Delta E_v$ to be 70 ± 5% of the bandgap difference.

The valence band offset in InP/GeSi has been calculated using an average bond energy model by Ke et al [38]. Continenza et al [39] have used local density theory to calculate band offsets in $(InAs)_n/(InP)_n$ (001) strained superlattices and Zhang et al [40] have calculated the offsets in InAlGaAs/InAlAs (001) lattice matched to InP.

## REFERENCES


[1] F. Capasso, G. Magaritondo (Eds.) [ *Heterojunction Band Discontinuities: Physics and Device Applications* (North-Holland, Amsterdam, 1987) ]

[2] C.G. Van de Walle [ *Phys. Rev. B (USA)* vol.39 no.3 (1989) p.1871-83 ]

[3] R.E. Cavicchi et al [ *Appl. Phys. Lett. (USA)* vol.54 no.8 (1989) p.739-41 ]

[4] D. Gershoni, H. Temkin, J.M. Vandenberg, S.N.G. Chu, R.A. Hamm, M.B. Panish [ *Phys. Rev. B (USA)* vol.60 no.5 (1988) p.448-51 ]

[5] L. Hrivnak [ *Phys. Status Solidi A (Germany)* vol.124 (1991) p.K111-6 ]

[6] D.V. Lang et al [ *J. Vac. Sci. Technol. B (USA)* vol.5 no.4 (1987) p.1215-20 ]

[7] J.P. Landesman et al [ *Appl. Phys. Lett. (USA)* vol.60 no.8 (1992) p.1241-3 ]

[8] W. Seidel, O. Krebs, P. Voisin, J.C. Harmand, F. Aristone, J.F. Palmier [ *Phys. Rev. B (USA)* vol.55 no.4 (1997) p.2274-9 ]

[9] J.R. Waldrop, E.A. Kraut, C.W. Farley, R.W. Grant [ *J. Appl. Phys. (USA)* vol.69 no.1 (1991) p.372-8 ]

[10] J. Almeida et al [ *J. Appl. Phys. (USA)* vol.78 no.5 (1995) p.3258-61 ]

[11] M.A. Haase, N. Pan, G.E. Stillman [ *Appl. Phys. Lett. (USA)* vol.54 no.15 (1989) p.1457-9 ]

[12] A.W. Higgs, H.J. Hutchinson, L.L. Taylor, N. Apsley, S.J. Bass [ *Semicond. Sci. Technol. (UK)* vol.5 (1990) p.581-5 ]

[13] M. Ogawa, M. Miyoshi, S. Shimizu, T. Murakami, T. Miyoshi [ *Jpn. J. Appl. Phys. (Japan)* vol.27 no.7 (1988) p.L1334-6 ]

[14] M. Leibovitch et al [ *Appl. Phys. Lett. (USA)* vol.69 no.17 (1996) p.2587-9 ]
[15] B.R. Nag, S. Mukhopadhyay [ *Appl. Phys. Lett. (USA)* vol.58 no.10 (1989) p.1056-8 ]
[16] M.S. Hybertsen [ *Appl. Phys. Lett. (USA)* vol.58 no.16 (1991) p.1759-61 ]
[17] S.R. Forrest, P.H. Schmidt, R.B. Wilson, M.L. Kaplan [ *Appl. Phys. Lett. (USA)* vol.45 no.11 (1984) p.1199-201 ]
[18] M. Ogura, M. Mitzuta, K. Onaka, H. Kukimoto [ *Jpn. J. Appl. Phys. (Japan)* vol.22 no.10 (1983) p.1502-9 ]
[19] S.R. Forrest, P.H. Schmidt, R.B. Wilson, M.L. Kaplan [ *J. Vac. Sci. Technol. B (USA)* vol.4 no.1 (1985) p.37-44 ]
[20] S.R. Forrest, O.K. Kim [ *J. Appl. Phys. (USA)* vol.53 no.8 (1982) p.5738-45 ]
[21] D.V. Lang et al [ *Appl. Phys. Lett. (USA)* vol.50 no.12 (1987) p.736-8 ]
[22] E.A. Montie, P.J.A. Thijs, G.W. 't Hooft [ *Appl. Phys. Lett. (USA)* vol.53 no.17 (1988) p.1611-3 ]
[23] M. Zachau et al [ *Superlattices Microstruct. (UK)* vol.5 no.1 (1989) p.19-22 ]
[24] P.E. Brunemeier, D.G. Deppe, N. Holonyak Jr. [ *Appl. Phys. Lett. (USA)* vol.46 no.8 (1985) p.755-7 ]
[25] E. Lugagne-Delpon, J.P. André, P. Voisin [ *Solid State Commun. (USA)* vol.86 no.1 (1993) p.1-6 ]
[26] Y. Foulon, C. Priester [ *Phys. Rev. B (USA)* vol.45 (1992) p.6259 ]
[27] R.G. Dandrea, C.B. Duke, A. Zunger Kaplan [ *J. Vac. Sci. Technol. B (USA)* vol.10 (1992) p.1744 ]
[28] J. Böhrer, A. Krost, T. Wolf, D. Bimberg [ *Phys. Rev. B (USA)* vol.47 no.11 (1993) p.6439-44 ]
[29] P. Abraham, M.A. Garcia Perez, T. Benyattou, G. Guillot, M. Sacilotti, X. Letartre [ *Semicond. Sci. Technol. (UK)* vol.10 (1995) p.1585-94 ]
[30] M. Quillec [ *Proc. SPIE - Int. Soc. Opt. Eng. (USA)* vol.1361 (1991) p.34 ]
[31] Y. Zhu, Y. Ishimaru, N. Takahashi, M. Shimizu [ *J. Appl. Phys. (USA)* vol.80 no.3 (1996) p.1617-22 ]
[32] J. Lammasniemi, K. Tappura, K. Smekalin [ *Appl. Phys. Lett. (USA)* vol.65 no.20 (1994) p.2574-5 ]
[33] M. Beaudoin et al [ *Phys. Rev. B (USA)* vol.53 no.4 (1996) p.1990-6 ]
[34] C. Monier, M.F. Milela, I. Serdiukova, A. Freundlich [ *Appl. Phys. Lett. (USA)* vol.72 no.13 (1998) p.1587-9 ]
[35] T. Anan, K. Nishi, A. Tomita, K. Tokutome, S. Sugou [ *Jpn. J. Appl. Phys. (Japan)* vol.37 no.7 (1998) p.3915-8 ]
[36] J. Hu et al [ *Appl. Phys. Lett. (USA)* vol.73 no.19 (1998) p.2799-801 ]
[37] J.R. Chang, Y.K. Su, Y.T. Lu, D.H. Jaw, H.P. Shiao, W. Lin [ *Appl. Phys. Lett. (USA)* vol.74 no.5 (1999) p.717-9 ]
[38] S.H. Ke, R.Z. Wang, M.C. Huang [ *J. Appl. Phys. (USA)* vol.80 no.2 (1996) p.872-9 ]
[39] A. Continenza, S. Massidda, A.J. Frreman [ *Phys. Rev. B (USA)* vol.42 no.6 (1990) p.3469-77 ]
[40] X.H. Zhang, S.J. Chua, S.J. Xu, W.J. Fan [ *J. Appl. Phys. (USA)* vol.83 no.11 (1998) p.5852-4 ]

CHAPTER 5

# OPTICAL PROPERTIES

# 5.1 Refractive index of InP as a function of photon energy

A.R. Forouhi and I. Bloomer

This Datareview appeared in *Properties of Indium Phosphide* (INSPEC, IEE, London, UK, 1991) and is reproduced here for the reader's convenience.

Room temperature n and k data of InP at photon energies in the static-IR, reststrahlen, transparent, bandgap, and visible-UV regions, have been determined by different investigators. In the 0.00124 to 0.0186 eV (static-IR) range, k ≅ 0. Also, in the 0.1127 to 1.2651 eV (transparent) range, k ≅ 0, but exact values of k are very sensitive to extrinsic parameters such as doping. At certain energies, more than one set of values of n and k, reported by different authors, are presented. The discrepancies between these different sets requires further investigation. At certain other energies, values of n or k are denoted by the '-' symbol. In such cases, interpolation of the closest given values for the optical constants may be used.

TABLE 1.

| Energy (eV) | Wavelength (μm) | n | | k | |
|---|---|---|---|---|---|
| 0.00124 | 1000.0 | 3.522 | [1] | | |
| 0.00248 | 500.0 | 3.523 | | | |
| 0.00496 | 250.0 | 3.529 | | | |
| 0.00744 | 166.7 | 3.538 | | | |
| 0.00992 | 125.0 | 3.551 | | | |
| 0.01240 | 100.0 | 3.57 | | | |
| 0.01860 | 66.67 | 3.65 | | | |
| 0.02790 | 44.48 | 3.35 | [1,2] | 0.214 | [1,2] |
| 0.02827 | 43.86 | 3.39 | | 0.234 | |
| 0.02864 | 43.29 | 3.39 | | 0.234 | |
| 0.02901 | 42.74 | 3.40 | | 0.288 | |
| 0.02938 | 42.19 | 3.43 | | 0.302 | |
| 0.02976 | 41.67 | 3.43 | | 0.316 | |
| | | 4.13 | [1] | 0.022 | [1] |
| 0.03013 | 41.15 | 3.44 | [1,2] | 0.320 | [1,2] |
| 0.03050 | 40.65 | 3.47 | | 0.324 | |
| 0.03087 | 40.16 | 3.47 | | 0.320 | |
| 0.03124 | 39.68 | 3.47 | | 0.309 | |
| 0.03162 | 39.22 | 3.47 | | 0.295 | |
| 0.03199 | 38.76 | 3.47 | | 0.302 | |
| 0.03236 | 38.31 | 3.51 | | 0.257 | |
| 0.03273 | 37.88 | 3.55 | | 0.282 | |
| 0.03310 | 37.45 | 3.55 | | 0.282 | |
| 0.03348 | 37.04 | 3.67 | | 0.263 | |
| | | 4.79 | [1] | 0.068 | [1] |
| 0.03385 | 36.63 | 3.72 | [1,2] | 0.257 | [1,2] |
| 0.03422 | 36.23 | 3.76 | | 0.282 | |
| 0.03459 | 35.84 | 3.80 | | 0.302 | |
| 0.03496 | 35.46 | 3.89 | | 0.282 | |
| 0.03534 | 35.09 | 3.98 | | 0.229 | |
| | | 5.73 | [1] | 0.183 | [1] |
| 0.03571 | 34.72 | 4.12 | [1,2] | 0.219 | [1,2] |
| 0.03608 | 34.35 | 4.37 | | 0.162 | |
| 0.03645 | 34.01 | 4.68 | | 0.120 | |
| | | 7.28 | [1] | 0.525 | [1] |
| 0.03682 | 33.67 | 5.01 | [1,2] | 0.162 | [1,2] |

TABLE 1 continued.

| Energy (eV) | Wavelength (μm) | n | | k | |
|---|---|---|---|---|---|
| 0.03720 | 33.33 | 6.17 | | 0.447 | |
| | | 10.4 | [1] | 2.13 | [1] |
| 0.03757 | 33.00 | 7.24 | [1,2] | 1.26 | [1,2] |
| | | 12.7 | [1] | 8.24 | [1] |
| 0.03794 | 32.68 | 7.41 | [1,2] | 3.31 | [1,2] |
| | | 4.00 | [1] | 11.1 | [1] |
| 0.03831 | 32.36 | 5.89 | [1,2] | 5.25 | [1,2] |
| 0.03868 | 30.05 | 4.57 | | 5.56 | |
| | | 0.822 | [1] | 6.26 | [1] |
| 0.03906 | 31.75 | 3.24 | [1,2] | 5.50 | [1,2] |
| 0.03943 | 31.45 | 2.09 | | 5.25 | |
| | | 0.409 | [1] | 4.37 | [1] |
| 0.03980 | 31.15 | 1.20 | [1,2] | 4.52 | [1,2] |
| 0.04017 | 30.86 | 0.759 | | 3.94 | |
| | | 0.275 | [1] | 3.24 | [1] |
| 0.04054 | 30.58 | 0.447 | [1,2[ | 3.31 | [1,2] |
| 0.04092 | 30.30 | 0.251 | | 2.75 | |
| | | 0.221 | [1] | 2.41 | [1] |
| 0.04129 | 30.03 | 0.170 | [1,2] | 2.29 | [1,2] |
| 0.04166 | 29.76 | 0.174 | | 1.86 | |
| | | 0.209 | [1] | 1.69 | [1] |
| 0.04203 | 29.59 | 0.224 | [1,2] | 1.48 | [1,2] |
| 0.04240 | 29.24 | 0.282 | | 1.12 | |
| | | 0.273 | [1] | 0.92 | [1] |
| 0.04277 | 28.99 | 0.355 | [1,2] | 0.794 | [1,2] |
| 0.04315 | 28.74 | 0.631 | | 0.398 | |
| | | 0.869 | [1] | 0.22 | [1] |
| 0.04352 | 28.49 | 1.00 | [1,2] | 0.240 | [1,2] |
| 0.04389 | 28.25 | 1.32 | | 0.324 | |
| 0.04401 | 28.17 | 1.418 | [1] | 0.099 | [1] |
| 0.04426 | 28.01 | 1.41 | [1,2] | 0.427 | [1,2] |
| 0.04463 | 27.78 | 1.41 | | 0.398 | |
| | | 1.653 | [1] | 0.07 | [1] |
| 0.04501 | 27.55 | 1.45 | [1,2] | 0.316 | [1,2] |
| 0.04538 | 27.32 | 1.51 | | 0.234 | |
| 0.04575 | 27.10 | 1.68 | | 0.224 | |
| 0.04612 | 26.88 | 1.70 | | 0.240 | |
| 0.04649 | 26.67 | 1.76 | | 0.234 | |
| | | 2.072 | [1] | 0.035 | [1] |
| 0.04687 | 26.46 | 1.84 | [1,2] | 0.234 | [1,2] |
| 0.04724 | 26.25 | 1.84 | | 0.229 | |
| 0.04761 | 26.04 | 1.86 | | 0.182 | |
| 0.04798 | 25.84 | 1.93 | | 0.129 | |
| 0.04835 | 25.64 | 2.00 | | 0.115 | |
| | | 2.305 | [1] | 0.021 | [1] |
| 0.04873 | 25.45 | 2.02 | [1,2] | 0.126 | [1,2] |
| 0.04910 | 25.25 | 2.04 | | 0.126 | |
| 0.04947 | 25.06 | 2.07 | | 0.129 | |
| 0.04984 | 24.88 | 2.09 | | 0.132 | |
| 0.05021 | 24.69 | 2.11 | | 0.120 | |
| 0.05059 | 24.51 | 2.12 | | 0.112 | |
| 0.05096 | 24.33 | 2.14 | | 0.123 | |
| 0.05133 | 24.15 | 2.14 | | 0.145 | |
| 0.05170 | 23.98 | 2.16 | | 0.135 | |
| 0.05207 | 23.81 | 2.19 | | 0.126 | |

TABLE 1 continued.

| Energy (eV) | Wavelength (μm) | n | | k | |
|---|---|---|---|---|---|
| 0.055 | 23.53 | 2.662 | [1,3] | 0.0115 | [1,4] |
| 0.06 | 20.66 | 2.780 | | 0.0146 | |
| 0.06027 | 20.57 | - | | 0.0130 | [1,6,7] |
| 0.06071 | 20.42 | - | | 0.0115 | |
| 0.06095 | 20.34 | - | | 0.0108 | |
| 0.06141 | 20.19 | - | | 0.00949 | |
| 0.06199 | 20.00 | - | | 0.00794 | |
| 0.06268 | 19.78 | - | | 0.00602 | |
| 0.06319 | 19.62 | - | | 0.00473 | |
| 0.06355 | 19.51 | - | | 0.00384 | |
| 0.06549 | 18.93 | - | | 0.00232 | |
| 0.06888 | 18.00 | - | | 0.00181 | |
| 0.07293 | 17.00 | - | | 0.00177 | |
| 0.07491 | 16.55 | - | | 0.00231 | |
| 0.07564 | 16.39 | - | | 0.00333 | |
| 0.07615 | 16.28 | - | | 0.00516 | |
| 0.07648 | 16.21 | - | | 0.00667 | |
| 0.07681 | 16.14 | - | | 0.00746 | |
| 0.07749 | 16.00 | - | | 0.00712 | |
| 0.07822 | 15.85 | - | | 0.00613 | |
| 0.07877 | 15.74 | - | | 0.00522 | |
| 0.07988 | 15.52 | - | | 0.00563 | |
| 0.08 | 15.49 | 2.956 | | 0.0039 | [4] |
| 0.08024 | 15.45 | - | | 0.00626 | [1,6,7] |
| 0.08093 | 15.32 | - | | 0.00617 | |
| 0.08135 | 15.24 | - | | 0.00525 | |
| 0.08265 | 15.0 | - | | 0.00371 | |
| 0.08349 | 14.85 | - | | 0.003 | |
| 0.085 | 14.58 | 2.975 | | 0.00148 | |
| 0.08600 | 14.40 | - | | 0.00128 | |
| 0.09478 | 13.08 | - | | 0.00667 | |
| 0.09537 | 13.0 | 3.0006 | [5] | - | |
| 0.10332 | 12.0 | 3.0153 | [5] | 0.000527 | [1,6,7] |
| 0.11271 | 11.0 | 3.0302 | | | |
| 0.12 | 10.33 | 3.038 | [1,3] | | |
| 0.12398 | 10.0 | 3.0464 | [5] | | |
| 0.13775 | 9.0 | 3.0532 | | | |
| 0.14 | 8.853 | 3.053 | [1,3] | | |
| 0.15497 | 8.0 | 3.0623 | [5] | | |
| 0.16 | 7.746 | 3.062 | [1,3] | | |
| 0.17711 | 7.0 | 3.0707 | [5] | | |
| 0.18 | 6.886 | 3.069 | [1,3] | | |
| 0.20 | 6.197 | 3.074 | | | |
| 0.20663 | 6.0 | 3.0783 | [5] | | |
| 0.22542 | 5.5 | 3.0819 | | | |
| 0.24796 | 5.0 | 3.0857 | | | |
| 0.25 | 4.958 | 3.083 | [1,3] | | |
| 0.27551 | 4.5 | 3.0894 | [5] | | |
| 0.30 | 4.131 | 3.089 | [1,3] | | |
| 0.30995 | 4.0 | 3.0986 | [5] | | |
| 0.35 | 3.541 | 3.095 | [1,3] | | |
| 0.35423 | 3.5 | 3.0986 | [5] | | |
| 0.40 | 3.099 | 3.101 | [1,3] | | |
| 0.41327 | 3.0 | 3.1049 | [5] | | |
| 0.45 | 2.754 | 3.107 | [1,3] | | |

TABLE 1 continued.

| Energy (eV) | Wavelength (μm) | n | | k | |
|---|---|---|---|---|---|
| 0.49592 | 2.5 | 3.1151 | [5] | | |
| 0.50 | 2.479 | 3.114 | [1,3] | | |
| 0.55 | 2.254 | 3.121 | | | |
| 0.60 | 2.066 | 3.129 | | | |
| 0.6199 | 2.0 | 3.1326 | [5] | | |
| | | 3.1341 | [8] | | |
| 0.65 | 1.907 | 3.137 | [1,3] | | |
| 0.65253 | 1.9 | 3.1390 | [8] | | |
| 0.68878 | 1.8 | 3.1448 | | | |
| 0.70 | 1.771 | 3.146 | [1,3] | | |
| 0.72929 | 1.7 | 3.1519 | [8] | | |
| 0.75 | 1.653 | 3.156 | [1,3] | | |
| 0.77487 | 1.6 | 3.1607 | [8] | | |
| 0.80 | 1.549 | 3.167 | [1,3] | | |
| 0.82653 | 1.5 | 3.1717 | [8] | | |
| 0.85 | 1.458 | 3.178 | [1,3] | | |
| 0.88557 | 1.40 | 3.1860 | [8] | | |
| 0.90 | 1.377 | 3.191 | [1,3] | | |
| 0.95 | 1.305 | 3.205 | | | |
| 0.95369 | 1.30 | 3.2048 | [5] | | |
| 1.00 | 1.239 | 3.220 | [8] | | |
| 1.03317 | 1.20 | 3.2293 | [1,3] | | |
| | | 3.2308 | [8] | | |
| 1.05 | 1.181 | 3.236 | [1,3] | | |
| 1.10 | 1.127 | 3.254 | | | |
| 1.12709 | 1.1 | 3.2658 | [5] | | |
| | | 3.2683 | [8] | | |
| 1.15 | 1.078 | 3.274 | [1,3] | | |
| 1.2 | 1.033 | 3.297 | | | |
| 1.21 | 1.024 | 3.302 | | | |
| 1.22 | 1.016 | 3.307 | | | |
| 1.23 | 1.008 | 3.312 | | | |
| 1.2398 | 1.0 | 3.3271 | [5] | | |
| | | 3.3265 | [8] | | |
| 1.25 | 0.991 | 3.324 | [1,3] | | |
| 1.2523 | 0.99 | 3.3361 | [5] | | |
| | | 3.3341 | [8] | | |
| 1.26 | 0.984 | 3.330 | [1,3] | | |
| 1.2651 | 0.98 | 3.3461 | [5] | | |
| | | 3.3421 | [8] | | |
| 1.27 | 0.976 | 3.337 | [1,3] | | |
| 1.272 | 0.975 | 3.3464 | [8] | 0.0000113 | [1,4,6] |
| 1.278 | 0.97 | 3.3572 | [5] | - | |
| 1.28 | 0.9686 | 3.345 | [1,3] | 0.0000246 | |
| 1.285 | 0.965 | 3.3551 | [8] | 0.0000392 | |
| 1.291 | 0.960 | 3.3596 | | 0.0000739 | |
| | | 3.3699 | [5] | - | |
| 1.30 | 0.9537 | 3.362 | [1,3] | 0.000281 | |
| 1.305 | 0.950 | 3.3691 | [8] | 0.000527 | |
| 1.312 | 0.945 | 3.3742 | | 0.000967 | |
| 1.319 | 0.940 | 3.3793 | | 0.00318 | |
| 1.326 | 0.935 | 3.3847 | | 0.0059 | |
| 1.333 | 0.930 | 3.3901 | | 0.0109 | |
| 1.340 | 0.925 | 3.3958 | | 0.0355 | |
| 1.345 | 0.922 | 3.3995 | | 0.0571 | |

TABLE 1 continued.

| Energy (eV) | Wavelength (μm) | n | | k | |
|---|---|---|---|---|---|
| 1.5 | 0.8265 | 3.456 | [1,9] | 0.203 | [1,9] |
| | | | | 0.140 | [6,10] |
| 1.6 | 0.7749 | 3.467 | | 0.218 | [1,9] |
| | | | | 0.163 | [6,10] |
| 1.7 | 0.7293 | 3.476 | | 0.242 | [1,9] |
| | | | | 0.176 | [6,10] |
| 1.8 | 0.6888 | 3.492 | | 0.270 | [1,9] |
| | | | | 0.206 | [6,10] |
| 1.9 | 0.6525 | 3.517 | | 0.298 | [1,9] |
| | | 3.410 | [6,10] | 0.253 | [6,10] |
| 2.0 | 0.6199 | 3.549 | [1,9] | 0.317 | [1,9] |
| | | 3.430 | [6,10] | 0.298 | [6,10] |
| 2.1 | 0.5904 | 3.585 | [1,9] | 0.347 | [1,9] |
| | | 3.450 | [6,10] | 0.334 | [6,10] |
| 2.2 | 0.5635 | 3.629 | [1,9] | 0.380 | [1,9] |
| | | 3.472 | [6,10] | 0.358 | [6,10] |
| 2.3 | 0.5390 | 3.682 | [1,9] | 0.416 | [1,9] |
| | | 3.521 | [6,10] | 0.389 | [6,10] |
| 2.4 | 0.5166 | 3.745 | [1,9] | 0.457 | [1,9] |
| | | 3.567 | [6,10] | 0.430 | [6,10] |
| 2.5 | 0.4959 | 3.818 | [1,9] | 0.511 | [1,9] |
| | | 3.621 | [6,10] | 0.480 | [6,10] |
| 2.6 | 0.4768 | 3.903 | [1,9] | 0.579 | [1,9] |
| | | 3.675 | [6,10] | 0.531 | [6,10] |
| 2.7 | 0.4592 | 3.004 | [1,9] | 0.667 | [1,9] |
| | | 3.754 | [6,10] | 0.599 | [6,10] |
| 2.8 | 0.4428 | 4.121 | [1,9] | 0.786 | [1,9] |
| | | 3.833 | [6,10] | 0.670 | [6,10] |
| 2.9 | 0.4275 | 4.256 | [1,9] | 0.964 | [1,9] |
| | | 3.982 | [6,10] | 0.816 | [6,10] |
| 3.0 | 0.4133 | 4.395 | [1,9] | 1.247 | [1,9] |
| | | 4.083 | [6,10] | 1.056 | [6,10] |
| 3.1 | 0.3999 | 4.415 | [1,9] | 1.735 | [1,9] |
| | | 4.100 | [6,10] | 1.439 | [6,10] |
| 3.2 | 0.3874 | 3.976 | [1,9] | 2.143 | [1,9] |
| | | 3.835 | [6,10] | 1.804 | [6,10] |
| 3.3 | 0.3757 | 3.576 | [1,9] | 2.209 | [1,9] |
| | | 3.441 | [6,10] | 1.857 | [6,10] |
| 3.4 | 0.3646 | 3.299 | [1,9] | 2.060 | [1,9] |
| | | 3.192 | [6,10] | 1.747 | [6,10] |
| 3.5 | 0.3542 | 3.193 | [1,9] | 1.948 | [1,9] |
| | | 3.082 | [6,10] | 1.622 | [6,10] |
| 3.6 | 0.3444 | 3.133 | [1,9] | 1.872 | [1,9] |
| | | 3.047 | [6,10] | 1.550 | [6,10] |
| 3.7 | 0.3351 | 3.103 | [1,9] | 1.816 | [1,9] |
| | | 3.024 | [6,10] | 1.489 | [6,10] |
| 3.8 | 0.3263 | 3.095 | [1,9] | 1.773 | [1,9] |
| | | 3.027 | [6,10] | 1.440 | [6,10] |
| 3.9 | 0.3179 | 3.108 | [1,9] | 1.744 | [1,9] |
| | | 3.054 | [6,10] | 1.401 | [6,10] |
| 4.0 | 0.3099 | 3.141 | [1,9] | 1.730 | [1,9] |
| | | 3.105 | [6,10] | 1.392 | [6,10] |
| 4.1 | 0.3024 | 3.196 | [1,9] | 1.735 | [1,9] |
| | | 3.162 | [6,10] | 1.389 | [6,10] |

TABLE 1 continued.

| Energy (eV) | Wavelength (μm) | n | | k | |
|---|---|---|---|---|---|
| 4.2 | 0.2952 | 3.275 | [1,9] | 1.762 | [1,9] |
| | | 3.248 | [6,10] | 1.415 | [6,10] |
| 4.3 | 0.2883 | 3.384 | [1,9] | 1.826 | [1,9] |
| | | 3.347 | [6,10] | 1.468 | [6,10] |
| 4.4 | 0.2818 | 3.527 | [1,9] | 1.948 | [1,9] |
| | | 3.473 | [6,10] | 1.549 | [6,10] |
| 4.5 | 0.2755 | 3.697 | [1,9] | 2.186 | [1,9] |
| | | 3.655 | [6,10] | 1.691 | [6,10] |
| 4.6 | 0.2695 | 3.800 | [1,9] | 2.637 | [1,9] |
| | | 3.849 | [6,10] | 2.117 | [6,10] |
| 4.7 | 0.2638 | 3.560 | [1,9] | 3.223 | [1,9] |
| | | 3.729 | [6,10] | 2.635 | [6,10] |
| 4.8 | 0.2583 | 2.984 | [1,9] | 3.517 | [1,9] |
| | | 3.335 | [6,10] | 3.039 | [6,10] |
| 4.9 | 0.2530 | 2.546 | [1,9] | 3.514 | [1,9] |
| | | 2.885 | [6,10] | 3.144 | [6,10] |
| 5.0 | 0.2480 | 2.131 | [1,9] | 3.495 | [1,9] |
| | | 2.451 | [6,10] | 3.166 | [6,10] |
| 5.1 | 0.2431 | 1.745 | [1,9] | 3.291 | [1,9] |
| | | 2.132 | [6,10] | 2.982 | [6,10] |
| 5.2 | 0.2431 | 1.558 | [1,9] | 3.016 | [1,9] |
| | | 1.960 | [6,10] | 2.801 | [6,10] |
| 5.3 | 0.2339 | 1.482 | [1,9] | 2.802 | [1,9] |
| | | 1.834 | [6,10] | 2.675 | [6,10] |
| 5.4 | 0.2296 | 1.455 | [1,9] | 2.652 | [1,9] |
| | | 1.737 | [6,10] | 2.553 | [6,10] |
| 5.5 | 0.2254 | 1.426 | [1,9] | 2.562 | [1,9] |
| | | 1.668 | [1,6,10] | 2.442 | [1,6,10] |
| 5.6 | 0.2214 | 1.375 | [1,9] | 2.484 | [1,9] |
| | | 1.616 | [1,6,10] | 2.349 | [1,6,10] |
| 5.7 | 0.2175 | 1.325 | [1,9] | 2.383 | [1,9] |
| | | 1.573 | [1,6,10] | 2.267 | [1,6,10] |
| 5.8 | 0.2137 | 1.299 | [1,9] | 2.280 | [1,9] |
| | | 1.544 | [1,6,10] | 2.191 | [1,6,10] |
| 5.9 | 0.2101 | 1.301 | [1,9] | 2.183 | [1,9] |
| | | 1.516 | [1,6,10] | 2.130 | [1,6,10] |
| 6.0 | 0.2066 | 1.336 | [1,9] | 2.113 | [1,9] |
| | | 1.500 | [1,6,10] | 2.063 | [1,6,10] |
| 6.1 | 0.2032 | 1.508 | | 2.005 | |
| 6.2 | 0.1999 | 1.525 | | 1.982 | |
| 6.3 | 0.1968 | 1.526 | | 1.991 | |
| 6.4 | 0.1937 | 1.496 | | 2.008 | |
| 6.5 | 0.1907 | 1.453 | | 2.010 | |
| 6.6 | 0.1878 | 1.402 | | 1.994 | |
| 6.7 | 0.1850 | 1.354 | | 1.986 | |
| 6.8 | 0.1823 | 1.307 | | 1.966 | |
| 6.9 | 0.1797 | 1.261 | | 1.941 | |
| 7.0 | 0.1771 | 1.215 | | 1.915 | |
| 8.0 | 0.1550 | 0.960 | | 1.566 | |
| 9.0 | 0.1378 | 0.872 | | 1.304 | |
| 10.0 | 0.1240 | 0.806 | | 1.154 | |
| 10.5 | 0.1181 | 0.797 | | 1.056 | |
| 11.0 | 0.1127 | 0.793 | | 0.996 | |
| 11.5 | 0.1078 | 0.781 | | 0.946 | |
| 12.0 | 0.1033 | 0.771 | | 0.899 | |

TABLE 1 continued.

| Energy (eV) | Wavelength (μm) | n | k |
|---|---|---|---|
| 12.5 | 0.0991 | 0.754 | 0.861 |
| 13.0 | 0.0954 | 0.726 | 0.820 |
| 13.5 | 0.0918 | 0.701 | 0.765 |
| 14.0 | 0.0885 | 0.688 | 0.706 |
| 14.5 | 0.0855 | 0.675 | 0.645 |
| 15.0 | 0.0826 | 0.695 | 0.574 |
| 15.5 | 0.0780 | 0.719 | 0.529 |
| 16.0 | 0.0775 | 0.742 | 0.491 |
| 16.5 | 0.0751 | 0.785 | 0.457 |
| 17.0 | 0.0729 | 0.824 | 0.454 |
| 17.5 | 0.0708 | 0.840 | 0.469 |
| 18.0 | 0.0689 | 0.846 | 0.477 |
| 18.5 | 0.0670 | 0.843 | 0.487 |
| 19.0 | 0.0652 | 0.834 | 0.493 |
| 19.5 | 0.0636 | 0.815 | 0.499 |
| 20.0 | 0.0619 | 0.793 | 0.494 |

Notes:

- The references given in brackets apply down the column until a new reference appears.
- In refs [1], [5], [6], and [9], the data are presented in tabular form.
- In refs [3] and [8], the data are presented in graphical form and particular dispersion relations that fit the data are given. These dispersion relations, and the parameters describing them, are discussed in [11].
- In refs [2], [4], [7], and [10], the data are presented in graphical form.

## REFERENCES

[1] O.J. Glembocki, H. Piller [ in *Handbook of Optical Constants of Solids* Ed. E.D. Palik (Academic Press, London, 1985) p.503-16 ]

[2] H. Jamshidi, T.J. Parker [ *Int. J. Infrared Millim. Waves (USA)* vol.4 (1983) p.1037-44 ]

[3] A.N. Pikhtin, A.D. Yas'kov [ *Sov. Phys.-Semicond. (USA)* vol.12 no.6 (1978) p.622-6 ]

[4] R. Newman [ *Phys. Rev. (USA)* vol.111 (1958) p.1518-21 ]

[5] V.B. Bogdanov, V.T. Prokopenko, A.D. Yas'kov [ *Opt. Spectrosc. (USA)* vol.60 (1986) p.68-9 ]

[6] B.O. Seraphin, H.E. Bennett [ *Semicond. Semimet. (USA)* vol.3 (Academic Press, New York, 1967) p.499-543 ]

[7] W.N. Reynolds, M.T. Lilburne, R.M. Dell [ *Proc. Phys. Soc. London (UK)* vol.71 (1958) p.416-21 ]

[8] G.D. Pettit, W.J. Turner [ *J. Appl. Phys. (USA)* vol.36 (1965) p.2081 ]

[9] D.E. Aspnes, A.A. Studna [ *Phys. Rev. B (USA)* vol.27 no.2 (1983) p.985-1009 ]

[10] M. Cardona [ *J. Appl. Phys. (USA)* vol.36 (1965) p.2181-6 ]

[11] A.R. Forouhi, I. Bloomer [ *Properties of Indium Phosphide* (INSPEC, IEE, London, UK, 1991) p.113-5 ]

# 5.2 Refractive index of InP: dependence on temperature, pressure and carrier injection

A.R. Forouhi and I. Bloomer

This Datareview appeared in *Properties of Indium Phosphide* (INSPEC, IEE, London, UK, 1991) and is reproduced here for the reader's convenience.

Extrinsic effects such as temperature and pressure, as well as injection of electrons and holes, produce changes in band structure which in turn affect optical functions. This Datareview is divided into three parts: Temperature Effects, Pressure Effects, and Carrier Induced Effects.

## A TEMPERATURE EFFECTS

Effects of temperature on optical properties of InP (and, in general, any semiconductor material) are significant throughout the static-IR to visible-UV spectral range.

In the static-IR regime, $E \cong 0$ to 0.02 eV, the real part of the dielectric function, $\varepsilon_1$, approaches the electrically measured static dielectric function $\varepsilon_0$. Meiners [1] performed low-frequency capacitance measurements to determine the static dielectric constant of InP from 100 to 300 K. A value of $12.56 \pm 0.2$ was obtained at 300 K and an extrapolated value of $11.93 \pm 0.2$ was obtained at 77 K. Below about 220 K the temperature dependence was expressed by the relationship $\varepsilon_0 = 11.76 (1 + 2.26 \times 10^{-4} T)$.

Seeger [2] used a microwave transmission interference method to determine the static dielectric constant of InP. At 77 and 300 K, $\varepsilon_0$ was found to be 12.2 and 12.5, respectively, in good agreement with Meiners [1].

In the reststrahlen regime, $E \cong 0.02$ to 0.1 eV, Jamshidi and Parker [3] determined optical constants of InP at 6 K and 300 K, using dispersive Fourier transform spectroscopy. They found that as the temperature decreases, values of both the transverse and longitudinal optic modes, $\omega_{TO}$ and $\omega_{LO}$, increase. At 300 K, $\omega_{TO}$ and $\omega_{LO}$ are 307.9 $cm^{-1}$ and 347.1 $cm^{-1}$, respectively, whereas at 6 K they are 310.2 $cm^{-1}$ and 349.0 $cm^{-1}$. Nevertheless, inspection of the graphs in [3] depicting k versus wavenumber reveals no significant change in the position of the major peak when temperature changes, while the maximum value of k increases from approximately 5.6 at 300 K, to approximately 7.9 at 6 K.

In the transparent regime, $E \cong 0.1$ to 1.27 eV ($\lambda \cong 12$ μm to 1 μm), $k \cong 0$. Bogdanov et al [4] measured the temperature coefficient of the refractive index, dn/dT, of InP as a function of the wavelength. Inspection of the graph presented by Bogdanov et al [4] shows that dn/dT rapidly drops from approximately $2.3 \times 10^{-4}$ at 1 μm to approximately $1.6 \times 10^{-4}$ at 2 μm, then remains almost constant at this value up to about 10 μm, and then increases to approximately $2.0 \times 10^{-4}$ as wavelength increases to 20 μm. Only a slight dependence of dn/dT on doping level was observed.

Pettit and Turner [5] presented measured n data at 77 and 298 K, in the 0.6 to 1.4 eV range. In this energy range, values of n at 298 K are consistently larger than values of n at 77 K: for example, at $E = 1.2398$ eV ($\lambda = 1$ μm), $n = 3.3265$ at 298 K and $n = 3.278$ at 77 K. At $E = 0.6199$ eV ($\lambda = 2$ μm), $n = 3.1341$ at 298 K and $n = 3.106$ at 77 K. This is consistent with the results of Bogdanov et al [4].

The value of the optical bandgap, $E_g$, represents the onset of the absorption edge and marks the boundary between transparent regime and the bandgap region. Reynolds et al [6] report that at room temperature, for undoped or lightly doped InP, $E_g = 1.27 \pm 0.01$ eV, whereas at 293 K, $E_g = 1.36$ eV. Reynolds et al [6] also report that this is in close agreement with Oswald's result [7], i.e. $E_g = 1.41 - 4.6 \times 10^{-4}$ T eV. However, Turner et al [8] argue that the temperature dependence of the energy gap above 100 K is $-2.9 \times 10^{-4}$ eV/K in contrast to the previously accepted value of $-4.6 \times 10^{-4}$ eV/K given by Oswald [7].

In the visible-UV range, E ≅ 2.5 to 8 eV, as the temperature decreases the energy at which peaks occur increases and the peaks themselves get sharper. Furthermore, in some cases single peaks split into doublets. The effect of temperature on band structure can be decomposed into a dilation effect and a phonon population effect. With respect to a dilation effect, this is interpreted as causing a change in the lattice constant. Temperature dilation is closely related to a change caused by hydrostatic stress. The effect of phonon population change is not so easily interpreted. It causes broadening of reflectivity spectra as T increases, a shift in the energy bands, and a change in exciton lifetimes, as discussed by Cohen and Chelikowsky [9].

Lautenschlager et al [10] have measured $\varepsilon_1(E)$ and $\varepsilon_2(E)$ at different temperatures ranging from 30 to 750 K, in the visible-UV range, using spectroscopic ellipsometry. From the experimental lineshape they analysed the temperature dependence of the critical point parameters, i.e. amplitude, threshold energy, broadening, and excitonic phase angle associated with various equations for $\varepsilon_2(E)$ when E is near a critical point transition energy. Their data (privately communicated to us by S. Zollner), ranging from E = 1.66 to 5.6 eV in intervals of 0.1 eV, are given in TABLE 1. It should be noted that this table represents an abbreviated version of their results which involve intervals of 0.01 eV.

Experimental results for n(E) and k(E) at different temperatures, such as those of Lautenschlager et al [10] presented in TABLE 1, can be fitted to the Forouhi-Bloomer [11] equations:

$$n(E) = n(\infty) + \sum_{i=1}^{q} \frac{G_i E + H_i}{E^2 - B_i E + C_i} \tag{1}$$

$$k(E) = \left[ \sum_{i=1}^{q} \frac{A_i}{E^2 - B_i E + C_i} \right] (E - E_g)^2 \tag{2}$$

The parameters $A_i$, $B_i$, $C_i$, $G_i$, $H_i$, $E_g$, and $n(\infty)$ are defined in [22]. However, in order to obtain as good a fit as in the room temperature case [11], measured results beyond 6 eV are needed so that the fourth major critical point transition peak is included. At present these measurements are unavailable.

Cardona's [12] reflectance measurements showed a doublet structure at around 3.15 eV at 90 K. This structure, not observable at higher temperatures, is attributed to spin-orbit splitting at $L_3$.

## B PRESSURE EFFECTS

At high pressures diamond and zinc-blende semiconductors may cross over from a direct energy bandgap to an indirect energy bandgap semiconductor (or vice versa). This is because the pressure coefficient of the Γ and X conduction minima, as measured relative to the valence band maximum, have opposite signs. So if the main gap is direct at low pressure, it becomes indirect at high pressure (or vice versa). The transition pressure is usually measured from the rapid decrease in the photoluminescence intensity [13].

TABLE 1 Optical functions obtained by Lautenschlager et al [10] at 30, 245, 300, 524 and 740 K via spectroscopic ellipsometry from 1.66 to 5.6 eV. This table represents an abbreviated version of the actual results in which the intervals are 0.01 eV.

| Energy (eV) | 30 K | | 245 K | | 300 K | | 524 K | | 740 K | |
|---|---|---|---|---|---|---|---|---|---|---|
| | n | k | n | k | n | k | n | k | n | k |
| 1.66 | 3.4476 | 0.29283 | 3.4335 | 0.32897 | 3.4829 | 0.28779 | 3.4822 | 0.35079 | 3.5386 | 0.34305 |
| 1.7 | 3.4544 | 0.28943 | 3.4386 | 0.34102 | 3.4925 | 0.30063 | 3.4916 | 0.36075 | 3.5483 | 0.35229 |
| 1.8 | 3.4665 | 0.31765 | 3.4525 | 0.36198 | 3.5091 | 0.32945 | 3.5085 | 0.38921 | 3.5739 | 0.37907 |
| 1.9 | 3.4890 | 0.33888 | 3.4740 | 0.38410 | 3.5362 | 0.35176 | 3.5356 | 0.40938 | 3.6090 | 0.39960 |
| 2.0 | 3.5136 | 0.36672 | 3.4981 | 0.41194 | 3.5672 | 0.38309 | 3.5677 | 0.43409 | 3.6495 | 0.42493 |
| 2.1 | 3.5477 | 0.39268 | 3.5304 | 0.43685 | 3.6040 | 0.41166 | 3.6062 | 0.46403 | 3.7025 | 0.45573 |
| 2.2 | 3.5861 | 0.41641 | 3.5678 | 0.46696 | 3.6474 | 0.44229 | 3.6545 | 0.49508 | 3.7645 | 0.49411 |
| 2.3 | 3.6326 | 0.45125 | 3.6143 | 0.49798 | 3.7013 | 0.47407 | 3.7116 | 0.53789 | 3.8388 | 0.54787 |
| 2.4 | 3.6857 | 0.48792 | 3.6687 | 0.53892 | 3.7633 | 0.52039 | 3.7809 | 0.58956 | 3.9207 | 0.62204 |
| 2.5 | 3.7493 | 0.53389 | 3.7335 | 0.58993 | 3.8359 | 0.57465 | 3.8606 | 0.65651 | 4.0111 | 0.72088 |
| 2.6 | 3.8218 | 0.58898 | 3.8083 | 0.65268 | 3.9194 | 0.64650 | 3.9483 | 0.74778 | 4.1072 | 0.85335 |
| 2.7 | 3.9066 | 0.65857 | 3.8956 | 0.73445 | 4.0177 | 0.73674 | 4.0474 | 0.87185 | 4.1992 | 1.0411 |
| 2.8 | 4.0082 | 0.74929 | 3.9949 | 0.84103 | 4.1330 | 0.86063 | 4.1513 | 1.0463 | 4.2471 | 1.3109 |
| 2.9 | 4.1285 | 0.87702 | 4.1141 | 0.99826 | 4.2636 | 1.0470 | 4.2331 | 1.3127 | 4.1553 | 1.6329 |
| 3.0 | 4.2684 | 1.0620 | 4.2336 | 1.2319 | 4.3879 | 1.3402 | 4.1841 | 1.6845 | 3.9147 | 1.8502 |
| 3.1 | 4.4178 | 1.3667 | 4.2996 | 1.6063 | 4.3770 | 1.8317 | 3.8987 | 1.9686 | 3.6661 | 1.9204 |
| 3.2 | 4.4511 | 1.9216 | 4.0484 | 2.0981 | 3.9271 | 2.1992 | 3.5914 | 2.0432 | 3.4811 | 1.9039 |
| 3.3 | 3.8051 | 2.3289 | 3.6028 | 2.2095 | 3.5248 | 2.2485 | 3.3651 | 1.9974 | 3.3633 | 1.8654 |
| 3.4 | 3.3723 | 2.4136 | 3.2654 | 2.1410 | 3.2660 | 2.0941 | 3.2380 | 1.9250 | 3.2898 | 1.8275 |
| 3.5 | 3.1248 | 2.1014 | 3.1095 | 1.9883 | 3.1616 | 1.9821 | 3.1637 | 1.8641 | 3.2433 | 1.7980 |
| 3.6 | 3.0530 | 1.9741 | 3.0436 | 1.8928 | 3.1021 | 1.9053 | 3.1247 | 1.8184 | 3.2213 | 1.7791 |
| 3.7 | 3.0061 | 1.8882 | 3.0086 | 1.8228 | 3.0705 | 1.8456 | 3.1097 | 1.7870 | 3.2184 | 1.7716 |
| 3.8 | 2.9838 | 1.8220 | 2.9992 | 1.7732 | 3.0606 | 1.8037 | 3.1136 | 1.7704 | 3.2284 | 1.7775 |
| 3.9 | 2.9829 | 1.7692 | 3.0062 | 1.7387 | 3.0736 | 1.7756 | 3.1336 | 1.7686 | 3.2528 | 1.8021 |

TABLE 1 continued.

| Energy (eV) | 30 K | | 245 K | | 300 K | | 524 K | | 740 K | |
|---|---|---|---|---|---|---|---|---|---|---|
| | n | k | n | k | n | k | n | k | n | k |
| 4.0 | 3.0047 | 1.7304 | 3.0358 | 1.7223 | 3.1037 | 1.7610 | 3.1678 | 1.7843 | 3.2908 | 1.8498 |
| 4.1 | 3.0489 | 1.7070 | 3.0852 | 1.7192 | 3.1566 | 1.7646 | 3.2216 | 1.8236 | 3.3398 | 1.9227 |
| 4.2 | 3.1118 | 1.6998 | 3.1547 | 1.7387 | 3.2303 | 1.7966 | 3.2910 | 1.8964 | 3.3904 | 2.0393 |
| 4.3 | 3.2047 | 1.7162 | 3.2462 | 1.7878 | 3.3311 | 1.8571 | 3.3685 | 2.0203 | 3.4316 | 2.2144 |
| 4.4 | 3.3417 | 1.7688 | 3.3729 | 1.8898 | 3.4649 | 1.9840 | 3.4363 | 2.2171 | 3.4159 | 2.4490 |
| 4.5 | 3.5346 | 1.8852 | 3.5287 | 2.0893 | 3.6086 | 2.2268 | 3.4407 | 2.5178 | 3.2862 | 2.7132 |
| 4.6 | 3.8044 | 2.1796 | 3.6464 | 2.4831 | 3.6758 | 2.6650 | 3.2812 | 2.8640 | 3.0488 | 2.9228 |
| 4.7 | 3.9090 | 2.8962 | 3.4399 | 3.0413 | 3.4185 | 3.1835 | 2.9452 | 3.1064 | 2.7585 | 3.0328 |
| 4.8 | 3.3352 | 3.4883 | 2.9299 | 3.3180 | 2.8789 | 3.4260 | 2.5777 | 3.1729 | 2.4715 | 3.0399 |
| 4.9 | 2.7041 | 3.5436 | 2.5166 | 3.3123 | 2.4789 | 3.4253 | 2.2578 | 3.1375 | 2.2324 | 2.9807 |
| 5.0 | 2.4317 | 3.5132 | 2.1700 | 3.3075 | 2.0765 | 3.3952 | 1.9945 | 3.0223 | 2.0504 | 2.8865 |
| 5.1 | 1.9202 | 3.6035 | 1.7921 | 3.1716 | 1.7293 | 3.1898 | 1.8093 | 2.8743 | 1.9106 | 2.7810 |
| 5.2 | 1.3964 | 3.2347 | 1.5605 | 2.9141 | 1.5738 | 2.9326 | 1.7012 | 2.7279 | 1.1841 | 2.6855 |
| 5.3 | 1.3351 | 2.8246 | 1.4992 | 2.6839 | 1.4874 | 2.7212 | 1.6351 | 2.6073 | 1.7389 | 2.5899 |
| 5.4 | 1.3564 | 2.5971 | 1.4870 | 2.5273 | 1.4916 | 2.5720 | 1.6000 | 2.5152 | 1.6735 | 2.5086 |
| 5.5 | 1.4350 | 2.4304 | 1.5043 | 2.4306 | 1.4910 | 2.4873 | 1.5586 | 2.4260 | 1.6442 | 2.4241 |
| 5.6 | 1.4626 | 2.3492 | 1.5037 | 2.3754 | 1.4536 | 2.3359 | 1.5516 | 2.3266 | 1.6598 | 2.3626 |

For some semiconductors this inversion is not observable due to transitions into another crystal phase before the inversion of the two conduction minima. In particular, III-V compounds can undergo a metallic first-order phase transition under high pressure. As a consequence the crystals turn opaque and the photoluminescence is totally quenched. The situation for InP is significant because the recent experimental data indicate that values of the crossover Γ-X pressure and phase transition pressure are very close to each other [13].

Gorczyca et al [13] examined the high pressure phase transition and the high pressure transition from the direct to indirect gap. They point out that there is conflicting experimental evidence as to whether the gap crossover occurs before the phase transition.

Gorczyca et al [13] used a self-consistent linear muffin tin orbital band structure calculation to investigate the optical and structural properties of InP under pressure. They determined the value of the pressure at the crossover from the direct to the indirect gap to be 104 ± 2 kbar whereas the transition to the metallic phase is determined to occur at 98 kbar. Thus, their calculation shows that the phase transition occurs at a pressure that is slightly below the pressure at which the gap in the zinc-blende structure would change from being direct (Γ) to indirect (Γ-X).

Gorczyca et al [13] derived the first and second order pressure coefficients $\alpha = dE/dP$ and $\beta = d^2E/dP^2$ via least squares fits to

$$E_g = E_g^0 + \alpha P + \beta P^2 \tag{3}$$

pertaining to the calculated values of the gaps at Γ, X, and L. They obtain:

$$E_{g,\Gamma}^0 = 1.42 \text{ eV}, \alpha_\Gamma = 7.8 \pm 0.2 \times 10^{-6} \text{ eV/bar}, \beta_\Gamma = -0.9 \times 10^{-11} \text{ eV/bar}^2;$$

$$E_{g,x}^0 = 2.24 \text{ eV}, \alpha_x = -1.1 \times 10^{-6} \text{ eV/bar}, \beta_x = -0.02 \times 10^{-11} \text{ eV/bar}^2;$$

$$E_{g,L}^0 = 1.94 \text{ eV}, \alpha_L = 3.8 \pm 0.1 \times 10^{-6} \text{ eV/bar}, \beta_L = -0.45 \times 10^{-11} \text{ eV/bar}^2.$$

Menoni et al [14] carried out photoluminescence measurements on n-type InP under hydrostatic pressure in a diamond anvil cell at 20 K. The photoluminescence peak corresponding to the direct bandgap transition was observed up to 120 kbar and was found to change sublinearly with pressure, similar to room temperature measurements. They fitted their data to EQN (3) to obtain:

$$E_{g,\Gamma}^0 = 1.393 \pm 0.005 \text{ eV}, \alpha_\Gamma = 7.5 \pm 0.2 \times 10^{-6} \text{ eV/bar}, \beta_\Gamma = -1.2 \pm 0.5 \times 10^{-11} \text{ eV/bar}^2.$$

Müller et al [15] measured the variation of the lowest direct absorption edge of InP with pressure at room temperature with a diamond anvil cell. Both transmission and luminescence techniques were used. They examined the linear and non-linear pressure coefficients of the $E_{g,\Gamma}$ gap by measuring absorption and luminescence on InP at room temperature up to the phase transition (101 kbar).

Müller et al [15] fitted their bandgap data to EQN (3) to obtain:

$$E_{g,\Gamma}^0 = 1.37 \text{ eV}, \alpha_\Gamma = 8.4 \pm 0.2 \times 10^{-6} \text{ eV/bar}, \beta_\Gamma = -1.8 \pm 0.3 \times 10^{-11} \text{ eV/bar}^2.$$

They observed that luminescence intensity remained nearly constant as a function of pressure up to 90 kbar. At this pressure the efficiency began to decrease thus signalling the crossing of the Γ and X conduction band minima. They found $\alpha_X = -3 \pm 1 \times 10^{-6}$ eV/bar.

Kobayashi et al [16] did a systematic study of the high-pressure optical absorption of n-type InP at 77 K in order to observe the effects of high doping on the absorption edge. With increasing pressure the absorption spectra shift towards the higher energies. This corresponds to a rise in the $\Gamma_1^c$ minimum of the conduction band relative to the top of the valence band: the direct gap increases. The absorption spectrum showed a low-energy tail starting at about 80 kbar for pure InP. This indicates the onset of the indirect optical transition. The pressure dependence of the absorption edge in heavily doped samples is different from that in undoped or moderately doped samples in the high pressure range. For pressures above 65 kbar the energy shift occurs at a lower rate than that in the low pressure range. The observed behaviour can be explained by considering the pressure dependence of the Fermi level.

Zurawsky et al [17] studied the effect of pressure on the bandgap $E_{g,\Gamma}$, and on the bottom of the conduction band, $E_c$, and top of the valence band, $E_v$, of n-type and p-type InP up to 10 kbar at 300 K. A value of $\alpha_\Gamma = 8.0 \times 10^{-6}$ eV/bar was obtained. It was determined that the increase of the energy gap of the direct transition is, to first order, due to an increase in energy of the conduction band edge.

Chang et al [18] used a pseudopotential method within the local density approximation to calculate the pressure coefficients at critical points for several diamond and zinc-blende semiconductors. For InP their calculation yields $\alpha_\Gamma = 8.0 \times 10^{-6}$ eV/bar (R), $\alpha_\Gamma = 7.4 \times 10^{-6}$ eV/bar, $\alpha_X = -2.1 \times 10^{-6}$ eV/bar, and $\alpha_L = 3.7 \times 10^{-6}$ eV/bar, where (R) stands for the relativistic calculation. They calculate the critical pressure for the inversion of the $L_1$ and $X_1$ minima to be approximately 105 kbar.

Camphausen et al [19] extended Van Vechten's model [10] of chemical bonds in tetrahedrally coordinated $A^NB^{8-N}$ compounds in order to calculate pressure coefficients as well as the pressure coefficient of the refractive index, d(ln(n))/dP. Applying their model to InP they calculate $\alpha_\Gamma = 9.5 \times 10^{-6}$ eV/bar, $\alpha_X = -1.0 \times 10^{-6}$ eV/bar, $\alpha_L = 3.4 \times 10^{-6}$ eV/bar, and d(ln(n))/dP = $-0.4 \pm 0.2 \times 10^{-6}$/bar.

## C CARRIER INDUCED EFFECTS

There are three principal carrier induced effects on optical functions: the plasma effect, the bandfilling effect, and a bandgap shrinkage effect due to electron-electron or hole-hole interactions [20].

Intraband free carrier absorption, also known as the plasma effect, affects the bandgap. According to the classical Drude interaction, the change in the absorption coefficient, $\Delta\alpha$, as well as the change in the refractive index, $\Delta n$, are directly proportional to $\lambda^2$. The sign in $\Delta n$ is always negative. Because of the $\lambda^2$ dependence, the plasma effect increases as the photon energy is decreased below the bandgap. However, just below the absorption edge, a $\lambda^3$ dependence is observed, which requires a non-classical, quantum mechanical description [21].

A shift of the fundamental absorption edge to higher photon energies has been observed for several doped semiconductors. The phenomenon, known as the Burstein Moss effect, has been explained by bandfilling. In the case of n-type semiconductors, the density of states in the conduction band is sufficiently low that a relatively small number of electrons can fill the band to an appreciable depth. With the lowest energy states in the conduction band filled, electrons from the valence band require energies greater than the nominal bandgap to be optically excited into the conduction band. The optical excitation corresponds to absorption; hence, there is a shift in the absorption edge to higher energies. The situation is similar for holes in p-type materials, but their larger effective mass means a higher density of states and hence a smaller bandfilling effect for a given carrier concentration.

In addition to bandfilling, the injection of excess carriers into an undoped semiconductor can cause a decrease in the bandgap because of electron-electron or hole-hole interactions [20].

Bennett and Soref [20] analysed the effects of these three principal carrier induced changes in the refractive index, $\Delta n$, and the absorption coefficient, $\Delta\alpha$, in the transparent region due to electrical injection of electrons and holes for $10^{15}$ to $10^{19}$ carriers/cm$^3$, from $E \cong 0.12$ to 1.34 eV. Their results also approximately apply to depletion of carriers from an initially doped region, although the sign of $\Delta n$ in depletion is opposite to that in injection. They showed that the bandfilling effect can produce a significant change in the index of refraction. In addition, bandgap shrinkage and plasma effects may be important, depending on carrier type and carrier concentration. The bandfilling and plasma effects both produce a negative $\Delta n$ for wavelengths in the transparent regime while bandgap shrinkage produces a positive $\Delta n$. Their results suggest that low-loss switches and phase modulators based upon carrier effects are feasible.

Jensen [21] describes the quantum theory of free carrier absorption, i.e. intraband absorption which is accompanied by scattering of optical or acoustical mode phonons or by charged impurities. Deviations from the quasi-classical Boltzmann equation or the Drude theory occur when photon energies become large compared with electron energies, i.e. when $E >> k_0T$ or $E >> E_f$ for non-degenerate or degenerate materials, respectively, where $k_0$ is the Boltzmann constant and $E_f$ the Fermi energy measured relative to the conduction band edge. The quantum result predicts a $\lambda^3$ wavelength dependence of $\alpha$, the absorption coefficient, for polar optical mode scattering and a $\lambda^{3.5}$ to $\lambda^4$ dependence for ionised impurity scattering. The low frequency $E \cong k_0T$ limit gives a good estimate for the DC mobility as a function of carrier concentration. At high frequencies, in lightly doped materials in which polar scattering dominates, $\varepsilon_1$ is proportional to $\lambda^3$, and $\varepsilon_2$ and k are proportional to $\lambda^4$. Furthermore, $\varepsilon_1$ is given approximately by the Drude theory expression and $n \cong (\varepsilon_\infty)^{1/2}$ for $\omega_p << \omega << E_g/\hbar$. As $\omega \rightarrow E_g/\hbar$, there is a small quantum correction to $\varepsilon_1$ and to n. Based on this model Jensen [22] calculates the optical functions of InP at 300 K in the range E = 0.01351 to 0.1361 eV ($\lambda$ = 91.8 to 9.18 μm) as a function of free carrier density. However, his results do not describe the reststrahlen region.

## REFERENCES


[1] L.G. Meiners [ *J. Appl. Phys. (USA)* vol.59 no.5 (1986) p.1611-3 ]

[2] K. Seeger [ *Appl. Phys. Lett. (USA)* vol.54 no.13 (1989) p.1268-9 ]

[3] H. Jamshidi, T.J. Parker [ *Int. J. Infrared Millim. Waves (USA)* vol.4 (1983) p.1037-44 ]

[4] V.B. Bogdanov, V.T. Prokopenko, A.D. Yas'kov [ *Opt. Spectrosc. (USA)* vol.60 (1986) p.68-9 ]

[5] G.D. Pettit, W.J. Turner [ *J. Appl. Phys. (USA)* vol.36 (1965) p.2081 ]

[6] W.N. Reynolds, M.T. Lilburne, R.M. Dell [ *Proc. Phys. Soc. London (UK)* vol.71 (1958) p.416-21 ]

[7] F. Oswald [ *Z. Nat.forsch. A (Germany)* vol.10a (1955) p.927-30 ]

[8] W.J. Turner, W.E. Reese, G.E. Pettit [ *Phys. Rev. (USA)* vol.136 (1964) p.A1467-70 ]

[9] M.L. Cohen, J.R. Chelikowsky [ *Electronic Structure and Optical Properties of Semiconductors* (Springer-Verlag, 1988) ]

[10] P. Lautenschlager, M. Garriga, M. Cardona [ *Phys. Rev. B (USA)* vol.36 no.9 (1987) p.4813-9 ]

[11] A.R. Forouhi, I. Bloomer [ *Phys. Rev. B (USA)* vol.38 no.3 (1988) p.1865-74 ]

[12] M. Cardona [ *J. Appl. Phys. (USA)* vol.32 (1961) p.958 ]

[13] L. Gorczyca, N.E. Christensen, M. Alouani [ *Phys. Rev. B (USA)* vol.39 no.11 (1989) p.7705-12 ]

[14] C.S. Menoni, H.D. Hochheimer, I.L. Spain [ *Phys. Rev. B (USA)* vol.33 no.8 (1986) p.5896-8 ]

[15] H. Müller, R. Trommer, M. Cardona, P. Vogl [ *Phys. Rev. B (USA)* vol.21 no.10 (1980) p.4879-83 ]

[16] T. Kobayashi, K. Aoki, K. Yamamoto [ *Physica B&C (Netherlands)* vol.139/140 (1986) p.537-40 ]

[17] W.P. Zurawsky, J.E. Littman, H.G. Drickamer [ *J. Appl. Phys. (USA)* vol.54 no.6 (1983) p.3216-9 ]

[18] K.J. Chang, S. Froyen, M.L. Cohen [ *Solid State Commun. (USA)* vol.50 no.2 (1984) p.105-7 ]

[19] D.L. Camphausen, G.A.N. Connell, W. Paul [ *Phys. Rev. Lett. (USA)* vol.26 (1971) p.184-8 ]

[20] B.R. Bennett, R.A. Soref [ *Proc. SPIE - Int. Soc. Opt. Eng. (USA)* vol.994 (1989) p.168-86 ]

[21] B. Jensen [ in *Handbook of Optical Constants of Solids* Ed. E.D. Palik (Academic Press, London, 1985) ch.9 p.169-88 ]

[22] A.R. Forouhi, I. Bloomer [ *Properties of Indium Phosphide* (INSPEC, IEE, London, UK, 1991) p.118-22 ]

# 5.3 Photoluminescence of InP

T.S. Kim and B.G. Streetman

July 1998

## A INTRODUCTION

Photoluminescence (PL) analysis is a very powerful technique for studying impurity and defect states in InP. PL is often sensitive enough to measure impurity concentrations in parts-per-trillion ranges. PL is affected by many factors, yielding information on both bulk and surface properties of InP. PL is used to identify acceptor impurities, to assess the crystal quality, to measure the surface band bending and surface recombination velocity, and to study the energy levels in quantum well structures. Recently, a significant level of effort has been spent on PL study of InP-based heterostructures, although this is not discussed here. As measurements can be made non-destructively, PL is also used for in situ or ex situ inspection of various semiconductor-processing steps [1-5]. First, a brief description of the PL process is given with an emphasis on the factors that need to be considered in analysing PL spectra. Lists of peaks that frequently appear in low temperature PL spectra of InP are given with brief discussions. PL peaks arising from shallow levels, deep-level impurities and complexes, processing-induced defects and surface effects are discussed.

## B PHOTOLUMINESCENCE PROCESS

As PL is sensitive to many factors, analysis of PL spectra requires good understanding of the following three primary processes [6]:

(a) creation of electron-hole pairs by absorption of incident photons,
(b) generation of photons by recombination of the electron-hole pairs, and
(c) escape of the generated photons from the sample.

In materials with high quantum efficiencies, other secondary processes such as stimulated emission, re-absorption and re-emission should also be considered [7,8]. During a PL process, the semiconductor is excited to a non-equilibrium state by absorbing incident photons whose energy is larger than the bandgap. The photogenerated electrons and holes quickly relax to the lowest available energy states near the conduction and valence bands before recombining [9]. Since the lowest available energy states depend on the Fermi level, PL spectra change with doping concentration. When the conduction band is partially filled due to heavy n-type doping, the PL peak shifts to higher energy with doping concentration, which is known as the Burstein-Moss shift [10-12]. In heavily doped p-type InP, the PL peak shifts to lower energy with doping concentration due to local fluctuations in the band edges [13]. The electron-hole pairs can recombine either radiatively or non-radiatively. PL measurements are designed to monitor photons emitted from the sample as a result of radiative recombination. Intensity of the luminescence is plotted as a function of either wavelength or energy, using either a conventional dispersive spectrometer or an interferometer with Fourier transformation capability. Thus, each radiative recombination process with different transition energy appears as a peak in a luminescence spectrum. Peaks broaden with sample temperature due to thermal energy [14,15]. Each recombination route competes against the others, making quantitative analysis of PL complicated. The presence of non-radiative recombination routes is manifested as a low overall PL intensity, which can be used to assess the crystal quality.

Analysis of PL spectra is further complicated by diffusion of the photogenerated carriers into the bulk. The presence of surface states results in a higher recombination rate at the surface than in the bulk, lowering the excess carrier concentration at the surface. The high recombination rate at the surface is described by a boundary condition known as the surface recombination velocity. Thus, surface recombination velocity, carrier mobility and carrier lifetime affect the spatial distribution of excess carriers. The excess carrier profile often extends a few microns into the bulk, while the absorption depth of excitation light is typically less than 100 nm. Diffusion of photogenerated carriers needs to be considered when a non-homogeneous structure is studied. For example, an implanted-and-annealed layer sometimes has a low quantum yield due to the radiation damage, in which case the PL spectrum would be dominated by the luminescence originating from the underlying region.

The small critical angle of InP for total internal reflection (~18 degrees) gives an additional complication. Only a small fraction (~2%) of the generated photons can escape directly from the front surface, the rest being reflected toward the back surface [16]. Multiple internal reflections increase the amount of luminescent light escaping through the front surface, but they also distort the PL spectra. As generated photons propagate through the sample, some of them are re-absorbed. When the quantum efficiency is high, new photons are then generated through radiative recombination of the electron-hole pairs (or excitons) created by re-absorption of the luminescence light. Such a cascade process, often called photon recycling, effectively increases the apparent carrier diffusion length and quantum efficiency [17]. Enhanced carrier diffusion by photon recycling is also known as radiative diffusion, and is especially important in low temperature PL of InP. Near the exciton energy, the absorption coefficient is a rapidly changing function of the wavelength and high-energy exciton peaks are preferentially quenched by re-absorption. As a result, a measured low-temperature PL spectrum is often distorted and does not reflect the true recombination rate of each transition [11].

## C ROOM-TEMPERATURE PHOTOLUMINESCENCE

Room-temperature PL is often used to assess the crystal quality by measuring the overall luminescence intensity. At room temperature the major radiative recombination process is the band-to-band transition. PL intensity is a measure of dominance of the radiative recombination over non-radiative recombination. The presence of non-radiative recombination centres generally indicates lattice imperfections and deep-level impurities. The room temperature PL measurement can be used, for instance, for mapping the polishing damage on an InP substrate [18]. Such effects as surface recombination velocity and surface depletion complicate analysis of the room-temperature PL intensity variation. The surface recombination velocity of InP is fairly low and very sensitive to the surface treatment. For this reason, room-temperature PL has been adopted to study InP surfaces exposed to gases [19-21] and chemical etches [22,23], effectiveness of a surface passivating layer [24], and interfaces within metal-insulator-InP structures [4,25,26]. An order of magnitude decrease in PL intensity is reported when PL measurement is done in an oxygen ambient instead of nitrogen [21]. Thus, variations in room-temperature PL intensity can be used to examine the quality of the surface or interface. Surface recombination velocity, hence PL intensity, also depends on the doping concentrations of InP [27]. The peak positions also vary depending on the doping concentration in degenerately doped InP [10-13], which may be used to non-destructively map the doping concentrations. Accurate analysis of the peak shape should take the self-absorption process into account [11].

## D LOW-TEMPERATURE PHOTOLUMINESCENCE

To gain further information on the impurity and defect states, PL is measured at cryogenic temperatures at which thermal energy is low enough that a variety of transitions can be resolved. For convenience, PL spectra at liquid He temperature only are discussed here. PL peaks are typically grouped into the three categories: band-edge transitions, shallow-impurity transitions and deep-level transitions.

## D1 Exciton Transition

The highest energy band in a low-temperature PL spectrum results from decay of excitons [28]. An exciton is an electron-hole pair bound by coulombic attraction, forming a hydrogen-like structure. In InP, recombination involving excitons bound to shallow impurities forms an overlapping band with free-exciton emission, and this band is frequently referred to as the band-edge (BE) peak. Since the binding energies of free excitons, donor-bound excitons and neutral acceptor-bound excitons are slightly different, the BE peak exhibits fine structure at low temperature. The bound-exciton peaks in pure InP are very sharp (FWHM < 0.1 meV) and normally do not show a shift with increasing excitation power density. The exciton peaks become broad as the impurity concentration of the crystal increases, due to the screening effect. At high doping concentrations the fine structure of the BE peak disappears. A list of exciton peaks in InP is given in TABLE 1.

TABLE 1 Exciton peaks in InP [29]. The exact peak position and assignment may differ from those given in other references [30,31].

| Exciton recombination mechanism | Peak position at 2 K (eV) |
|---|---|
| Free | 1.4188 |
| Bound to neutral donor (n = 5) | 1.4185 |
| Bound to neutral donor (n = 4) | 1.4183 |
| Bound to neutral donor (n = 3) | 1.4178 |
| Bound to neutral donor (n = 2) | 1.4176 |
| Bound to neutral donor (n = 1) | 1.4173 |
| Neutral donor to hole | 1.4168 |
| Bound to ionised donor | 1.4166 |
| Bound to neutral acceptor | 1.4144 |
| Bound to neutral acceptor | 1.4142 |

## D2 Shallow Acceptor Transition

At wavelengths longer than for the BE peak are 'conduction band to neutral acceptor' (e-A) and 'neutral donor to neutral acceptor' (D-A) transitions. These peaks appear at lower energies than the BE peak since the ionisation energies of impurities are larger than the exciton binding energies. Longitudinal optical (LO) phonon replicas of these peaks also appear at 43 meV lower in energy than the fundamental peaks. Donor-to-valence band transitions appear within the BE-peak, since the ionisation energy of shallow donors in InP is only about 7 meV. Shallow donors and acceptors are like hydrogen atoms imbedded in the semiconductor with electrons and holes having effective masses. Actual energy levels are shifted from the ideal hydrogen-like energy levels due to the effect of the core potentials of the donor or acceptor impurities [32], which is called the chemical shift or central cell correction. The chemical shift of shallow donors in InP is very small due to the small electron effective mass. A small effective mass gives a large Bohr radius, reducing the effect of the core potential. For this reason, the shallow levels of different donor species are not resolved in usual PL spectra of InP. In contrast, acceptors exhibit a strong chemical shift due to the larger effective mass of holes. Since the chemical shift is not easy to calculate, ionisation energies of acceptor species are measured empirically by, for example, low-temperature PL. Identification of an acceptor species with a particular PL peak is often done by intentional doping. If an acceptor-related peak intensity increases with the doping concentration of an acceptor impurity, the peak can generally be assigned to that acceptor impurity. Once the ionisation energies of acceptors are established by intentional doping experiments, low temperature PL can be used to identify the acceptor species in a sample. A list of the peak positions of typical acceptor species in InP is given in TABLE 2. The exact peak position of a D-A transition cannot be given, since a D-A peak shows a blue shift with temperature or excitation power density. The blue shift is due to the coulombic potential between ionised donors and acceptors which must be overcome by an electron making a transition from a neutral donor to a neutral acceptor [33]. The coulombic potential depends on the distance between the donor and acceptor impurities.

The transition probability also depends on the donor-acceptor distance, with closer donor-acceptor pairs having a higher transition rate than remote pairs. In an actual PL spectrum of InP, a D-A peak appears as a broad peak that is an unresolved superposition of many pair transitions having slightly different transition energies. With increasing temperature or excitation power density, faster transitions become favourable, causing a blue shift (a shift toward high energy). A shift of as much as 4 meV may be observed due to these effects. The 1.378 eV peak in TABLE 2, which is frequently referred to as the 'A1' peak, frequently appears in InP grown by a variety of methods. It used to be assigned to a carbon acceptor, but is probably induced by either Be or Mg [34]. Assigning a peak to an acceptor species by intentional doping experiments is sometimes very difficult.

TABLE 2 Shallow-impurity related PL transitions of InP.

| Peak (eV) | Impurity | Transition | Ref |
|---|---|---|---|
| 1.40 | Surface state | | [35] |
| | Unidentified | D-A | [36] |
| 1.3785 | Mg | D-A | [34] |
| 1.3783 | 'A1' | D-A | [37] |
| 1.3782 | Be | D-A | [34] |
| 1.378 | Ca | D-A | [38] |
| 1.3755 | C | D-A | [34] |
| 1.3732 | Zn | D-A | [34] |
| 1.35 | Defect induced | | [39] |
| | Cu | | [40] |
| 1.369 | Cd | | [37] |

## D3 Deep-Level Transition

There is no clear demarcation energy at which transitions can be classified into shallow- and deep-level transitions. As the ionisation energy of an impurity or defect becomes large, transitions tend to become non-radiative. Nevertheless there are several PL peaks that frequently appear in the energy range of 1.3 to 0.6 eV. The impurities and defects that induce deep levels also tend to distort the lattice substantially, enhancing the interactions with phonons and even generating localised phonons. Typically, deep-level peaks are broad due to strong interaction with the local vibrational modes and sometimes show fine structure originating from phonon replicas. A good example of fine structure due to phonon replicas is the Cu-related peak in InP [41]. Copper gives rise to a distinct PL band in InP which consists of a sharp zero-phonon line at 1.2889 eV with many sharp phonon replicas at lower energy superimposed on a broad background originating from interaction with the local vibrational modes. In addition to the LO phonons with energy of 43 meV, localised phonons with energy of 33.3 meV and seven other localised phonons are induced to give rise to many sharp peaks. Some deep-level impurities may create more than one level within the bandgap, inducing a transition between these levels [42,43]. Such intracentre transitions also typically show sharp zero-phonon lines with broad bands. The identification of the exact chemical origin and energy states of the impurities and defects responsible for deep-level transitions is often very difficult (TABLE 3).

TABLE 3 Deep-level PL transitions in InP. The assignments of the peaks are often tentative and disagreements exist in the literature regarding some assignments.

| Peak (eV) | Assignment | Ref |
|---|---|---|
| 1.2889 | Cu | [41] |
| 1.24 | Yb intracentre | [44] |
| 1.21 | Acceptor, V(In) or Si | [45] |
| 1.14 | Acceptor, V(In) impurity | [45] |
| | Mn | [46,47] |
| 1.10 | Ge, Fe | [48,49] |
| 1.08 | Acceptor, D-A complex | [45] |
| 0.99 | Donor, V(P) impurity | [45] |
| 0.80 - 0.79 | Er intracentre | [50,51] |
| 0.75 | Unknown | [45] |
| 0.35 | Fe intracentre | [52] |

## D4 Processing-Induced Defect Transition

After certain processing steps, several peaks have been reported to appear in InP that cannot be easily attributed to any impurity species (TABLE 4). Several excitonic peaks are reported to appear after heat treatment or electron irradiation of InP [53-56]. The origins of these peaks are still not well established. A mechanically damaged surface of InP is reported to give rise to a peak at 1.306 eV [57]. Further examination of the 1.306 eV peak may prove useful in understanding the surface states of InP. Radiation damage induced by (Si) ion implantation gives rise to a transition at 1.35 eV. A deep-donor level with an ionisation energy of 33 meV is believed to be induced by the radiation damage [39]. The 1.35 eV peak also appears after Cu is diffused into InP. Consequently, the 1.35 eV peak has been attributed to Si (possibly a Si acceptor) [58], a Si-defect complex [59], and a Cu acceptor [60]. The 1.35 eV peak shows a blue shift of 1.4 meV per decade increase in excitation power density, causing a wide variation in the estimated ionisation energy. Hydrogenation does not induce a new transition, but affects PL spectra as it neutralises deep-level defects and ionised impurities [61].

TABLE 4 InP PL peaks that appear after certain processing.

| Peak (eV) | Assignment (processing) | Ref |
|---|---|---|
| 1.409 | Exciton bound to unidentified deep-level acceptor (heat treatment, electron irradiation) | [53,54] |
| 1.407 | Unknown (heat treatment) | [55] |
| 1.403 - 1.402 | D-A pair transition, possibly e-A of 1.396 eV peak (heat treatment, electron irradiation) | [53,54] |
| 1.399 | Complexes involving In-on-P antisite defect (alpha particle irradiation) | [62] |
| 1.398 | Exciton bound to deep level (heat treatment, electron irradiation) | [53,54] |
| 1.396 | D-A pair transition involving an unidentified shallow acceptor (heat treatment, electron irradiation) | [53,54] |
| 1.394 | Exciton bound to an unidentified deep-level acceptor (heat treatment, electron irradiation, epitaxy) | [36] |
| 1.404 - 1.392 | Surface-related transition or D-A pair transition involving an unidentified shallow acceptor (surface treatment, epitaxy) | [36,63] |
| 1.306 | Saw-damage induced peak (sawing, polishing) | [57] |
| 1.35 | Radiation-damage induced peak (ion implantation, Cu diffusion) | [39] |

### D5 Surface Effect and Surface State-Induced Transition

Surface treatment with acid or DI water is known to affect low-temperature PL spectra [63,64]. After dipping in HF, a peak appears near 1.4 eV. DI-water rinse, for example, makes the 1.4 eV peak disappear, indicating that it arises from or near the surface. The peak position typically ranges from 1.392 to 1.404 eV, showing a blue shift with excitation power density. The 1.4 eV emission band is attributed to recombination from the bulk conduction band and bulk neutral donors to surface states distributed from the valence band to at least 300 meV into the bandgap [63]. The 1.4 eV peak was earlier attributed to an unidentified D-A transition [36]. Substantial distortion of exciton peaks after surface treatment with acids has also been reported [64]. The extent of this distortion varies depending on the excitation power density and surface preparation. This distortion is mainly due to the re-absorption of band-edge luminescence as discussed in Section B. The distortion of the BE luminescence results in the appearance of artifactual peaks in the low-energy tail of the band-edge region which sometimes dominate the spectra and can lead to misinterpretation of spectra.

## E OTHER RELATED MEASUREMENTS

Information regarding the band structures beyond the conduction band minimum and valence band maximum is typically not available from conventional PL analysis. Since the excitation light has energy larger than the bandgap, a small fraction of the photogenerated electrons and holes recombine radiatively while they are still in high-energy states. Hot carrier recombination gives rise to a very weak, high-energy tail in PL which is called hot photoluminescence (HPL). Information on the band structure and kinetics of relaxation processes in the femtosecond range can be obtained by analysing HPL [65]. The kinetics of a radiative recombination process are studied by time-resolved PL measurements. In time-resolved PL measurements, a pulsed excitation is used, and PL is measured at various time delays from the excitation pulse. Time-resolved measurements are essential for measuring lifetimes and transition times of recombination processes. The identification of effective-mass donors in InP cannot be made using conventional PL due to small chemical shifts. In magneto-photoluminescence (MPL) measurements, an additional chemical shift is induced by a magnetic field [66]. MPL is also used for studying potential profiles across heterostructures [67].

PL measurements are often made at high pressures to study the changes in the band structure and impurity levels of a semiconductor with pressure [68,69]. At sufficiently high pressures, InP becomes an indirect bandgap material as Γ and L minima increase while X minima decrease in energy with pressure. At low pressures, PL intensity remains more or less the same. At high pressures PL intensity decreases exponentially with pressure as X approaches Γ. From the changes in PL intensity with pressure, the crossover pressure at which the InP bandgap becomes indirect is estimated. Although measured and calculated cross-over pressure ranges from 7 to 10.5 GPa, the PL peak due to direct band-to-band transitions does not disappear even at 12 GPa [70-72].

## F SUMMARY

PL is sensitive to many factors, making it useful for studying surface/interface properties and bulk impurity and defect states. As PL measurements are non-destructive, scanning PL measurements are often used for inspecting substrates and epilayers. The PL process is somewhat complex, requiring good understanding of factors that affect PL spectra. For example, PL may vary significantly depending on surface treatment alone, mainly due to the low surface recombination velocity and high quantum efficiency of InP. Better understanding of the surface-related phenomena in InP and their effects on PL are desired. In the literature, disagreement exists on many peak assignments, and the origins of many peaks have yet to be identified. Identifying the origins of deep-level transitions is an especially difficult task and more work is needed. As InP-based heterostructures are now widely used for many device applications, the focus of PL study appears to be shifting to heterostructures.

## REFERENCES


[1] M.F. Lambrinas et al [ *J. Electrochem. Soc. (USA)* vol.144 no.6 (1997) p.2086-95 ]

[2] W. Jantz et al [ *Mater. Res. Soc. Symp. Proc. (USA)* vol.325 (1994) p.409-18 ]

[3] H. Hasegawa et al [ *Proc. 4th Int. Conf. on InP and Related Materials* Newport, RI, USA, 21-24 Apr. 1992 (IEEE, NY, USA, 1992) p.24-7 ]

[4] F. Kiel, W. Kulisch, R. Kassing [ *Proc. 4th Int. Conf. on InP and Related Materials* Newport, RI, USA, 21-24 Apr. 1992 (IEEE, NY, USA, 1992) p.347-50 ]

[5] G.E. Carver, R.W. Heebner, R.D. Moore [ *Proc. 16th State-of-the-Art Program on Compound Semiconductors* St. Louis, MO, USA, 1992 (Electrochem. Soc., Pennington, NJ, USA, 1992) p.239-50 ]

[6] H.B. Bebb, E.W. Williams [ *Semicond. Semimet. (USA)* vol.8 (1975) ch.4 p.181-320 ]

[7] D.Z. Garbuzov [ in *Semiconductor Physics* Eds. V.M. Tuchkevich, V.Ya. Frenkel (Consultants Bureau, New York, 1986) p.53-86 ]

[8] S.M. Durbin et al [ *Proc. IEEE Photovoltaic Specialists Conf. (USA)* vol.23 (1993) p.628-32 ]

[9] G.R. Allan, H.M. van Driel [ *Proc. Hot Carriers in Semiconductors* Eds. K. Hess, J.P. Leburton, U. Ravaioli (Plenum Press, New York, NY, USA, 1995) p.121-4 ]

[10] Y. Morishita et al [ *J. Cryst. Growth (Netherlands)* vol.104 no.2 (1990) p. 457-62 ]

[11] R.M. Sieg, S.A. Ringel [ *J. Appl. Phys. (USA)* vol.80 no.1 (1996) p.448-58 ]

[12] R. Schwabe et al [ *Solid State Commun. (USA)* vol.58 no.7 (1986) p.485-91 ]

[13] C. Blaauw et al [ *J. Electron. Mater. (USA)* vol.18 no.5 (1989) p.567-72 ]

[14] R. Benzaquen et al [ *Phys. Rev. B (USA)* vol.52 no.3 (1995) p.1485-8 ]

[15] R. Benzaquen et al [ *Phys. Rev. B (USA)* vol.52 no.4 (1995) p.R2273-6 ]

[16] C.A. Warwick [ *J. Appl. Phys. (USA)* vol.59 no.12 (1986) p.4182-4 ]

[17] W.P. Dumke [ *Phys. Rev. (USA)* vol.105 (1957) p.139 ]

[18] Z. Laczik, G.R. Booker, A. Mowbray [ *Mater. Sci. Eng. B (Switzerland)* vol.42 no.1-3 (1996) p.217-24 ]

[19] H. Nagai, Y. Noguchi [ *Appl. Phys. Lett. (USA)* vol.33 no.4 (1978) p.312-4 ]

[20] T. Suzuki, M. Ogawa [ *Appl. Phys. Lett. (USA)* vol.34 no.7 (1979) p.447-9 ]

[21] S.D. Lester, T.S. Kim, B.G. Streetman [ *J. Appl. Phys. (USA)* vol.60 no.12 (1986) p.4209-14 ]

[22] H. Nagai, Y. Noguchi [ *J. Appl. Phys. (USA)* vol.50 no.3 (1979) p.1544-5 ]

[23] S.K. Krawczyk, G. Hollinger [ *Appl. Phys. Lett. (USA)* vol.45 no.8 (1984) p.870-2 ]

[24] I.K. Han et al [ *J. Appl. Phys. (USA)* vol.81 no.10 (1997) p.6986-91 ]

[25] B. Sautreuil, P. Viktorovitvh, R. Blanchet [ *J. Appl. Phys. (USA)* vol.57 no.6 (1985) p.2322-4 ]

[26] M. Ochiai et al [ *Electron. Lett. (UK)* vol.29 no.6 (1993) p.568-9 ]

[27] D.E. Aspnes [ *Surf. Sci. (Netherlands)* vol.132 no.1-3 (1983) p.406-21 ]

[28] P.J. Dean [ *NATO Adv. Study Inst. Ser. B, Phys. (USA)* vol.88 (1983) p.247-315 ]

[29] M. Razeghi [ *The MOCVD Challenge* vol.1 (Adam Hilger, Philadelphia, USA, 1989) p.56-7 ]

[30] L.D. Zhu et al [ *J. Appl. Phys. (USA)* vol.57 no.12 (1985) p.5486-92 ]

[31] T. Inoue et al [ *J. Appl. Phys. (USA)* vol.67 no.11 (1990) p.7165-8 ]

[32] N.W. Ashcroft, N.D. Mermin [ *Solid State Physics* (Saunders College, Philadelphia, USA, 1976) ch.28 ]

[33] R. Dingle [ *Phys. Rev. (USA)* vol.184 no.3 (1969) p.788-96 ]

[34] B.J. Skromme et al [ *J. Electron. Mater. (USA)* vol.13 no.3 (1984) p.463-91 ]

[35] T.S. Kim, S.D. Lester, B.G. Streetman [ *J. Appl. Phys. (USA)* vol.61 no.5 (1987) p.2072-4 ]

[36] J.U. Fischbach et al [ *Solid State Commun. (USA)* vol.11 no.5 (1972) p.725-7 ]

[37] K. Hess, N. Stath, K.W. Benz [ *J. Electrochem. Soc. (USA)* vol.121 no.9 (1974) p.1208-12 ]

[38] E. Kubota, A. Katsui, K. Sugii [ *J. Appl. Phys. (USA)* vol.59 no.11 (1986) p.3841-6 ]

[39] T.S. Kim, S.D. Lester, B.G. Streetman [ *J. Appl. Phys. (USA)* vol.62 no.4 (1987) p.1363-7 ]

[40] E.W. Williams et al [ *J. Electrochem. Soc. (USA)* vol.120 no.12 (1973) p.1741-9 ]

[41] M.S. Skolnick et al [ *J. Phys. C (UK)* vol.16 no.10 (1983) p.1967-85 ]

[42] W. Koerber et al [ *Inst. Phys. Conf. Ser. (UK)* no.91 (1987) p.303-6 ]

[43] See, for example, papers in [ *Mater. Res. Soc. Symp. Proc. (USA)* vol.301 (1993) and vol.422 (1996) ]

[44] H. Ennen, G. Pomrenke, A. Axmann [ *J. Appl. Phys. (USA)* vol.57 no.6 (1985) p.2182-5 ]

[45] H. Temkin et al [ *J. Appl. Phys. (USA)* vol.53 no.11 (1982) p.7526-35 ]

[46] Phil Won Yu [ *Solid State Commun. (USA)* vol.34 no.3 (1980) p.183-6 ]

[47] A.M. Loshinskii et al [ *Sov. Phys.-Semicond. (USA)* vol.19 no.11 (1985) p.1223-6 ]

[48] A.M. White, P.J. Dean, B. Day [ *Proc. Int. Conf. on Phys. of Semicond.* Rome, 1976, Ed. F.G. Fumi (North Holland, Amsterdam, 1976) p.1057 ]

[49] L. Eaves et al [ *Proc. Conf. on Semi-insulating III-V Materials* Eds. S. Makram-Ebeid, B. Tuck (Shiva, Nantwick, UK, 1982) p.199-207 ]

[50] X. Zhao, K. Hirakawa, T. Ikoma [ *Appl. Phys. Lett. (USA)* vol.54 no.8 (1989) p.712-4 ]

[51] Y. Fujiwara et al [ *Mater. Sci. Forum (Switzerland)* vol.196-201 pt.2 (1995) p.621-6 ]

[52] B.V. Shanabrook et al [ *Proc. Conf. on Semi-insulating III-V Materials* Eds. S. Makram-Ebeid, B. Tuck (Shiva, Nantwick, UK, 1982) p.310-3 ]

[53] J. Bandet et al [ *Solid State Commun. (USA)* vol.54 no.9 (1985) p.767-70 ]

[54] J. Frandon et al [ *J. Appl. Phys. (USA)* vol.59 no.5 (1986) p.1627-32 ]

[55] T.S. Kim, S.D. Lester, B.G. Streetman [ *J. Appl. Phys. (USA)* vol.61 no.9 (1987) p.4598-602 ]

[56] K. Pressel et al [ *Mater. Res. Soc. Symp. Proc. (USA)* vol.262 (1992) p.695-700 ]

[57] J.D. Oberstar, B.G. Streetman [ *Surf. Sci. (Netherlands)* vol.108 no.3 (1981) p.L470-6 ]

[58] P.K. Bhattacharya, W.H. Goodman, M.V. Rao [ *J. Appl. Phys. (USA)* vol.55 no.2 (1984) p.509-14 ]

[59] M.V. Rao [ *Appl. Phys. Lett. (USA)* vol.48 no.22 (1986) p.1522-4 ]

[60] J.-C. Lee, A.G. Milnes, T.E. Schlesinger [ *Phys. Rev. B (USA)* vol.34 no.10 (1986) p.7385-7 ]

[61] S. Balasubramanian et al [ *J. Appl. Phys. (USA)* vol.77 no.10 (1995) p.5398-405 ]

[62] M.L. Favaro et al [ *Phys. Status Solidi A (Germany)* vol.156 no.2 (1996) p.523-32 ]

[63] S.D. Lester, T.S. Kim, B.G. Streetman [ *J. Appl. Phys. (USA)* vol.62 no.7 (1987) p.2950-4 ]

[64] S.D. Lester, T.S. Kim, B.G. Streetman [ *J. Appl. Phys. (USA)* vol.63 no.3 (1988) p.853-7 ]

[65] M.A. Alekseev et al [ *Sov. Phys.-Semicond. (USA)* vol.23 no.5 (1989) p.479-90 ]

[66] M.S. Skolnick et al [ *Appl. Phys. Lett. (USA)* vol.44 no.9 (1984) p.881-3 ]

[67] T. Lundstrom et al [ *Proc. 23rd Int. Conf. on Phys. of Semicon.* Berlin, Germany, 21-26 July 1996 (World Scientific, Singapore, 1996) vol.3 p.1827-30 ]

[68] D.L. Camphausen, G.A. Neville Connell, W. Paul [ *Phys. Rev. Lett. (USA)* vol.26 (1971) p.184-8 ]

[69] K. Takarabe [ *Phys. Status Solidi B (Germany)* vol.198 no.1 (1996) p.211-22 ]

[70] K. Kobayashi et al [ *J. Lumin. (Netherlands)* vol.28 (1981) p.347 ]

[71] C.S. Menoni, H.D. Hochheimer, I.L Spain [ *Phys. Rev. B (USA)* vol.33 no.8 (1986) p.5896-8 ]

[72] K.J. Chang et al [ *Solid State Commun. (USA)* vol.50 no.2 (1984) p.105-7 ]

# 5.4 Intersubband transitions in InP-based quantum wells

C. Sirtori

September 1998

## A INTRODUCTION

In a semiconductor, the optical transitions between electronic states of the same band (conduction or valence) are currently defined as intersubband transitions (ISBTs) [1,2]. These states originate from size quantisation in attractive potential wells and they were first investigated (in the 1960s) in a two-dimensional electron gas at the silicon/silicon dioxide interface [3]. These investigations were subsequently extended to heterojunction quantum wells [4] and finally to coupled-quantum well quasi-molecules [5,6]. In these structures it has been shown that by judicious control of quantum well widths and tunnelling barrier thickness one can conceive artificial potentials such that energy level separations, dipole matrix elements and lifetimes are dependent on the characteristic potential design. Therefore, within the framework which is imposed by the physical parameters of the constituent materials, using ISBTs it is possible to tailor the optical and electrical properties of the specific heterostructure and eventually obtain the desired 'materials by design' [7]. This has caused a major impetus to the engineering of new photonic materials and related optoelectronic devices in the mid-infrared region of the electromagnetic spectrum (wavelength 3 - 15 μm) [1,2,6], leading to the invention of the quantum cascade (QC) laser [8] and the quantum well infrared photodetector [9] (QWIP). It was at the beginning of the development of QWIPs that great attention was focused on ISBTs between valence-band states in order to fabricate detectors operating with light at normal incidence with respect to the wafer surface [9]. The selection rules for transitions in the conduction band, in fact, forbid absorption of light propagating normal to the layers [10]. Nowadays technology has overcome this problem by making two-dimensional gratings on the detector surface and ISBTs in the valence band are, at present, of little interest [9]. Therefore, the core of this Datareview is essentially on transitions connecting conduction-band states.

There are only a few material systems where optical ISBTs have been experimentally observed: GaAs/GaInAs/AlGaAs, InP/GaInAs/AlInAs, GaSb/AlGaSb/InGaAs and Si/SiGe. For all these heterostructures, different material compositions (lattice matched and strained) were used in order to vary the depth of the quantum wells. Particular attention has been paid to heterostructures with large values of the conduction-band offset ($\Delta E_c \geq 300$ meV) due to the possibility of achieving higher transition energies [9,11,12]. It is important to notice that ISBTs are typically observed in materials with very abrupt heterojunction interfaces and where the thickness is controlled with atomic precision. The growth technique which best complies with these requirements is with no doubt molecular beam epitaxy (MBE). This is why more than 90% of the research on intersubband transitions has been carried out using heterostructures grown by this technique, the few other reports referring to materials grown with chemical beam epitaxy (CBE) or other metal-organic epitaxy techniques.

## B OPTICAL PROPERTIES OF ISBTS IN $Ga_{0.47}In_{0.53}As/Al_{0.47}In_{0.53}As/InP$

Among all the heterostructures $Ga_{0.47}In_{0.53}As/Al_{0.48}In_{0.52}As$ grown lattice matched on InP is probably the most indicated to study ISBTs in quantum wells, and is the one which has received more attention up to date. It offers a higher direct energy band offset than the GaAs/AlGaAs system and, with respect to the Sb based materials, can count on a more mature growth technique which has been developed for high speed electronics and light wave communications. Heterostructures made of $Ga_{0.47}In_{0.53}As/InP$, where

InP is used as the barrier material, have also been investigated. However, the weak conduction band offset (~230 meV) [9] strongly limits the range of accessible energy separations between the subbands and makes this heterostructure much less attractive than that where AlInAs is used as the barrier material. Other material combinations based on InP have been studied to tailor the spectral response of QWIPs: InGaAsP/InP for $\lambda$ = 8 - 12 μm, $GaAs/Ga_{0.5}In_{0.5}P$ for $\lambda$ ~ 8 μm and $GaAs/Al_{0.5}In_{0.5}P$ for $\lambda$ ~ 3 μm [9]. Note that the last two heterostructures are grown lattice matched on GaAs substrates.

One of the interesting peculiarities of ISBTs is their relatively simple physical description. Theoretical predictions of the electronic energies and of the optical transition strengths have been mainly based on standard one-band envelope-function Hamiltonians [5]. However, in $Ga_{0.47}In_{0.53}As/Al_{0.48}In_{0.52}As$ the intersubband energy can be a relevant fraction of the energy gap of the constituent materials. In this situation the interaction between the conduction- and the full-valence-band cannot be neglected and a more rigorous description has to be taken into account. Fortunately, as long as electron (conduction) states are considered, the four-band Kane Hamiltonian can be reduced to an 'effective' two band model, which can in turn be recast into a one-band model, provided an appropriate energy dependence is given to the effective mass (non-parabolicity effects). The complete theoretical treatment is given in [13]. In the following table all the physical constants that characterise this heterostructure are given.

TABLE 1 Physical parameters of the $Ga_{0.47}In_{0.53}As/Al_{0.47}In_{0.53}As$ heterostructure.

| Physical constant | Value | Physical constant | Value |
|---|---|---|---|
| $\Delta E_c$ | 0.52 eV | $\gamma$ (non-parabolicity coefficient) | $1.13 \times 10^{-18}$ $m^{-2}$ |
| m* ($Ga_{0.47}In_{0.53}As$) | 0.043 $m_e$ | $E_{gap}$ ($Ga_{0.47}In_{0.53}As$) | 0.81 eV |
| m* ($Al_{0.48}In_{0.52}As$) | 0.078 $m_e$ | $E_{gap}$ ($Al_{0.48}In_{0.52}As$) | 1.63 eV |

## B1 Absorption

The first observation of ISBTs in $Ga_{0.47}In_{0.53}As/Al_{0.48}In_{0.52}As$ was made by Levine et al in 1988, by measuring the absorption peak at 230 meV ($\lambda$ = 5 μm) induced by electrons making optical transitions between two confined subbands in a quantum well [14]. Absorption measurements are a convenient way to measure ISBTs and characterise coupled-quantum well semiconductors. From the analysis of the spectra one can not only individuate the positions of the resonances, and therefore the energy separations between the states, but also measure the areas under the peaks, which are related to the oscillator strengths of the transitions. This provides a direct determination of the dipole matrix elements, i.e. of the optical coupling between the subbands. It is evident that in order to observe absorption, electrons must be added in the conduction band. This is typically obtained by doping the structures in the wells with a concentration such that the sheet density per quantum well or coupled-quantum well ranges from ~$10^{10}$ to $10^{12}$. Higher doping concentrations are not suitable since they strongly broaden the intersubband transitions due to scattering with impurities and the coupling with the plasma resonance. In FIGURE 1 the intersubband absorption spectrum, at 5 K, between two bound states in a single quantum well 10 nm wide is reported [15]. The line shape can be fitted over more than a decade of absorbance with a Lorenzian curve of full-width-at-half-maximum (FWHM) = 5.1 meV. It is plausible that in this specific case the dominant broadening mechanism is homogeneous and corresponds to elastic scattering of in-plane long wavelength electrons against the very short pace roughness at the interface, with a dephasing time of 130 fs. This is not valid as soon as we explore more complicated structures made of coupled quantum wells. In the case, for instance, of diagonal transitions, i.e. when the wavefunctions are spatially separated in different quantum wells and the optical coupling is provided only by the tails of the probability distributions, the FWHM can easily be 20 meV and the absorption line shape fits a Gaussian curve, or does not have any well defined shape [16]. The highest transition energy that can be obtained in a single quantum well is about 300 meV but can easily be extended to 360 meV in engineered coupled-quantum wells having electron Bragg reflectors as barriers [17]. In this case a continuum resonance can be strongly localised in the well at energies above the barrier height. Note that this case is very different from the bound to continuum transitions normally used for QWIPs [9].

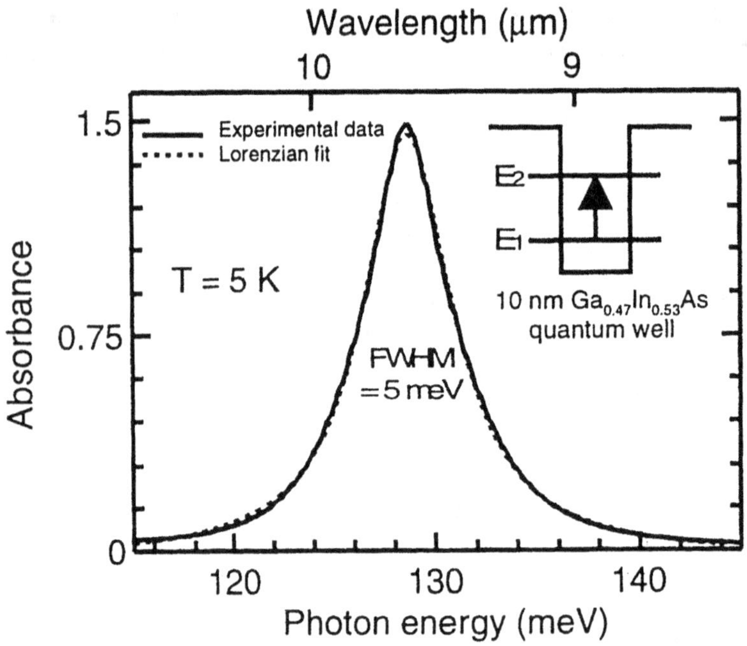


FIGURE 1 Intersubband absorption spectrum between bound states in a quantum well.

Formation of energy bands in superlattices made of short period $Ga_{0.47}In_{0.53}As/Al_{0.48}In_{0.52}As$ n-doped layers has also been investigated using ISBT absorption measurements. In this case it can be shown that the shape of the absorption spectrum is directly related to the energy dispersion along the superlattice axis [18].

## B2 Emission

Electroluminescence associated with ISBTs at energies greater than the optical phonon energy was observed for the first time by Faist et al [19] in a $Ga_{0.47}In_{0.53}As/Al_{0.48}In_{0.52}As$ highly engineered heterostructure based on the cascade principle. The mid infrared emission was generated by electron injection into the excited state of a coupled-quantum well region via a $Ga_{0.47}In_{0.53}As/Al_{0.48}In_{0.52}As$ pseudoquaternary graded gap alloy. FIGURE 2 illustrates the conduction band diagram (and the corresponding moduli squared of the relevant wavefunctions) of a portion of the luminescent structure under an applied electric field ~$6 \times 10^4$ V/cm$^2$ corresponding to the potential drop necessary to line up the ground state of the injector with the excited state of the optical transition ($3 \rightarrow 2$, indicated by the wavy arrow). The dashed lines are the effective conduction band edges of the digitally graded electron-injecting region. In the specific case illustrated in FIGURE 2 electrons pass from the injector ground state through a 4.4 nm AlInAs barrier into the n = 3 energy level of the active region. The latter includes 3.1 nm, 7.2 nm and 5.4 nm thick GaInAs wells separated by 2.0 nm and 1.5 nm AlInAs barriers respectively. Electrons escape from this region in the next adjacent injector through a 2.2 nm barrier [19]. Sandwiched between cladding layers for the optical waveguide and using standard processing for semiconductor ridge lasers, this structure is the base of the QC laser [8]. It is worth mentioning that active region designs different from that of FIGURE 2 have been used to fabricate QC lasers: in particular, lasers based on transitions between minibands [20] and with strongly diagonal transitions have been demonstrated [21].

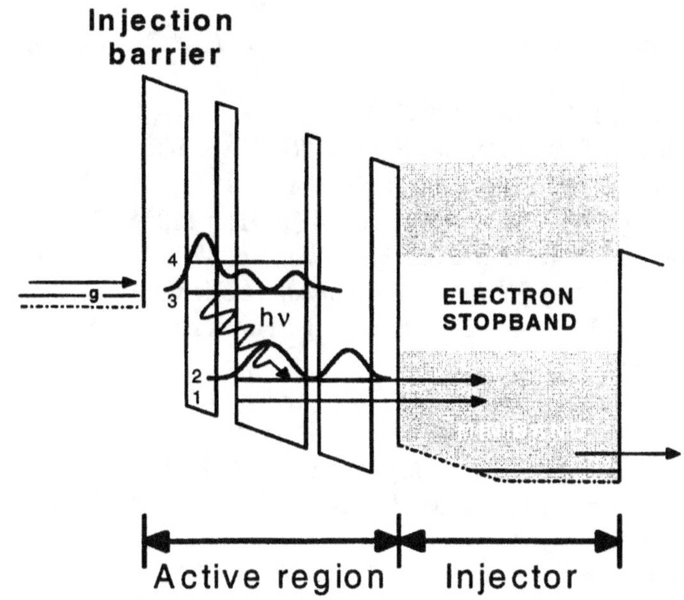


FIGURE 2 Schematic band diagram of a quantum cascade laser showing also the moduli squared of the relevant wavefunctions.

One of the main problems solved by QC laser design is the population inversion between subbands. The difficulty is caused by the extremely short lifetimes $\tau_i$ of the excited states, which are typically in the order of 1 ps for vertical transitions and for an energy separation between the subbands in the order of a few (4 - 5) longitudinal optical phonon energies ($\hbar\omega_{lo}$ = 36 meV in this material system) [15,22]. This short lifetime is dictated by the electron-LO phonon interaction which is the dominant energy relaxation mechanism for electrons in the excited subbands [23]. The characteristic time of this interaction is associated with the overlap between the wavefunctions and with the momentum transfer $\Delta k$ that electrons exchange with the lattice. The larger $\Delta k$ the longer is $\tau_i$, which, far from resonance, has roughly a quadratic dependence on k ($\tau_i \sim \Delta k^2$). On the contrary, when the separation between the subbands is of the order of $\hbar\omega_{lo}$, $\tau$ is exceptionally short in the 0.2 - 0.3 ps range. Recent investigations on the energy relaxation dynamics of electrons resonantly promoted to excited subbands can be found in [24].

Very large non-linear optical susceptibilities have been measured in $Ga_{0.47}In_{0.53}As/Al_{0.48}In_{0.52}As$ asymmetric quantum wells due to resonant intersubband transitions [25]. Second and third harmonic generation, at $\lambda$ = 5 μm and $\lambda$ = 3.3 μm respectively, have been observed by shining a $CO_2$ laser onto suitably engineered quantum well semiconductors [16]. Due to their resonant nature these non-linearities have also shown a strong dependence on an external electric bias, which may allow the non-linear effects to be enhanced or suppressed. Nevertheless, they are practically of no technological interest, as their conversion efficiency is very low and at most can reach 1%.

## C CONCLUSION

The main characteristics of ISBTs in InP and related compounds are summarised in this Datareview. Among all these heterostructures $Ga_{0.47}In_{0.53}As/Al_{0.48}In_{0.52}As$ grown lattice matched on InP is the one that has drawn most attention and is also the material system where the QC laser has been developed.

**REFERENCES**


[1] H.C. Liu, B.F. Levine, J.Y. Anderson [ *NATO ASI Ser. E, Appl. Sci. (Netherlands)* vol.270 (1994) ]

[2] E. Rosencher, B. Vinter, B.F. Levine [ *NATO ASI Ser. B, Phys. (USA)* vol.288 (1992) ]

[3] T. Ando, A.B. Fowler, F. Stern [ *Rev. Mod. Phys. (USA)* vol.54 (1982) p.437-672 ]

[4] L.C. West, S.J. Eglash [ *Appl. Phys. Lett. (USA)* vol.46 (1985) p.1156-9 ]

[5] E. Rosencher, P. Bois [ *Phys. Rev. B (USA)* vol.44 (1991) p.11315-27 ]

[6] F. Capasso, J. Faist, C. Sirtori [ *J. Math. Phys. (USA)* vol.37 (1996) p.4775-92 ]

[7] F. Capasso, J. Faist, C. Sirtori [ *Solid State Commun. (USA)* vol.102 (1997) p.231-42 ]

[8] J. Faist, F. Capasso, D.L. Sivco, C. Sirtori, A.L. Hutchinson, A.Y. Cho [ *Science (USA)* vol.264 (1994) p.553-6 ]

[9] B.F. Levine [ *J. Appl. Phys. (USA)* vol.74 (1993) p.R1-81 ]

[10] H.C. Liu, M. Buchanan, Z.R. Wasilewski [ *Appl. Phys. Lett. (USA)* vol.72 (1998) p.1682-4 ]

[11] H. Asai, Y. Kawamura [ *Appl. Phys. Lett. (USA)* vol.56 (1990) p.746-8 ]

[12] H.C. Chui, S.M. Lord, E. Martinet, M.M. Fejer, J.S. Harris Jr. [ *Appl. Phys. Lett. (USA)* vol.63 (1993) p.364-6 ]

[13] C. Sirtori, F. Capasso, J. Faist, S. Scandolo [ *Phys. Rev. B (USA)* vol.50 (1994) p.8663-74 ]

[14] B.F. Levine, A.Y. Cho, J. Walker, R.J. Malik, D.A. Kleinman, D.L. Sivco [ *Appl. Phys. Lett. (USA)* vol.52 (1988) p.1481-3 ]

[15] J. Faist et al [ *Appl. Phys. Lett. (USA)* vol.63 (1993) p.1354-6 ]

[16] F. Capasso, C. Sirtori, A.Y. Cho [ *IEEE J. Quantum Electron. (USA)* vol.30 (1994) p.1313-26 ]

[17] C. Sirtori, F. Capasso, J. Faist, D.L. Sivco, S.N.G. Chu, A.Y. Cho [ *Appl. Phys. Lett. (USA)* vol.61 (1992) p.898-900 ]

[18] C. Sirtori, F. Capasso, D.L. Sivco, A.Y. Cho [ *Appl. Phys. Lett. (USA)* vol.64 (1994) p.2982-4 ]

[19] J. Faist et al [ *Appl. Phys. Lett. (USA)* vol.64 (1994) p.1144-6 ]

[20] G. Scamarcio et al [ *Science (USA)* vol.276 (1997) p.773-6 ]

[21] J. Faist, F. Capasso, C. Sirtori, D.L. Sivco, A.L. Hutchinson, A.Y. Cho [ *Nature (UK)* vol.387 (1997) p.777-82 ]

[22] A. Bonvalet, J. Nagle, V. Berger, A. Migus, J.-L. Martin, M. Joffre [ *Phys. Rev. Lett. (USA)* vol.76 (1996) p.4392-5 ]

[23] R. Ferreira, G. Bastard [ *Phys. Rev. B (USA)* vol.40 (1989) p.1074-88 ]

[24] S. Lutgen et al [ *Phys. Rev. Lett. (USA)* vol.77 (1996) p.3657-60 ]

[25] C. Sirtori, F. Capasso, J. Faist, L.N. Pfeiffer, K.W. West [ *Appl. Phys. Lett. (USA)* vol.65 (1994) p.445-7 ]

# 5.5 Raman spectra of InP

G. Irmer

February 1998

## A INTRODUCTION

Raman spectra of InP are useful in characterising the bulk (excitation with infrared laser lines: InP is transparent to the Nd:YAG laser radiation) and the surface region (excitation with visible laser lines: typical sampling depths of light scattering ~100 nm). From the phonon and free carrier excitation spectra, information about the lattice perfection, carrier concentration and mobility are obtained. Raman scattering has demonstrated its sensitivity for local and non-destructive investigation of perturbations induced in the perfect crystal in several ways: e.g. mechanical alteration, stress, incorporation of impurities, carriers, alloying and ion implantation.

## B PHONONS

The first order Raman spectrum of undoped semi-insulating InP consists of two narrow lines of the LO(Γ) and TO(Γ) phonons having a frequency of 344.5 and 303.7 $cm^{-1}$ [1,2] at 300 K, respectively. The Raman spectra of the phonons were shown to depend on hydrostatic pressure [1], uniaxial pressure [4] and temperature [3-5]. From the scattering intensity ratio of the LO(Γ) and TO(Γ) phonons, the Faust-Henry coefficient was determined: $C = -0.53$ at $\lambda = 1.06$ μm [6] and $C = -0.46$ at $\lambda = 476.5$ nm [4].

The features in the second order Raman spectrum [7-10] correspond to singularities in the two-phonon density of states which occur when the dispersion curves are parallel or one is horizontal, particularly at the critical points of the Brillouin zone. Phonon frequencies [9] derived from the second-order Raman spectrum are given in TABLE 1. They are in good accord with two-phonon IR absorption [11] and neutron scattering measurements [12]. The latter measurements are generally less precise than optical measurements but give access to phonons of all wavevectors and could be used to resolve the ambiguities of the optical data (see TABLE 1). Raman scattering was studied up to fifth-order under non-resonant conditions [13].

## C CARRIERS

Collective excitations of free carriers and LO-phonons interact via their macroscopic electric field. As a result in the Raman spectra coupled plasmon LO-phonon modes are observed (see [16] for a review). Concentrations and mobilities of free carriers in n-InP were obtained in agreement with IR-reflectivity measurements from their bandwidths and frequency shifts using the Nd:YAG laser excitation at 1.06 μm wavelength [2]. The Raman spectra excited with different visible laser lines and the dispersion of the modes were successfully described by evaluating the Lindhard-Mermin dielectric function including the non-parabolicity of the band structure up to $n \sim 2 \times 10^{19}$ $cm^{-3}$ [17]. The coupled modes of heavily doped n-type InP were measured under hydrostatic pressure near the Γ-X crossover [18].

TABLE 1 Phonon frequencies (in $cm^{-1}$) derived from second-order Raman scattering [9] in comparison with those derived from two-phonon absorption [11] and neutron scattering measurements [12].

| Critical point | Mode | Bedel et al [9] | Ulrici and Jahne [11] | Borcherds et al [12] |
|---|---|---|---|---|
| Γ | LO | 348.5 ± 0.5 | 345 ± 1 | |
| | TO | 306.0 ± 0.5 | 305 ± 3 | 307 ± 7 |
| X | LO | 334 ± 2 | 336 ± 3 | 332 ± 3 |
| | TO | 320 ± 2 | 317 ± 1 | 324 ± 7 |
| | LA | 190 ± 4 | 191 ± 2 | 193 ± 10 |
| | TA | 68 ± 0.5 | 64 ± 3 | 68 ± 3 |
| L | LO | 344 ± 2 | 341 ± 1 | 340 ± 10 |
| | TO | 312 ± 2 | 313 ± 1 | 317 ± 5 |
| | LA | 168 ± 2 | 168 ± 2 | 167 ± 3 |
| | TA | 55 ± 0.5 | 55 ± 3 | 55 ± 1 |
| W | W6 | 344 ± 2 | 340 ± 5 | |
| | W5 | 311 ± 2 | 325 ± 5 | |
| | W4 | 311 ± 2 | 312 ± 5 | |
| | W3 | 155 ± 4 | 125 ± 5 | |
| | W2 | 96 ± 1 | 80 ± 5 | |
| | W1 | 77 ± 1 | 63 ± 5 | |
| K | K6 | 336 ± 2 | 335 ± 5 | |
| | K5 | 311 ± 2 | 316 ± 5 | |
| | K4 | 311 ± 2 | 312 ± 5 | |
| | K3 | 155 ± 4 | 166 ± 5 | |
| | K2 | 97.5 ± 0.5 | 70 ± 5 | |
| | K1 | 68 ± 0.5 | 65 ± 5 | |

See [11,15] for impurity modes.

The single-particle electron excitations and the line-shapes of their Raman spectra have been studied in samples with concentrations in the range n ~ $10^8$ - $10^{18}$ $cm^{-3}$ [19].

Electronic Raman scattering has been used to study bound carrier transitions of shallow acceptors [20].

With time-resolved Raman scattering photoexcited plasmas in InP were observed with spatial and temporal resolution [21-26]. Plasma densities ranging from n ~ $10^{16}$ - 5 × $10^{17}$ $cm^{-3}$ were deduced from fitting the Raman spectra with the plasmon LO-phonon scattering theory. For an initial injection plasma density of n ~ 2 × $10^{17}$ $cm^{-3}$ and at 300 K the photoexcited plasma expands laterally along the surface of the crystal at a velocity of v ~ 5 × $10^4$ cm/s and perpendicularly into the crystal at a velocity of v ~ 1.5 × $10^5$ cm/s [26].

## D SURFACE INVESTIGATIONS

The investigations were performed with visible laser lines; different depths of the substance may be probed using different wavelengths of the incident light.

### D1 Stress

The long-wavelength optical phonon frequencies, linewidths and polarisations are sensitive to surface damage and strain. Using Raman scattering from the LO-phonon excited by various lines of an Ar-ion laser polish-induced strain and its depth dependence were investigated [27,28]. Raman measurements of radiation defect induced strain [29] and sputter-etched surfaces [30,31] are reported. Strain-induced frequency shifts caused by uniaxial stress were measured and phonon deformation potentials have been obtained from the data [4].

**D2 Ion Implantation**

Raman scattering can give sensitive information about the appearance of defects induced in a crystal during the ion implantation process or about their removal during annealing. Studies were made of implantation with ions of Zn [32], Ga [33], Se [34,35], Si [35-37], Be [38], Mg [39], P[40], Yb [41] and He [42].

As the translational symmetry of the lattice is progressively lost in the process of implantation, more and more phonons around the centre of the Brillouin zone participate in the first order Raman scattering. In the Raman spectra with increasing fluences the following features are observed:

(i) softening of the second-order spectrum, which is more sensitive to long-range order than the first-order spectrum;

(ii) broad peaks assigned to the disorder-activated transverse acoustic (DATA) mode around 50 $cm^{-1}$ and to the disorder-activated longitudinal acoustic (DALA) mode around 130 $cm^{-1}$ appear and coupled plasmon LO-phonon bands disappear;

(iii) disorder-activation of transverse (DATO) and longitudinal (DALO) optical modes which appear as shoulders on the high- and low-energy sides of the TO(Γ) and LO(Γ) peaks, respectively;

(iv) the two optical peaks (DATO, DALO) merge into a single feature located around 300 $cm^{-1}$ and a band at 420 $cm^{-1}$ appears, perhaps due to phosphorus-phosphorus bond vibrations [43]. The Raman spectrum resembles that obtained for a-InP [44], for which correlation lengths of only a few interatomic distances are reached.

**D3 Annealing**

(i) At about 600 K there is an increase in correlation length, the optical band splits into two bands (DATO and DALO) and the appearance of second order scattering shows a polycrystalline state.

(ii) At about 800 K - 900 K the recovery of long range order starts, the monocrystalline state is progressively reconstructed and the bands DALO and DATO are transformed into the TO(Γ) and LO(Γ) bands.

With Raman polarisation measurements of the TO(Γ) and LO(Γ) phonons the random or preferential orientation of crystallites is analysed in laser annealed InP [34].

Raman scattering has been used to study the thermal oxidation of InP [45,46]. Elemental red phosphorus and oxides are detected in films grown between 650 K and 850 K [45].

**D4 Schottky Barrier Formation**

Investigation of the Raman scattering by LO-phonons in the surface space charge layer and the coupled plasmon LO-phonon modes of the bulk can be used to obtain quantitative information on the band bending at the surface or at interfaces in InP [47-50]. The experiments can also be carried out on surfaces covered with partially transparent layers such as metallic films [47,51].

## E RESONANT RAMAN SCATTERING

Resonant Raman scattering provides a powerful tool for obtaining band-structure information.

From the resonance in the two-LO-phonon scattering, the position of the $E_0 + \Delta_0$ gap and its Lorentzian broadening are determined at 100 K to be $1.509 \pm 0.003$ eV and $5.0 \pm 0.5$ meV, respectively. An electron-phonon deformation potential $d_0 = 14$ eV is derived which is much smaller than usually assumed [52].

The measured lineshapes of the 2LO Raman peak excited at different laser energies are described in terms of an excitonic model in the low energy range, and within a band picture for higher energies [53].

The efficiency of Raman scattering by LO phonons in a magnetic field exhibits resonances which can be associated with interband magneto-optical transitions between Landau levels [54,55].

Spin-flip single-particle and collective intersubband excitations were investigated in MIS accumulation layers [56].

## REFERENCES


[1] R. Trommer, H. Muller, M. Cardona, P. Vogl [ *Phys. Rev. B (USA)* vol.21 no.10 (1980) p.4869-78 ]

[2] B.H. Bairamov et al [ *Phys. Rev. B (USA)* vol.38 no.8 (1988) p.5722-5 ]

[3] G. Irmer, M. Wenzel, J. Monecke [ *Phys. Status Solidi (Germany)* vol.195 no.1 (1996) p.85-95 ]

[4] E. Anastassakis, Y.S. Raptis, M. Hunermann, W. Richter, M. Cardona [ *Phys. Rev. B (USA)* vol.38 no.11 (1988) p.7702-9 ]

[5] J.R. Shealy, G.W. Wicks [ *Appl. Phys. Lett. (USA)* vol.50 no.17 (1987) p.1173-5 ]

[6] N. Suzuki, K. Tada [ *Jpn. J. Appl. Phys. 1 (Japan)* vol.23 no.3 (1984) p.291-5 ]

[7] G.F. Alfrey, P.H. Borcherds [ *J. Phys. C (UK)* vol.5 no.20 (1972) p.L275-8 ]

[8] M. Sinyukov, R. Trommer, M. Cardona [ *Phys. Status Solidi B (Germany)* vol.86 no.2 (1978) p.563-8 ]

[9] E. Bedel, G. Landa, R. Carles, J.P. Redoules, J.B. Renucci [ *J. Phys. C (UK)* vol.19 no.10 (1986) p.1471-80 ]

[10] E.T.M. Kernohan, R.T. Phillips, B.H. Bairamov, D.A. Ritchie, M.Y. Simmons [ *Solid State Commun. (USA)* vol.100 no.4 (1996) p.263-7 ]

[11] B. Ulrici, E. Jahne [ *Phys. Status Solidi B (Germany)* vol.74 no.2 (1976) p.601-7 ]

[12] P.H. Borcherds, G.F. Alfrey, D.H. Saunderson, A.D.B. Woods [ *J. Phys. C (UK)* vol.8 no.13 (1975) p.2022-30 ]

[13] L. Artus, R. Cusco, J.M. Martin, G. Gonzales-Diaz [ *Phys. Rev. B (USA)* vol.50 no.16 (1994) p.11552-5 ]

[14] D.N. Talwar, M. Vandevyver, M. Zigone [ *J. Phys. C (UK)* vol.13 no.19 (1980) p.3775-93 ]

[15] E. Jahne, P. Kleinert, B.Kh. Bairamov, V.V. Toporov [ *Phys. Status Solidi B (Germany)* vol.104 no.2 (1981) p.531-9 ]

[16] G. Abstreiter, M. Cardona, A. Pinczuk [ in *Light Scattering in Solids IV* Eds M. Cardona, G. Guntherodt (Springer-Verlag, Berlin, 1984) p.5 ]

[17] W. Richter, U. Nowak, H. Jurgensen, U. Rossler [ *Solid State Commun. (USA)* vol.67 no.3 (1988) p.199-204 ]

[18] S. Ernst, A.R. Goñi, K. Syassen, M. Cardona [ *Phys. Rev. B (USA)* vol.53 no.3 (1996) p.1287-93 ]

[19] B.H. Bairamov, V.A. Voitenko, I.P. Ipatova, V.V. Toporov [ *Proc. Int. Conf. Raman Spectroscopy*, Calcutta, India, 2-6 Nov. 1988, Eds S.B. Banerjee, S.S. Jha (World Scientific, Singapore, 1989) p.386 ]

[20] M. Wenzel, G. Irmer, J. Monecke [ *Solid State Commun. (USA)* vol.104 no.6 (1997) p.371-4 ]

[21] A.R.B. de Castro, R.S. Turtelli [ *Solid State Commun. (USA)* vol.47 no.6 (1983) p.475-8 ]

[22] T. Nakamura, T. Katoda [ *J. Appl. Phys. (USA)* vol.55 no.8 (1984) p.3064-7 ]

[23] J.F. Young, K. Wan [ *Phys. Rev. B (USA)* vol.35 no.5 (1987) p.2544-7 ]

[24] J.F. Young, K. Wan, A.J. Spring Thorpe, P. Mandeville [ *Phys. Rev. B (USA)* vol.36 no.2 (1987) p.1316-9 ]

[25] K.T. Tsen, G. Halama, O.F. Sankey, S.-C.Y. Tsen [ *Solid-State Electron. (UK)* vol.32 no.12 (1989) p.1331-6 ]

[26] K.T. Tsen, G. Halama, O.F. Sankey, S.-C.Y. Tsen, H. Morkoc [ *Phys. Rev. B (USA)* vol.40 no.11 (1989) p.8103-6 ]

[27] H. Shen, F.H. Pollak [ *Appl. Phys. Lett. (USA)* vol.45 no.6 (1984) p.692-4 ]

[28] Z. Hang, H. Shen, F.H. Pollak [ *J. Appl. Phys. (USA)* vol.64 no.6 (1988) p.3233-42 ]

[29] C.R. Wie et al [ *Proc. Beam-Solid Interactions and Transient Processes Symposium*, Boston, USA, 1-4 Dec 1986 (Mater. Res. Soc., Pittsburgh, USA, 1987) p.517 ]

[30] D.J. Olego, R. Schachter, M. Viscogliosi, L.A. Bunz [ *Appl. Phys. Lett. (USA)* vol.49 no.12 (1986) p.719-21 ]

[31] N. Bouadma, P. Devoldere, B. Jusserand, P. Ossart [ *Appl. Phys. Lett. (USA)* vol.48 no.19 (1986) p.1285-7 ]

[32] E. Bedel, G. Landa, R. Carles, J.B. Renucci, J.M. Roquais, P.N. Favennec [ *J. Appl. Phys. (USA)* vol.60 no.6 (1986) p.1980-4 ]

[33] S.J. Yu, H. Asahi, S. Emura, H. Sumida, S. Gonda, H. Tanone [ *J. Appl. Phys. (USA)* vol.66 no.2 (1989) p.856-60 ]

[34] I.I. Reshina [ *Phys. Status Solidi A (Germany)* vol.106 no.1 (1988) p.261-70 ]

[35] D. Kirillov, J.L. Merz [ *Mater. Res. Soc. Symp. Proc. (USA)* vol.23 (1984) p.707-12 ]

[36] L.L. Abels, S. Sundaram, R.L. Schmidt, J. Comas [ *Appl. Surf. Sci. (Netherlands)* vol.9 (1981) p.2-13 ]

[37] R. Cusco, G. Talamas, L. Artus, J.M. Martin, G. Gonzales-Diaz [ *J. Appl. Phys. (USA)* vol.79 no.8 (1996) p.3927-9 ]

[38] C.S. Rama Rao, S. Sundaram, R.L. Schmidt, J. Comas [ *J. Appl. Phys. (USA)* vol.54 no.4 (1983) p.1808-15 ]

[39] A. Yamada, Y. Makita, H. Asakura, T. Iida, S. Kimura, T. Matsumori [ *Appl. Phys. A, Solids Surf. (Germany)* vol.53 no.2 (1991) p.102-8 ]

[40] Bo-Rong Shi, N. Cue, Tian-Bing Xu, S. Au [ *J. Appl. Phys. (USA)* vol.80 no.4 (1996) p.2127-31 ]

[41] H. Katsumata, S.-I. Uekusa, H. Sai, M. Kumagai [ *J. Appl. Phys. (USA)* vol.80 no.4 (1996) p.2383-7 ]

[42] I.M. Tiginyanu, A.I. Terletsky, V.V. Ursaki [ *Solid State Commun. (USA)* vol.96 no.10 (1995) p.789-92 ]

[43] M. Wihl, M. Cardona, J. Tauc [ *J. Non-Cryst. Solids (Netherlands)* vol.8-10 (1972) p.172 ]

[44] J.S. Lannin [ in *Tetrahedrally Bonded Amorphous Semiconductors* Eds M.H. Brodsky, H. Kirkpatrick, D.A. Weaire (AIP, New York, 1974) p.260 ]

[45] G.P. Schwartz, W.A. Sunder, J.E. Griffiths [ *Appl. Phys. Lett. (USA)* vol.37 no.10 (1980) p.925-7 ]

[46] G.P. Schwartz, G.J. Gualtieri, L.H. Dubois, W.A. Bonner, A.A. Ballman [ *J. Electrochem. Soc. (USA)* vol.131 no.7 (1984) p.1716-20 ]

[47] A. Pinczuk, A.A. Ballman, R.E. Nahory, M.A. Pollak, J.M. Worlock [ *J. Vac. Sci. Technol. (USA)* vol.16 no.5 (1979) p.1168-70 ]

[48] D.N. Bose, Y. Rampraksah, S. Basu [ *Mater. Lett. (Netherlands)* vol.8 no.9 (1989) p.364-8 ]

[49] X. Chen, X. Si, V. Malhotra [ *J. Electrochem. Soc. (USA)* vol.140 no.7 (1993) p.2085-8 ]

[50] A.N. Obraztsov, A.A. Gomanyuk, A.V. Mikulenok, F.S. Terra [ *Semiconductors (USA)* vol.29 no.11 (1995) p.2082-7 ]

[51] N. Esser, U. Resch, U. Rettweiler, W. Richter, D.R.T. Zahn, R.H. Williams [ *Vacuum (UK)* vol.38 no.4/5 (1988) p.229-32 ]

[52] W. Kauschke, M. Cardona [ *Phys. Rev. B (USA)* vol.33 no.8 (1986) p.5473-81 ]

[53] E.T.M. Kernohan, R.T. Phillips [ *Solid State Commun. (USA)* vol.100 no.4 (1996) p.245-9 ]

[54] T. Ruf et al [ *Phys. Rev. B (USA)* vol.39 no.18 (1989) p.1378-88 ]

[55] V. López, F. Comas, C. Trallero-Giner, T. Ruf, M. Cardona [ *Phys. Rev. B (USA)* vol.54 no.15 (1996) p.10502-7 ]

[56] G. Abstreiter, R. Huber, G. Trankle, B. Vinter [ *Solid State Commun. (USA)* vol.47 no.8 (1983) p.651-4 ]

# 5.6 IR reflection spectra of InP

Q.H. Hua

This Datareview appeared in *Properties of Indium Phosphide* (INSPEC, IEE, London, UK, 1991) and is reproduced here for the reader's convenience.

## A INTRODUCTION

IR-reflection spectra of InP (especially of doped bulk crystal) are often used for assessment of free-carrier properties [1-4]. In recent years, InP has become more important for use in transferred electron devices, operating at higher frequencies than GaAs devices, and in optoelectronic integrated circuits. Compared with Hall measurements, reflectivity measurements have the advantages of being non-destructive and not requiring either contacts or specially shaped samples. Thus it is much easier to assess the qualities of InP with acceptable errors both for bulk materials and for epitaxial layers.

## B BULK MATERIAL

The reflectivity of bulk InP has already been studied by Kohl [5] for carrier density higher than $10^{18}$ $cm^{-3}$. Later works extended the measurable range of carrier density from $10^{16}$ $cm^{-3}$ to $10^{20}$ $cm^{-3}$ with a relative error of 15% and of mobility with a relative area of 25%. The equation relating carrier density n with plasma frequency ω(P) used in many papers is as follows:

$$\omega(P) = \left[\frac{ne^2}{m^*\varepsilon(\infty)\varepsilon(0)}\right] \qquad (1)$$

where e is the electron charge, m* the electron effective mass, and ε(∞) and ε(0) are the permittivities at infinite and zero frequencies, respectively. The reflection spectra of InP in a wide range of $\nu = 4000$ $cm^{-1}$ to 30 $cm^{-1}$ have been studied and exhibited obvious frequency-dependence. At high frequencies R is almost constant (40% as for the intrinsic semiconductor) and has no relation to n. When the frequency is decreased, two minima due to coupled plasmon-LO phonons appear. It can be seen that the two minima are sensitive to the carrier density over different ranges. At low n, the lower minimum (below 300 $cm^{-1}$) is due to a plasmon-like mode and varies with n while the upper minimum is phonon-like being constant near ω(LO); at high n, the upper minimum is plasmon-like and varies with n while the lower minimum is constant near ω(TO), where ω(LO) and ω(TO) are longitudinal and transverse optical-phonon frequencies, respectively. The sharp rise on the low frequency side of the plasmon-like minimum is called the plasma edge. Between ω(TO) and ω(LO), the reflectivity (as high as 90%) is the reststrahlen peak due to the permittivity becoming a pure image producing strong reflection. Difference values of ω(TO) and ω(LO) of InP in the literature are listed in TABLE 1.

TABLE 1.

| ω(TO) ($cm^{-1}$) | ω(LO) ($cm^{-1}$) | Ref |
|---|---|---|
| 302.94 | 345.06 | [10] |
| 303.7 | 344.5 | [9] |
| 303.7 | 350.6 | [7] |
| 307 | 351 | [2,3,11] |
| 307.9 | 347.1 | [6] |

## C EPITAXIAL LAYER

A comprehensive study of the reflection spectra of epitaxial layers on semi-insulating and conductive substrates has been made by Pickering [1]. The spectra have similar behaviour to that of the bulk crystal but may be affected by the substrate, depending on the thickness of the epilayers. Due to increasing penetration depth with increasing n, thicker layers are needed for accurate measurements on SI-substrate at lower n. The data presented in [1] are listed in TABLE 2.

TABLE 2.

| n ($cm^{-3}$) | Thickness (micron) | Error (%) |
|---|---|---|
| $10^{18}$ | 1 | |
| $10^{17}$ | | ±10 |
| $10^{16}$ | 10 | ±20 |
| $5 \times 10^{15}$ | | ±40 |

The spectra of layers on conducting substrates exhibit interference effects for low n, unless the n of the layer is about the same or greater than that of the substrate.

The reflectivity at 77 K has also been measured, the results indicating that the polar optical scattering is still important at low temperature, which makes the optical mobility much lower than the Hall mobility.

## D OTHER ASPECTS

Jamshidi et al [6] determined the amplitude and phase reflection spectra of InP at 6 and 300 K and used them to calculate optical constants (refractive index n and extinction coefficient k) and dielectric parameters. Anastassakis et al studied the effect of uniaxial stress on long-wavelength optical phonons of InP by far-infrared reflection spectra to obtain the phonon deformation potential; the results are tabulated in [7]. Chen et al [8] investigated the reflection spectra of InP at liquid helium temperature under hydrostatic pressure up to 7 kbar. They resolved n = 1 and n = 2 excitonic levels and deduced a change in Rydberg energy.

## REFERENCES


[1] C. Pickering [ *J. Phys. D (UK)* vol.16 no.2 (1983) p.213-23 ]

[2] O.K. Kim, W.A. Bonner [ *J. Electron. Mater. (USA)* vol.12 no.5 (1983) p.827-36 ]

[3] Q.H. Hua, G.P. Li, X.K. He, Q. Wang [ *Mater. Lett. (Netherlands)* vol.3 no.3 (1985) p.93-7 ]

[4] M.A. Il'in, P.Yu. Karasw, N.A. Denisova, A.V. Rezvov, S.V. Tyurina [ *Izv. Vyssh. Uchebn. Zaved. Fiz. (USSR)* vol.31 (1988) p.22 ]

[5] F. Kohl [ *J. Phys. Chem. Solids (UK)* vol.32 (1971) p.2481 ]

[6] H. Jamshidi, T.J. Parker [ *Int. J. Infrared Millim. Waves (USA)* vol.4 (1983) p.1037 ]

[7] E. Anastassakis, Y.S. Raptis, M. Hunermann, W. Richter, M. Cardona [ *Phys. Rev. B (USA)* vol.38 no.11 (1988) p.7702-9 ]

[8] Y. Chen, B. Gil, A. Kadri, J. Allegre, J. Camassel, H. Mathieu [ *Physica B&C (Netherlands)* vol.139&140 (1986) p.491-4 ]

[9] B.H. Bairamov et al [ *Phys. Rev. B (USA)* vol.38 no.8 (1988) p.5722-5 ]

[10] J.M. Xu, L.Y. Shao, Y.T. Wang, C. Lin [ *Chin. J. Infrared Res. (China)* vol.6A (1987) p.271 ]

[11] M. Hass, B.W. Henvis [ *J. Phys. Chem. Solids (UK)* vol.23 (1962) p.1099 ]

# 5.7 Photoconductivity spectra due to impurities in InP

R.A. Stradling (revised and updated by M.R. Brozel)

June 1999

## A INTRODUCTION

IR photoconductivity is due to the photoionisation of either donors or acceptors to produce free carriers. This makes it a valuable tool to investigate the structures of impurity states and also to aid chemical analysis of the semiconductor.

## B SHALLOW STATES

Only donors will be considered, firstly because these are the most important residual impurities in InP and secondly because acceptors are much deeper in the bandgap, have distinct ionisation energies and are best investigated using photoluminescence (Datareview 5.3). For shallow donors, far-infrared photoexcitation produces an internal electron transition (usually 1s - 2p). However, the subsequent thermal excitation of the 2p electron into the conduction band produces the free electron that results in photoconductivity. This is the so-called 'photothermal ionisation' process [1]. As a result of this mechanism the photoconductivity spectrum consists of sharp lines.

### B1 Introduction

As with GaAs, shallow donor states in InP are well described by effective mass theory. Even in the 1s ground state the chemical shifts for different donor species amount only to a few percent of the total donor binding energy. Generally 'central cell structure' due to the presence of two or more residual contaminants can be observed as fine structure on the 1s - 2p lines. However, the line widths are such that only a small amount of back-doping with specific dopants can be tolerated before the broadening becomes greater than the chemical shifts, making the identification of the individual contaminants a difficult exercise. Furthermore, the prime cause of the broadening is the interaction with the Coulomb fields from randomly-distributed neighbouring impurities. This broadening is usually 'inhomogeneous', producing an asymmetric line-shape which causes the peak positions to shift with increasing line width.

In the low concentration limit this broadening can be treated by the theory of Larsen [2]. However, with total impurity concentrations greater than about $10^{15}$ $cm^{-3}$ a number of satellite peaks or shoulders appear on the side of the main 1s - 2p line with both InP and GaAs [3,4]. The relative intensities of these peaks increase with increasing impurity concentration but the peak positions are concentration independent. This behaviour is explained by the formation of 'molecular' orbitals where the electronic wavefunctions are distributed over two or more donor sites [3,4].

A complicating factor in the identification of components of the central line is the distortion of the line shape that can arise when the absorption becomes strong. A characteristic 'notched' line shape which can occur with less pure or intentionally-doped material was first observed and discussed by Stillman et al [5].

As with GaAs, photoconductive techniques are frequently used with InP epitaxial films in order to obtain high sensitivity. The photothermal process [1] usually generates a strong photoconductive signal provided that the temperature is not reduced below 4.2 K. However, the temperature and electric field dependence of the line intensities [6] shows that hopping between adjacent donor sites can become the dominant mechanism for generating the photosignal for impurity concentrations of the order of $10^{15}$ $cm^{-3}$. In this case the enhancement of the contributions from closely-spaced molecular pairs can cause additional systematic shifts in the peak positions from those for isolated donors [7], restricting the range of back-doping which can be employed to identify the residual contaminants.

A number of techniques can be employed to narrow the donor lines, such as the application of a magnetic field, the use of bandgap illumination, the choice of optimum temperature and the application of hydrostatic pressure. These techniques, together with the general principles involved in the use of FIR photoconductivity for the chemical identification of contaminants, are reviewed in a Datareview on GaAs [8].

The first far-infrared (FIR) spectroscopic observation of shallow donors in InP was by Chamberlain et al [9]. At that time the effective mass of the Γ-conduction band of InP was thought to be smaller than that of GaAs because the direct bandgap of InP was smaller. However, the shallow donor binding energy found by FIR spectroscopy was ~7.4 meV compared with 5.8 meV in GaAs. A substantially higher band-edge mass of $0.0810 \pm 0.0005$ $m_o$ was found from the Zeeman splitting of the donor lines [9] and this value was confirmed by cyclotron resonance experiments ($0.0810 \pm 0.0003$ $m_o$ [10]) and by the magnetophonon effect ($0.0820 \pm 0.0010$ $m_o$ [11]). A more recent measurement using Zeeman splitting of donor states in gas-source MOCVD gives a band-edge mass of conduction band electrons in InP of $0.0817 \pm 0.0002$ $m_o$ [42]. The anomalously low effective mass of GaAs can be explained using a k.p. treatment (Hermann and Weisbuch [12]).

At this time the peak low temperature mobilities available were of the order of 50,000 $cm^2/V$ s (in VPE material). As a result of improvements in growth technology, particularly in the purification of the trimethyl indium precursor used in MOCVD growth, low temperature mobilities exceeding 400,000 $cm^2/V$ s have now been reported [13]. As a result, the shallow donor magneto-optical spectrum of InP [14,15] is even richer than that of GaAs [16-18], enabling twelve lines grouped in energy beneath the N = 1 Landau level to be detected in the field range 3 - 8 T as well as eight lines beneath the N = 2 Landau level and six below the N = 3 level. Despite a dramatic increase in the number of spectral lines observed with increasing mobility, the 1s - 2p lines did not narrow by more than a factor of two compared with the earlier measurements [9,14]. This observation is suggestive of anomalous broadening associated with a non-photothermal mechanism for the generation of the photosignal. Theory can predict the observed line positions to an accuracy of about 0.3%. The particular interest in such a comparison between theory and experiment is that the magnetic field range covers the awkward intermediate field regime where Coulombic and Zeeman terms are comparable as signified by the dimensionless magnetic field variable being of the order unity ($\gamma = \frac{1}{2}\hbar\Omega_c/R^*$ where $\Omega_c$ is the cyclotron frequency and $R^*$ is zero-field effective mass binding energy). Particularly useful in checking the accuracy of the variational calculation in this intermediate field region ($\gamma \sim 1$) is the measurement of 'interexcited state transitions', i.e. transitions not involving the ground state [19], as these lines do not involve any significant chemical shifts.

The high purity of the currently available samples results in the appearance of strong lines from negatively charged donor ions ($D^-$ states) when a non-equilibrium carrier distribution is produced either by illumination with bandgap radiation or by the application of a non-ohmic electric field bias (~1 V/cm) [15,20,21].

In general, FIR photoconductivity studies of the central cell structure show that the improvement in the electron mobility of both VPE and MOCVD has been accompanied by a reduction in the proportion of sulphur donors present.

It is also worthy of note that a DC electrical measurement (the magneto-impurity effect) can produce comparable resolution to the infrared measurements in detecting the 1% differences in energy between the individual components of the central cell structure [22,23].

The good quality far infrared spectra observable have enabled time resolved studies of the carrier capture dynamics to be undertaken [24,25].

### B2 VPE InP

As with GaAs, epitaxial layers of InP produced by both the hydride and halide VPE processes have very similar sets of residual donor contaminants [9,26-30]. Generally, two contaminants account for most of the oscillator strength in the observed central cell spectrum. FIR and photoluminescence studies with samples lightly back-doped with sulphur and with silicon have established that these two donors are the most likely candidates for the residual contaminants [28-31]. With earlier samples of InP, sulphur was generally the dominant contaminant [9,26] accounting for about 70% of the total concentration of all the shallow donors. However, when low-temperature mobilities of the order of 150,000 $cm^2/V$ s were achieved with the In-$PCl_3$-$H_2$ process, sulphur and silicon were found at comparable concentration, indicating that the mobility improvement resulted in part from the reduction in sulphur content by the halide process. A large number of weak residual donor lines were observed when bandgap illumination was used to narrow the lines [27,28]. The change in the relative intensities of the Si and S lines found on reducing the thickness of the film by etching and also from a comparison with photoluminescence data from the same samples, indicated that the silicon donors aggregate close to the epilayer/substrate interface [28]. In general the results from FIR photoconductivity and 'two-electron' photoluminescence measurements are in agreement regarding the identities and proportions of the residual contaminants [31,32].

Experiments were also performed with samples back-doped with Sn and Ge and the magnitude of the chemical shifts determined [29,31]. The values for the individual chemical shifts estimated for zero magnetic field are: Si = +0.10 meV (0.77 $cm^{-1}$); S = +0.15 meV (1.20 $cm^{-1}$); Sn = +0.17 meV (1.40 $cm^{-1}$); Ge = +0.21 meV (1.67 $cm^{-1}$). The results quoted above are in agreement with photoluminescence (PL) measurements on the same samples except that in PL the position of the Sn donor could not be distinguished from S giving a chemical shift of Sn of +0.15 ± 0.02 meV. As can be seen from TABLE 1, the chemical shifts require that the effective mass value for the binding energy (effective Rydberg) be known accurately. The assumed value is 7.31 ± 0.05 meV as deduced from a study of interexcited state transitions [33]. The differences in chemical shifts achievable with high purity samples can be estimated to ±0.02 meV with linewidths of ~0.1 $cm^{-1}$ and therefore can be obtained to higher accuracy than their absolute values.

TABLE 1.

| Donor | GaAs Chemical shift (meV) | InP Chemical shift (meV) |
|---|---|---|
| $X_o$ | - | -0.11 |
| $X_1$ | - | 0.00 |
| Pb | 0.03 | |
| Si | 0.05 | 0.10 |
| Se | 0.06 | |
| Sn | 0.07 | 0.17 |
| S | 0.10 | 0.15 |
| Ge | 0.18 | 0.21 |

FIR photoconductivity spectra from the shallow residual donors in InP films co-doped with rare earths have been reported [10,11]. The residual donor concentration was rather high in the reactor concerned (~$10^{16}$ $cm^{-3}$) and without doping with rare earths no shallow donor spectra could be observed. When doped with Yb, Gd or Sm (≤0.1 at.%), fairly sharp donor lines could be observed, with the best result being obtained with 0.1 at.% Yb where the full width of the 1s - 2p line at zero field was ~0.32 meV. The donor lines could only be observed on illumination with bandgap radiation and the line widths were too great for central cell structure to be detectable, indicating large compensation.

## B3 MOCVD InP

Although the improvement in the low temperature mobility of MOCVD InP has been linked with the reduction in the silicon content of the trimethyl indium [13,36], the FIR photoconductivity spectra for all MOCVD samples with 77 K mobilities in excess of 120,000 $cm^2/V$ s show quite high concentrations of silicon with sulphur as a minority contaminant. For example, some high mobility samples show Si as the dominant residual donor with 70% of the total concentration of the shallow donors [14,15]. Sulphur was present at around 20% of the total abundance and traces of Ge and another donor with a chemical shift intermediate between those for S and Si were also detected [14] (FIGURE 1). Other high purity samples showed Si as the most abundant donor with a substantial proportion of S and with also a fourth as yet unidentified donor ($X_o$) which may have a negative chemical shift. Remarkably similar spectra showing S and Si with approximately the same relative intensity and line widths and probably the same impurity with the intermediate chemical shift were found by Lee et al [37]. They also detected a weak signal due to $X_o$.

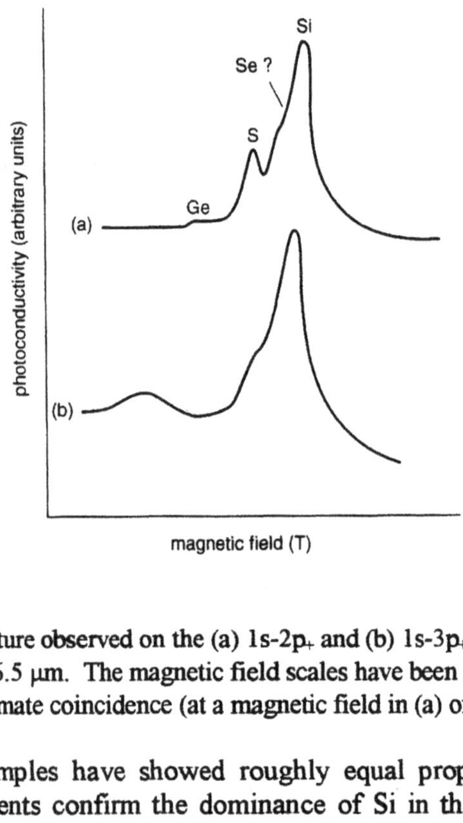


FIGURE 1 The central line structure observed on the (a) 1s-2p₊ and (b) 1s-3p₊ lines for an MOCVD-grown InP sample at a wavelength of 96.5 μm. The magnetic field scales have been shifted to bring the strongest line (Si) into approximate coincidence (at a magnetic field in (a) of ~3.7 T (from [14])).

Somewhat lower mobility samples have showed roughly equal proportions of S and Si [27,28]. Photoluminescence measurements confirm the dominance of Si in the high mobility samples [31]. Thus, although the new preparative techniques for trimethyl indium may have succeeded in reducing the silicon content, the sulphur contamination has apparently been reduced even more. In a more recent publication, the use of hydrogen to passivate a fraction of residual donors and acceptors was investigated by photoconductivity. Passivation was found to reduce the carrier concentration and increase the 77 K mobility. However, no change in the photoconductivity was observed. Clearly only partial passivation was attained [49].

## B4 Bulk InP

Sulphur is present at 20 - 40% of the total donor abundance for the four LEC samples studied in [28]. However, an unidentified donor ($X_1$) is as abundant with a very small chemical shift. In two of the four samples a significant amount of a third donor ($X_o$) was also observed. Donor ($X_o$) is also found in MOCVD material and is estimated to have a negative chemical shift. Only a small amount of Si was present in these LEC samples. Photoluminescence experiments confirm the dominance of sulphur and $X_1$ as the majority contaminants in LEC material [38].

### B5 LPE and MBE InP

The low temperature mobilities achieved with liquid phase and molecular beam epitaxies are inferior to those which can be obtained by VPE or MOCVD. No far infrared photoconductivity spectra have been reported for MBE material, although in gas-source MBE both Si and S were identified as being the major donor contaminants [42]. There is one report of spectra for LPE InP [30]. S was the dominant residual donor and Si was also present [30,39].

### B6 Neutron Transmutation Doped InP

On exposure to thermal neutrons In is transmuted to Sn and P to S. However, because of the giant capture cross-section of In, Sn is by far the most dominant of the neutron induced donors with S only accounting for 0.1% of the total.

The first far infrared photoconductivity results for thermal neutron irradiated InP were reported by Lee et al [37] who found an increase in the integrated absorption intensity in the region of the sulphur donor line after introducing an additional tin donor concentration of $6 \times 10^{14}$ $cm^{-3}$ but could not resolve a separate peak. This result is in agreement with photoluminescence measurements of Skolnick et al [32] who concluded that the chemical shift of tin was very similar to that of sulphur. However, Najda et al [40] were able to resolve an additional line 0.02 meV deeper in energy than sulphur with samples where the additional donor concentration arising from neutron transmutation was $2 \times 10^{14}$ $cm^{-3}$.

### B7 DX Centre in InP

By applying a hydrostatic pressure of greater than 82 kbar to a sample of S-doped InP, persistent photoconductivity due to the creation of a DX centre can be produced. The measured optical ionisation energy is between 0.86 and 1.02 eV. By extrapolating the energy back to zero pressure, it is deduced that the energy of this DX centre lies at $0.51 \pm 0.07$ eV above the $\Gamma$ minimum [43]. The creation of DX centres in InP by applying hydrostatic pressure is, therefore, similar to the case of GaAs.

### B8 Summary

The chemical shifts are quite strongly dependent on magnetic field. The results quoted in TABLE 1 for InP are a compilation of the best data as discussed in detail in Section B of this Datareview. The GaAs results are taken from [8]. All the results are calculated for zero magnetic field according to the theory of [41]. With the exception of the values for Sn, the shifts for InP are consistently larger than for GaAs as would be expected from the larger effective mass for InP. GaAs also showed a weak peak with a negative central cell correction [28].

## C SELECTED DEEP LEVELS

Optical excitation of deep electronic levels in the bandgap can give rise to direct (rather than photothermal) photoionisation. The transition from the level to either band occurs at a much greater photon energy than those considered in Section B and it follows that these measurements need not be made at liquid helium temperatures as thermal ionisation is not so important. However, the resultant spectra are relatively broad unless an internal transition is also necessary for ionisation to take place.

Most work has been performed on Fe-doped InP because of its commercial importance as a substrate material. The photoconductivity spectrum of a very deep acceptor such as $Fe_{Ga}$, which is only partially ionised in SI InP, occurring as $Fe_{Ga}^{3+}$ and $Fe_{Ga}^{2+}$ consists of at least two components corresponding to:

- $e_{VB} + Fe^{3+} \rightarrow Fe^{2+}$ (capture by a neutral Fe atom of an electron from the valence band leaving a free hole), and
- $Fe^{2+} \rightarrow Fe^{3+} + e_{CB}$ (electron emission from an ionised Fe atom into the conduction band).

The optical cross-sections for these two major transitions, $\sigma_p^o$ and $\sigma_n^o$, are complicated because of the different regions of the valence band and conduction band, respectively, which can be involved in the transitions [44] (FIGURE 2). Although this is true for all photoconductive transitions, this effect is particularly pronounced for levels near the middle of the bandgap as $\sigma_n^o$ and $\sigma_p^o$ can overlap considerably under these circumstances. However, there is often a 'magic wavelength' (where $\sigma_n^o = \sigma_p^o$) where photoconductive measurements can give an indication of the total Fe concentration.

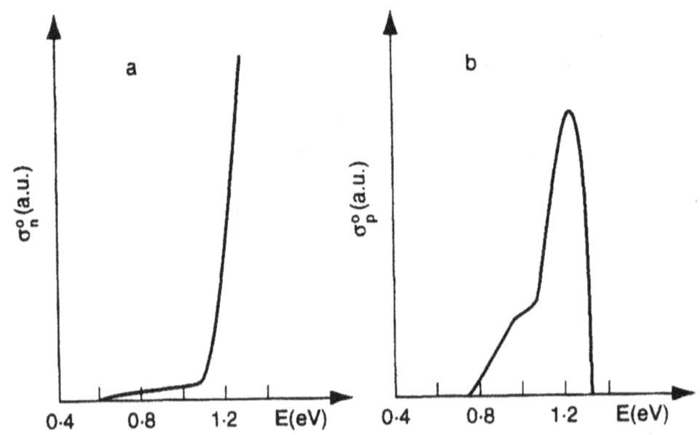


FIGURE 2 Optical photoionisation cross-sections of the $Fe_{In}$ deep level in InP: (a) $\sigma_n^o$ (hν) (250 K), (b) $\sigma_p^o$ (hν) (258 K) (from [44]).

Because these measurements can be made at room temperature and are restricted to the near-infrared they can be automated. This can then result in the presentation of spatial maps of Fe concentration in SI InP. Typical data obtained on wafers that have undergone different heat treatments to homogenise electrical characteristics are presented in [45,46]. Photocurrents last only a short time after the photoexcitation is extinguished. Analysis of the decay can give information about carrier dynamics. This technique has been applied to Fe-doped InP [47].

Of recent interest is the incorporation of erbium (Er) into semiconductors in an attempt to produce light of 1.55 μm wavelength without having to grow alloys of extremely precise composition. The photoconductive properties of Er-doped InP grown by atmospheric pressure MOCVD have been determined [48]. Two peaks, the dominant at 1.27 eV associated with a free to bound transition of an electron at a centre involving an $Er^{3+}$ atom and a smaller one at 0.807 eV involving the intracentre 4f transition, were observed. However, the relationship of either of these features to the required luminescence behaviour is uncertain. Other rare earths have been investigated in [10,11].

**REFERENCES**


[1] G.E. Stillman, C.M. Wolfe, J.O. Dimmock [ *Semicond. Semimet. (USA)* vol.12 (Academic Press, New York, 1977) ch.4 p.169-290 ]

[2] D.M. Larsen [ *Phys. Rev. B (USA)* vol.8 no.2 (1973) p.535-53 and vol.11 no.10 (1975) p.3904-9 ]

[3] K.K. Bajaj et al [ *J. Phys. C (UK)* vol.8 no.4 (1975) p.530-40 ]

[4] T. Inoshita, N. Iwata [ *Semicond. Sci. Technol. (UK)* vol.4 no.1 (1989) p.28-34 ]

[5] G.E. Stillman, T.S. Low, B. Lee [ *Solid State Commun. (USA)* vol.53 no.12 (1985) p.1041-8 ]

[6] C.J. Armistead, R.A. Stradling [ unpublished ]

[7] A.C. Carter, G.P. Carver, R.J. Nicholas, J.C. Portal, R.A. Stradling [ *Solid State Commun. (USA)* vol.24 no.1 (1977) p.55-60 ]

[8] R.A. Stradling [ in *Properties of Gallium Arsenide* third edition (INSPEC, IEE, UK, 1996) ch.8 p.262-70 ]

[9] J.M. Chamberlain, H.B. Ergun, K.A. Gehring, R.A. Stradling [ *Solid State Commun. (USA)* vol.9 (1971) p.1563-6 ]

[10] J.M. Chamberlain, P.E. Simmonds, R.A. Stradling, C.C. Bradley [ *J. Phys. C (UK)* vol.4 (1971) p.L37-41 ]

[11] L. Eaves, R.A. Stradling, S. Askenazy, J. Leotin, J.C. Portal, J.P. Ulmet [ *J. Phys. C (UK)* vol.4 (1971) p.L42-7 ]

[12] C. Hermann, C. Weisbuch [ *Phys. Rev. B (USA)* vol.15 no.2 (1977) p.816-22 and p.823-33 ]

[13] J.M. Boud, M.A. Fisher, D. Lancefield, A.R. Adams, E.J. Thrush, C.G. Cureton [ *Inst. Phys. Conf. Ser. (UK)* no.91 (1987) p.801-4 ]

[14] S.N. Holmes, P.D. Wang, R.A. Stradling [ *Semicond. Sci. Technol. (UK)* vol.5 no.2 (1990) p.143-9 ]

[15] P.D. Wang, S.N. Holmes, R.A. Stradling [ *Proc. 4th Int. Conf. on Shallow Impurities in Semiconductors* London, UK, 31 July-2 Aug. 1990 (Trans Tech Publications, Switzerland, 1990) p.375-80 ]

[16] C.J. Armistead, R.A. Stradling, Z. Wasilewski [ *Semicond. Sci. Technol. (UK)* vol.4 no.7 (1989) p.557-64 ]

[17] R.T. Grimes, M.B. Stanaway, J.M. Chamberlain, M. Henini, O.H. Hughes [ *Semicond. Sci. Technol. (UK)* vol.4 no.7 (1989) p.548-52 ]

[18] A.V. Klarenbosch, T.O. Klaassen, W.T. Wenckebach, C.T. Foxon [ *J. Appl. Phys. (USA)* vol.67 no.10 (1990) p.6323-8 ]

[19] C.J. Armistead, P. Makado, S.P. Najda, R.A. Stradling [ *J. Phys. C (UK)* vol.19 no.30 (1986) p.6023-38 ]

[20] C.J. Armistead, S.P. Najda, R.A. Stradling, J.C. Maan [ *Solid State Commun. (USA)* vol.53 no.12 (1985) p.1109-14 ]

[21] S.P. Najda, C.J. Armistead, C. Trager, R.A. Stradling [ *Semicond. Sci. Technol. (UK)* vol.4 no.6 (1989) p.439-54 ]

[22] L. Eaves et al [ *J. Phys. C (UK)* vol.7 no.11 (1974) p.1999-2007 ]

[23] S.N. Holmes, P.D. Wang, D.A. Cowan, C. Trader, R.A. Stradling [ *Semicond. Sci. Technol. (UK)* vol.5 no.2 (1990) p.150-8 ]

[24] T. Ohyama, E. Otsuka, S. Yamada, T. Fukui, N. Kobayashi [ *Jpn. J. Appl. Phys. 2 (Japan)* vol.22 no.11 (1983) p.L742-4 ]

[25] G.L.J.A. Rikken, P. Wyder, J.M. Chamberlain, G.A. Toombs [ *Inst. Phys. Conf. Ser. (UK)* no.95 (1989) p.243-8 ]

[26] R.A. Stradling, L. Eaves, R.A. Hoult, N. Miura, P.E. Simmonds [ *Inst. Phys. Conf. Ser. (UK)* no.17 (1972) p.65 ]

[27] C.J. Armistead et al [ *Lect. Notes Phys. (West Germany)* no.177 (1983) p.289-92 ]

[28] C.J. Armistead, P. Knowles, S.P. Najda, R.A. Stradling [ *J. Phys. C (UK)* vol.17 (1984) p.6415-34 ]

[29] B.J. Skromme, T.S. Low, T.J. Roth, G.E. Stillman, J.K. Kennedy, J.K. Abrokwah [ *J. Electron. Mater. (USA)* vol.12 no.2 (1983) p.433-57 ]

[30] L.W. Cook, M.M. Tashima, N. Tabatabaie, T.S. Low, G.E. Stillman [ *J. Cryst. Growth (Netherlands)* vol.56 no.2 (1982) p.476-84 ]

[31] P.J. Dean, M.S. Skolnick, L.L. Taylor [ *J. Appl. Phys. (USA)* vol.55 no.4 (1984) p.957-63 ]

[32] M.S. Skolnick et al [ *Appl. Phys. Lett. (USA)* vol.44 no.9 (1984) p.881-3 ]

[33] P.E. Simmonds, J.M. Chamberlain, R.A. Hoult, R.A. Stradling, C.C. Bradley [ *J. Phys. C (UK)* vol.7 no.22 (1974) p.4164-84 ]

[34] L.V. Berman, A.T. Gorelenok, A.G. Zhukov, V.V. Mamutin [ *Sov. Phys.-Semicond. (USA)* vol.19 no.8 (1985) p.857-9 ]

[35] V.G. Golubev, A.T. Gorelenok, V.I. Ivanov-Omskii, V.V. Mamutin, I.G. Minervin, A.V. Osutin [ *Sov. Tech. Phys. Lett. (USA)* vol.11 no.3 (1985) p.143-4 ]

[36] J.O. Williams [ in *Growth and Characterisation of Semiconductors* Eds. R.A. Stradling, P.C. Klipstein (Hilger, 1990) p.22 ]

[37] B. Lee, N. Pan, G.E. Stillman, K.L. Hess [ *J. Appl. Phys. (USA)* vol.62 no.3 (1987) p.1129-32 ]

[38] P.J. Dean, M.S. Skolnick, B. Cockayne, W.R. MacEwan, G.W. Iseler [ *J. Cryst. Growth (Netherlands)* vol.6 (1984) p.486 ]

[39] M.S. Skolnick, P.J. Dean, S.H. Groves, E. Kuphal [ *Appl. Phys. Lett. (USA)* vol.45 (1984) p.962-4 ]

[40] S.P. Najda, S.N. Holmes, R.A. Stradling, F. Kuchar [ *Semicond. Sci. Technol. (UK)* vol.4 no.9 (1989) p.791-6 ]

[41] D. Cabib, E. Fabri, G. Fiori [ *Solid State Commun. (USA)* vol.9 no.17 (1971) p.1517-20 ]

[42] X.H. Shi, P.L. Liu, S.C. Shen, J.X. Chen, H.P. Xin, A.Z. Li [ *J. Appl. Phys. (USA)* vol.80 no.8 (1996) p.4491-4 ]

[43] J.A. Wolk, W. Walukiewicz, M.L.W. Thebalt, E.E. Haller [ *Phys. Rev. Lett. (USA)* vol.68 no.24 (1992) p.3619-22 ]

[44] P. Leyral, F. Litty, G. Bremond, A. Nonillat, G. Guillot [ *Semi-Insulating III-V Materials* (Shiva, Nantwich, UK, 1982) p.192-8 ]

[45] A. Alvarez, M. Avella, J. Jimenez, M.A. Gonzalez, R. Fornari [ *Semicond. Sci. Technol. (UK)* vol.11 (1996) p.941 ]

[46] M. Avella et al [ *Inst. Phys. Conf. Ser. (UK)* no.160 (1997) p.157-60 ]

[47] R.S. Miranda, H.W.K. Tom, A.M. Johnson, T.J. Bridges, G.D. Aumiller [ *Appl. Phys. Lett. (USA)* vol.60 no.9 (1992) p.1105-7 ]

[48] X.Z. Wang, A.J. Neuhalfen, B.W. Wessels [ *Appl. Phys. Lett. (USA)* vol.64 no.4 (1994) p.466-8 ]

[49] R.W. Glew et al [ *Semicond. Sci. Technol. (UK)* vol.6 no.11 (1991) p.1088-92 ]

CHAPTER 6

# DEFECTS, DEEP LEVELS AND THEIR DETECTION

# 6.1 Extended defects in InP

L. Haji (revised by A.R. Peaker)

June 1998

## A INTRODUCTION

In bulk InP crystals, grown by the liquid encapsulation Czochralski (LEC) method, several kinds of defect have been detected and extensively investigated by various analytical techniques. Chemical etching [1] is a common tool to reveal some of these defects. However, due to its high resolving power and ability to provide detailed structural information, transmission electron microscopy (TEM) is especially suitable for the study of extended defects in InP.

The principal extended defects in InP are dislocation lines, inclusion-like defects, dislocation loops and twins. The special case of micro-precipitates is dealt with in detail in Datareview 6.2 in this book.

## B GROWN-IN DEFECTS

The grown-in defects include all crystalline defects which are generated during the growth of InP crystals. The two major kinds of defect usually detected in bulk InP are threading dislocation lines and defect clusters.

### B1 Dislocations

Dislocations are line defects induced by easy glide in the {111}<110> slip system as a result of stresses due to thermal gradients present during the growth process. The dislocation density in undoped LEC crystals is generally high and can reach $10^5$ $cm^{-2}$ [2]. This density can be reduced to below $10^3$ $cm^{-2}$ for undoped large diameter crystals grown by the dynamic gradient freeze (DGF) technique [3].

It is well known, in crystal growth, that the incorporation of suitable impurities at concentrations higher than $10^{18}$ atoms/$cm^3$ facilitates the growth of dislocation-free crystals. The reduction of the dislocation density is explained by a hardening effect produced by impurities in the crystals. Macroscopically dislocation-free LEC InP crystals have been obtained by Seki et al [4], Matsui et al [5] and Kamajima et al [6]. On the other hand, Feldman and Ballman [7] have shown that the double doping of InP with sulphur and zinc is also effective in allowing low dislocation densities to be achieved with independent control of the net free carrier concentration. No detectable precipitates were seen in this case, and it was noted that the double doping appears to increase the Zn solubility with regard to that reported when Zn is used alone [7]. The authors do not specify the homogeneity of these properties along the entire length of the ingots, but it is probable that there is inhomogeneity due to non-unity distribution coefficients.

The addition of isovalent impurities, such as gallium and antimony, has been shown to be effective for the reduction of the dislocation density [8], without change of the electrical properties of the material. Haji et al [9] obtained nearly dislocation-free LEC grown InP crystals double-doped with Ga and Fe. However, increasing the doping levels generally leads to the formation of a high density of microdefects. Microdefects accompanying single dislocations were observed for heavily doped crystals. Their formation near climbing dislocations is explained by a process which involves interactions between thermal point defects and impurities [10,11].

Solid phase precipitates could be created in InP for high concentrations of some kinds of impurities. Rumsby et al [2] have identified $FeP_2$ precipitates in LEC Fe-doped semi-insulating InP when 0.1 wt % of iron is added to the melt. Monberg et al [12] have found TiP precipitates in their (Zn + Ti) double doped InP crystals grown by the DGF technique.

### B2 Inclusion-Like Defects

Unusual large defect clusters or inclusion-like defects called grappes, which seem to be typical of InP, have been detected in either undoped or doped crystals at densities of $10^2$ to $10^4/cm^2$. The extent of these defects, as revealed by X-ray diffraction topography, ranges from 5 to 200 μm.

Al-Jassim et al [13], Mahajan and Chin [14], Elliott and Regnault [15], Augustus et al [16] and Franzosi et al [17] extensively investigated the inclusion-like defects in either undoped or doped InP crystals. Augustus et al [16] gave a detailed description of such a defect in Ge-doped LEC InP by using TEM combined with energy dispersive X-ray analysis (EDX) and other microanalytical means. The defect clusters consist of a central region or core which is surrounded by prismatic dislocation loops arranged in arrays along the <110> axes [16,17]. The shape of the core could be triangular with a network of close packed dislocations or spherical with a dense and random distribution of dislocation tangles.

It is known that the generation of prismatic dislocation loops takes place in crystals at inclusions or precipitates [18]. Thus, impurities must be associated with the core of the defect. However, attempts to reveal the chemical nature of the core of these defects have not succeeded, but it was found that the In/P ratio increases in the core of clusters [16].

Whilst it had been assumed that gaseous inclusions are the most probable cause of these defects, no direct evidence of the presence of precipitates or inclusions had been found within the core of such defects [15,16]. However, all the authors concur in the belief that the moisture in $B_2O_3$, used as an encapsulant in the LEC method, is responsible for the occurrence of these defect clusters as first suggested by Cockayne et al [19].

### B3 Dislocation Loops

Brown et al [20] discussed in detail the nature of prismatic dislocation loops occurring in undoped InP crystals grown from indium-rich melts. These loops were found to be interstitial in character and were shown to be lying in a {110} plane.

TEM observations showed that the loop plane was decorated with micro-precipitates and small loops which act as strain centres [16,20]. Similar results were found by Wagner et al [21] for Sn-doped InP and by Haji et al [9] for (Ga,Fe)-double doped InP. Warwick and Booker [22] studied the luminescence properties of dislocation loops which occur in Fe-doped crystals grown from In-rich melt. These loops showed a non-radiative behaviour.

EDX analyses have shown that the planes of the prismatic dislocation loops, contained in undoped crystals grown from a non-stoichiometric melt (In excess), are indium-rich [20]. A model based on the condensation of indium interstitials onto dislocation lines causing climb was proposed to explain the formation of these dislocation loops [16,20].

The effect of annealing on these defects was studied by Brown et al [23] for undoped and Fe-doped InP and by Haji et al [9] for Fe and (Ga,Fe)-doped InP. The results show that the prismatic dislocation loops are affected by annealing at temperatures higher than 750°C, that is, the small loops and microprecipitates decorating the plane of the loops in unannealed material disappear completely after annealing.

### B4 Twins

Twins occur frequently in InP crystals due to their low stacking fault energy. Twinning probability in LEC grown InP could be decreased only by a drastic control of the growth conditions.

Tohno et al [24] gave a detailed description of twins in undoped LEC InP grown along the <111> direction by means of X-ray diffraction topography. Correlations between the twin formation and growth conditions were established and the effects of twinning on the polycrystallisation process and the dislocation behaviour were discussed.

Recently, Holt et al [25] have reported on transmission cathodoluminescence (TCL) contrast of twin boundaries in Cr and Fe doped LEC InP. Their observations together with X-ray microanalysis have shown the presence of Cr segregation onto twinning dislocations which increases from seed to tail end of the ingot, whereas for Fe doped InP no segregation of Fe was observed, enabling better observation of the TCL contrast of twins.

## C PROCESS-INDUCED DEFECTS

The diffusion of dopants and implantation of energetic ions into InP materials can cause defect generation.

### C1 Diffusion-Induced Defects

Ueda et al [26] gave a detailed description of the defects induced by Cd-diffusion in Sn-doped LEC InP substrates. Two types of defect were observed by TEM: (i) bulk defects which consist of a high density of microdefects (100 - 1000 Å in diameter) uniformly distributed in the diffused region regardless of the phosphorus pressure during the diffusion process; (ii) surface defects which consist of precipitate-like defects presumably caused by Cd or Cd-alloy precipitation and dislocation networks considered to be due to a thermal stress relaxation and expected to be eliminated by improving the chemical treatment before diffusion [26].

The surface defects induced by Zn diffusion in S-doped InP substrates were studied by Eger et al [27]. These are composed of crystalline inclusions associated with In-rich droplets. The inclusions are identified by SEM/EDX and X-ray diffraction as $Zn_3P_2$ which have an epitaxial relationship with the InP substrate. The formation of the $Zn_3P_2$ inclusions is interpreted in terms of a chemical reaction between the Zn and the InP substrate. Eger et al [27] proposed a new metallic diffusion course, which is based on Zn-Ga-In-Cd mixtures, for the elimination of the surface defects created during the diffusion process.

### C2 Ion Implantation-Induced Defects

It has been established that ion implantation induces structural disorder in materials. The nature and distribution of the disorder depend on the type of the implanted ions, implant energy, fluences and substrate temperature. Small amorphous clusters or amorphous layers (buried or continuous layers) could be created after ion implantation [28].

Duhamel et al [29] investigated the structural damage induced by silicon implantation into semi-insulating Fe-doped LEC InP. They show that high Si implantation doses ($>1 \times 10^{14}$ cm$^{-2}$) create continuous amorphous layers extending to the surface. After conventional furnace annealing or rapid thermal annealing, used to activate the Si dopant, recrystallisation of the amorphous layers occurs with the production of high densities of dislocation loops.

Damage induced by high energy (MeV) ion implantation of nitrogen into Sn-doped InP wafers was investigated by Fulin Xiong et al [30] using various techniques. The implanted doses ranged from $5 \times 10^{13}$ to $5 \times 10^{16}$ ions $cm^{-2}$ and the beam flux was limited to 1 $\mu A\ cm^{-2}$. The high energy ion implantations were shown to result in the generation of high densities of point defect clusters, for samples implanted with low doses ($<5 \times 10^{14}$ $cm^{-2}$), while for samples implanted at higher doses ($>1 \times 10^{15}$ $cm^{-2}$) amorphisation occurs in a depth of about 4 μm. Post-implantation annealing at about 500°C under $H_2$ gas flow produces a reduction in the surface damage and causes a recrystallisation of the amorphous layer which is accompanied by the creation of micro-twins and stacking faults at the interface between the recrystallised layer and the substrate [30].

The issue of point defects introduced by ion-implantation, proton damage and electron irradiation is dealt with in more detail in Datareview 6.10 in this book.

## C3 Plastic Deformation

Intentionally induced plastic deformation has been extensively used to study dislocations in many materials. Studies of plastic deformation in InP reveal that a considerable number of vacancies are formed during the glide of dislocations primarily because of the dragging of jogs along screw dislocations. Leipner et al [31] show that these vacancies form secondary defects such as stable vacancy clusters.

## REFERENCES


[1] J.L. Wegher, R. Fornari, T. Gorog, J.J. Kelly [ *J. Cryst. Growth (Netherlands)* vol.141 (1994) p.57-67 ]

[2] D. Rumsby, R.M. Ware, M. Whitker [ *J. Cryst. Growth (Netherlands)* vol.54 no.1 (1981) p.32-40 ]

[3] E.M. Monberg, H. Brown, C.E. Bonner [ *J. Cryst. Growth (Netherlands)* vol.94 no.1 (1989) p.109-14 ]

[4] Y. Seki, J. Matsui, H. Watanabe [ *J. Appl. Phys. (USA)* vol.47 no.7 (1976) p.3374-6 ]

[5] J. Matsui, H. Watanabe, Y. Seki [ *J. Cryst. Growth (Netherlands)* vol.46 (1979) p.563 ]

[6] T. Kamajima, J. Matsui, Y. Seki, H. Watanabe [ *J. Appl. Phys. (USA)* vol.50 no.5 (1979) p.3312-21 ]

[7] R.D. Feldman, A.A. Ballman [ *Mater. Lett. (Netherlands)* vol.1 no.2 (1982) p.77-9 ]

[8] S. Tohno, A. Katsui [ *J. Cryst. Growth (Netherlands)* vol.78 no.2 (1986) p.249-56 ]

[9] L. Haji, R. Coquille, M. Gauneau, Y. Toudic [ *J. Cryst. Growth (Netherlands)* vol.82 no.3 (1987) p.487-94 ]

[10] P.M. Petroff, L.C. Kimerling [ *Appl. Phys. Lett. (USA)* vol.29 no.8 (1976) p.461-3 ]

[11] G. Wagner, V. Gottschalch [ *Philos. Mag. A (UK)* vol.52 no.3 (1985) p.395-406 ]

[12] E.M. Monberg, R.L. Barns, P.M. Bridenbaugh, A. Carter [ *Mater. Sci. Eng. B (Switzerland)* vol.6 (1990) p.29-32 ]

[13] M.M. Al-Jassim, C.A. Warwick, G.R. Booker [ *Inst. Phys. Conf. Ser. (UK)* no.60 (1981) p.357-64 ]

[14] S. Mahajan, A.K. Chin [ *J. Cryst. Growth (Netherlands)* vol.54 no.1 (1981) p.138-42 ]

[15] C.R. Elliot, J.C. Regnault [ *J. Electrochem. Soc. (USA)* vol.128 (1981) p.113 ]

[16] P.D. Augustus, D.J. Stirland, M. Yates [ *J. Cryst. Growth (Netherlands)* vol.64 no.1 (1983) p.121-8 ]

[17] P. Franzosi, G. Salviati, M. Cocito, F. Taiariol, G. Ghezzi [ *J. Cryst. Growth (Netherlands)* vol.69 no.2/3 (1984) p.388-98 ]

[18] J.W. Matthews [ *Scr. Metall. (USA)* vol.5 (1971) p.1053 ]

[19] B. Cockayne, G.T. Brown, W.R. MacEwan [ *J. Cryst. Growth (Netherlands)* vol.51 no.3 (1981) p.461-9 ]

[20] G.T. Brown, B. Cockayne, W.R. MacEwan [ *J. Electron. Mater. (USA)* vol.12 no.1 (1983) p.93-106 ]

[21] G. Wagner, V. Gottschalch [ *Cryst. Res. Technol. (East Germany)* vol.21 no.7 (1986) p.881-90 ]

[22] C.A. Warwick, G.R. Booker [ *J. Cryst. Growth (Netherlands)* vol.79 (1986) p.503-7 ]

[23] G.T. Brown, C.R. Whitehouse, J. Woodward [ *J. Cryst. Growth (Netherlands)* vol.64 no.1 (1983) p.129-36 ]

[24] S. Tohno, A. Katsui [ *J. Cryst. Growth (Netherlands)* vol.74 no.2 (1986) p.362-74 ]

[25] D.B. Holt, G. Salviati [ *J. Cryst. Growth (Netherlands)* vol.100 no.3 (1990) p.497-507 ]

[26] O. Ueda, H. Ishikawa, I. Umebu [ *Jpn. J. Appl. Phys. (Japan)* vol.23 (1984) p.1551-4 ]

[27] D. Eger, A.J. Springthorpe, A. Margittai, F.R. Shepherd, R.A. Bruce, G.M. Smith [ *J. Electron. Mater. (USA)* vol.16 no.3 (1987) p.163-7 ]

[28] D.K. Sadana [ *Mater. Res. Soc. Symp. Proc. (USA)* vol.92 (1987) p.319-28 ]

[29] N. Duhamel, B. Descouts, P. Krauz, K. Rao, J. Dangla, P. Henoc [ *J. Cryst. Growth (Netherlands)* vol.79 (1986) p.443-8 ]

[30] Fulin Xiong, C.W. Nieh, T.A. Tombrello, D.N. Jamieson, T. Vreeland Jr. [ *Vacuum (UK)* vol.39 no.2-4 (1989) p.177-82 ]

[31] H.S. Leipner, R. Krauserehberg, C. Huber [ *Mater. Sci. Forum (Switzerland)* vol.196 (1995) p.1267-71 ]

# 6.2 Structure of microprecipitates in SI InP

J.P. Fillard and J.M. Lussert (revised by A.R. Peaker)

June 1998

## A INTRODUCTION

Among the various defects [1] which affect III-V semiconductors and more especially InP, microprecipitates (MP) are frequently observed and have been shown to be of importance in relation to the performance of devices [2,3].

The sizes of the various particles which have been discovered in undoped or weakly doped materials are in the submicron range. In general three experimental methods have been used to investigate such flaws: (i) chemical etching of the surface, (ii) transmission electron microscopy, and (iii) laser scanning tomography.

InP ingots are usually grown in either the <001> or the <111> direction, the wafer always being a [100] plane. This means that wafers have to be cut perpendicular to the growth axis in the former case and at a 37.5° angle in the latter case. Wafers originating from <111> ingots are reportedly more homogeneous whereas the <001> ones give W shaped etch pit density (EPD) profiles.

## B ETCHING

From the initial work of Abrahams and Buiocchi [4] the 'AB' solution as well as the 'H' solution [5] are known to reveal structural defects at the surface of GaAs; this technique has been extended to InP (dislocations, stacking faults, grain boundaries, S-pits, etc.) where a selectivity exists [1,5,6] in the etching rates for the different kinds of defect. Also Weyher has shown that the DSL (dilute Sirtl and light) photo-etching technique [7,8] selectively revealed chemical defects such as striations and microprecipitates (dislocation decoration or matrix precipitates). This makes it possible to separate and to count different kinds of particle.

Usually [9-12] the density of MPs observed in selective etching experiments is smaller than the density of revealed dislocations ($10^2$ to $5 \times 10^4$ $cm^{-2}$). Recent results show that the size, the density and the texture of MPs are related to the dopant and to the stoichiometry of the material.

## C TRANSMISSION ELECTRON MICROSCOPY

Microprecipitates have been found using TEM by Kamejima et al [13] among the various kinds of lattice defect in InP. An extensive study has been performed by correlated TEM and ECAX analysis by Cockayne and co-workers [14-21,30] on InP materials of diversified dopings. In this case the doping level is very high (1 part over 1000 in weight) and the microprecipitates essentially originate in the doping species as second phase inclusions. The corresponding aggregates have a large size (1 to 100 μm) and a well defined chemical and crystallographical definition. It is not proven if extrapolation of the conclusion can be extended to much smaller particles revealed in purer materials.

Transition elements have a low solubility and often precipitate if introduced into the InP matrix. This has been observed for the case of Ni, Ti, Cr, V, Mn, Fe, Co and Cu. The condensed precipitates take the typical shapes of: hexagonal platelets (TiP, $Ni_2P$, VP); cuboids ($CoP_X$); prism-like whiskers ($FeP_2$, MnP); hollow centred needles ($FeP_2$, MnP, $Ni_2P$, $Ni_5P_2$); rods (CrP, $CoP_3$); and eutectic mixtures at grain boundaries ($Cu_3P$ + $Cu_{1.7}In$).

## D LASER SCANNING TOMOGRAPHY

This technique [22-28] is non-destructive and gives a direct image of the internal defects; it is capable of 3-dimensional bulk analysis [22,25,26]. A laser light beam is directed at the wafer and penetrating light scattered by local inhomogeneities is collected by a camera; scanning of the sample is performed and the image of the virtual plane is computer assembled.

It has been demonstrated after comparison with etching figures that the observed point defects are microprecipitates. This technique has been available since the late 1980s and it is apparent that major improvements have been effected in the growth technique during the last ten years. It is also possible to conclude that the improvements in the MP texture are directly related to the improvements in the electrical homogeneity (i.e. the ability to fabricate good ICs). It is also worth noting that these improvements are independent of the dislocation (EPD) density.

The influence of the dopants Zn, Fe, Sn, S and Cd on the MP texture is not dramatic but it can be said that (within similar process technology) Fe preferentially gives large but few MPs whereas Sn gives more abundant MPs than S doping; Cd and Zn give small particles.

Standard production before 1987 (any doping or process) showed an LST texture of bright points (density $10^5$ - $10^7$ $cm^{-3}$) which indicate large MPs (under 1 μm). Smaller particles are hardly observable because of the high intensity emitted by the largest which limits the detectivity of the camera. We have observed that these MPs are arranged along dislocations as decorations similarly to the case for GaAs [27,28] leading to a similar cell structure.

Gradually but regularly these scatterers have disappeared from the more recent materials. This enables observation of smaller and smaller particles with higher and higher densities (up to $10^{12}$ $cm^{-3}$) for increasing material quality.

<111> grown materials are more homogeneous than <100> which show W-type profiles of MP density; nevertheless the density of MPs increases from seed to tail [29].

## REFERENCES

[1] E.V. Mikryukova, A.N. Morozov, A.V. Berkova, A.Ya. Nashel'skii, S.V. Yakobson [ *Sov. Phys.-Crystallogr. (USA)* vol.33 (1988) p.722 ]

[2] H. Yamamoto, O. Oda, M. Seiwa, M. Taniguchi, H. Nakata, M. Ejima [ *J. Electrochem. Soc. (USA)* vol.136 no.10 (1989) p.3098-102 ]

[3] J.P. Fillard, M. Castagne, P. Gall, J. Bonnafe [ *Proc. GAAS-90 Application Symposium* Roma, 1990 ]

[4] M.S. Abrahams, C. Buiocchi [ *J. Appl. Phys. (USA)* vol.36 (1965) p.2855 ]

[5] A. Huber, N.T. Linh [ *J. Cryst. Growth (Netherlands)* vol.29 no.1 (1975) p.80-4 ]

[6] D.J. Stirland, B.W. Straughan [ *Thin Solid Films (Switzerland)* vol.31 (1976) p.139-70 ]

[7] J.L. Weyher, J. Van de Ven [ *J. Cryst. Growth (Netherlands)* vol.63 no.2 (1983) p.285-91 ]

[8] J.L. Weyher, L.J. Giling [ *J. Appl. Phys. (USA)* vol.58 (1985) p.219-22 ]

[9] S.N.G. Chu, C.M. Jodlank [ *J. Electrochem. Soc. (USA)* vol.129 no.2 (1982) p.352-4 ]

[10] J.L. Weyher, R. Fornari, J. Gorog [ *Mater. Sci. Eng. B (Switzerland)* vol.28 (1994) p.488-92 ]

[11] V. Gottschalch, R. Smanek [ *J. Mater. Sci. Lett. (UK)* vol.1 no.8 (1982) p.358-63 ]

[12] P.L. Giles, D.J. Stirland, P.D. Augustus, M.C. Hales, N.B. Hasdell, P. Davies [ *Inst. Phys. Conf. Ser. (UK)* no.56 (1980) p.669-78 ]

[13] T. Kamejima, J. Matsui, Y. Seki, H. Watanabe [ *J. Appl. Phys. (USA)* vol.50 no.5 (1979) p.3312-4 ]

[14] B. Cockayne, W.R. MacEwan, S.J. Courtney, I.R. Harris, N.A. Smith [ *J. Cryst. Growth (Netherlands)* vol.69 no.2-3 (1984) p.610-2 ]

[15] B. Cockayne, W.R. MacEwan, I.R. Harris, N.A. Smith [ *J. Mater. Sci. Lett. (UK)* vol.3 (1984) p.990 ]

[16] N.A. Smith, I.R. Harris, B. Cockayne, W.R. MacEwan [ *J. Cryst. Growth (Netherlands)* vol.68 (1984) p.517 ]

[17] I.R. Harris, N.A. Smith, B. Cockayne, W.R. MacEwan [ *J. Cryst. Growth (Netherlands)* vol.68 (1983) p.115-20 ]

[18] N.A. Smith, A.W. Nicholls, I.R. Harris, B. Cockayne, W.R. MacEwan [ *J. Cryst. Growth (Netherlands)* vol.69 (1984) p.269-74 ]

[19] I.R. Harris, N.A. Smith, B. Cockayne, W.R. MacEwan [ *J. Cryst. Growth (Netherlands)* vol.73 no.2 (1985) p.233-41 ]

[20] B. Cockayne, W.R. MacEwan, I.R. Harris, N.A. Smith [ *J. Cryst. Growth (Netherlands)* vol.76 no.2 (1986) p.251-8 ]

[21] B. Cockayne, W.R. MacEwan, I.R. Harris, N.A. Smith [ *J. Cryst. Growth (Netherlands)* vol.73 (1985) p.637-40 ]

[22] J.P. Fillard, P. Gall, M. Castagne, J. Bonnafe [ *Gadest'89* Ed. M. Kittler (Sci. Tech. Publications, 1989) p.403 ]

[23] J.P. Fillard, P.C. Montgomery, P. Gall, M. Castagne, J. Bonnafe [ *Proc. DRIP III Conf., J. Cryst. Growth (Netherlands)* vol.103 no.1-4 (June 1990) p.120-5 ]

[24] P. Gall, J.M. Lussert, J.P. Fillard, E.V.K. Rao [ *Proc. Conf. on Defects in Semiconductors* Yokohama, Japan, 1989 ]

[25] J.P. Fillard [ *J. Phys. Colloq. (France)* vol.49 no.C4 (1988) p.463-70 ]

[26] K. Moriya, T. Ogawa [ *Jpn. J. Appl. Phys. 2 (Japan)* vol.22 no.4 (1983) p.L207-9 ]

[27] J.P. Fillard, A. Baroudi, P. Gall, J. Bonnafe, P.N. Favennec, R. Coquille [ *Solid State Commun. (USA)* vol.64 no.9 (1987) p.1243-6 ]

[28] J.P. Fillard, P. Gall, A. Baroudi, A. George, J. Bonnafe [ *Jpn. J. Appl. Phys. 2 (Japan)* vol.26 no.8 (1987) p.L1255-7 ]

[29] H. L'Haridon et al [ *J. Cryst. Growth (Netherlands)* vol.113 no.1-2 (Aug. 1991) p.39-44 ]

[30] B. Cockayne, W.R. MacEwan, G.T. Brown [ *J. Cryst. Growth (Netherlands)* vol.55 (1981) p.263-7 ]

# 6.3 ESR and ODMR detection of paramagnetic defects in InP

B. Lambert and H.J. Von Bardeleben

This Datareview appeared in *Properties of Indium Phosphide* (INSPEC, IEE, London, UK, 1991) and is reproduced here for the reader's convenience.

## A INTRODUCTION

Over 1980 - 1990 numerous review papers have been devoted to the study of defects in III-V materials [1-6]. The present Datareview is devoted exclusively to ESR and ODMR defect properties in InP regardless of whether these defects are native or induced, but we emphasise defect aspects which are important to InP materials science and/or technologically significant.

Magnetic resonance studies have been performed on four classes of magnetic defects: effective mass like donors, 3d transition metal and 4f rare earth ions, and intrinsic defects. Both classical ESR and optically detected magnetic resonance (ODMR) techniques have been applied.

## B SHALLOW DONORS AND ACCEPTORS

The shallow acceptor resonance has not been observed, but could be if uniaxial stress were applied [7]. In contrast the delocalised electron resonance has been observed at g = 1.20 both in ODMR [8] and in ESR [9] at liquid He temperatures. Due to the motional narrowing the lineshape is greatly reduced to $\Delta H = 2$ G for samples of donor concentration $10^{16}$ $cm^{-3}$ [10].

## C 3d TRANSITION METALS

Seven transition-metal (3d) elements have been studied [9,11-17], including Fe, Cr and Ti, which have, or might have, importance for semi-insulating InP. The results of the ESR studies are summarised in TABLE 1. Note that the $^5T_2$ ($Cr^{2+}$) site symmetry in InP is reduced to tetragonal ($D_{2d}$) by a Jahn-Teller distortion. A review paper [6] is devoted to the transition-metal impurities in III-V compounds. Quantitative measurements of $Fe^{3+}$ [18], $Cr^{2+}$ and $Cr^{4+}$ [19] concentrations have been performed.

## D 4f RARE EARTH METALS

The most spectacular result on this system concerns the processing of luminescent devices (see a review by Ennen and Schneider [5]); in particular Er gives rise to an $Er^{3+}$ luminescence in the 1.54 μm region. From EPR investigations, the ground state spectra of $Gd^{3+}$ [20], $Er^{3+}$ [21] and $Yb^{3+}$ [22] in InP have been identified; europium in the charge state 2+ has been observed [23]. The results are summarised in TABLE 1. In these cases (except for Eu), rather narrow ($\Delta H$ 10 - 20 G) ESR linewidths could be observed, because inhomogeneous line broadening is strongly suppressed through the screening of the unpaired 4f electrons by the outer-lying closed $5s^2$ and $4p^6$ shells. This is remarkable since in general ESR of intrinsic and extrinsic lattice defects in InP is greatly hampered by a large linewidth, usually exceeding 100 G. The large linewidth related to $Eu^{2+}$ is attributed [23] to spin-spin interaction between the Eu centres and to the unresolved hyperfine structures.

ODMR in the optically excited $^2F_{5/2}$ ($\Gamma_8$) state of $Yb^{3+}$ has been observed [24] and confirms a Lande g-value ($g_{5/2}$ = 0.578) also observed in Zeeman studies [25].

TABLE 1.

| Dopant | Charge state | T (K) | Configuration | g | ΔH (G) | A ($\times 10^4$ $cm^{-1}$) | Ref |
|---|---|---|---|---|---|---|---|
| Ti | 3+ | 4 | $3d^1$,$2_E$ | 1.94 | 450 | | [11] |
| V | 3+ | 4 | $3d^2$,$^3A_2$ | 1.96 | 400 | $A^{50}$ = 55 | [9] |
| Cr | 2+ | 4 | $3d^4$,$^5T_2$ | $g_{\parallel}$ = 1.981 | | | [12] |
| | | | | $g_{\perp}$ = 2.010 | | | |
| | 4+ | 4 | $3d^2$,$^3A_2$ | 1.999 | 350 | | [13] |
| Mn | 2+ | 2.3 | $3d^5$,$^6A_1$ | 2.01 | 330 | $A^{55}$ = 56 | [14] |
| Fe | 3+ | 4.5 | $3d^5$,$^6A_1$ | 2.204 | 120 | | [15] |
| Co | 2+ | 3.5 | $3d^7$,$^4A_2$ | 2.192 | 150 | | [16] |
| Ni | 3+ | 20 | $3d^7$,$^4A_2$ | 2.098 | 97 | | [17] |
| Eu | 2+ | 17 | $4f^7$,$^8A_1$ | 1.90 | 800 | | [23] |
| Gd | 3+ | 300 | $4f^7$,$^8A_1$ | 1.985 | 11 | | [20] |
| Er | 3+ | 4 | $4f^{11}$ $^4I_{15/2}(\Gamma_7)$ | 5.699 | | $A^{167}$ = 210 | [21] |
| Yb | 3+ | 4 | $4f^{13}$ $^2F_{7/2}(\Gamma_7)$ | 3.29 | 12 | $A^{171}$ = 544<br>$A^{173}$ = 149 | [22] |

## E INTRINSIC DEFECTS IN InP

Of the basic intrinsic defects, vacancies, interstitials and anti-sites, the indium vacancy $V_{In}$, the phosphorus vacancy $V_P$ and the phosphorus anti-site $P_{In}$ are expected to have multiple charge states in the gap [26]. They are thus electrically active and paramagnetic in particular charge states: $V_{In}^{2+,0,2-}$, $V_P^{0/2+}$, $P_{In}^{+}$. The concentrations of these native intrinsic defects will strongly depend on the sample growth conditions. All published ESR and ODMR results have been obtained in LEC grown InP. In this case the dominant native defect is the $P_{In}$ defect. Intrinsic defects at much higher concentrations ($10^6$ - $10^{18}$ $cm^{-3}$) have been produced by particle irradiation ($e^-$, $n^o$, ions) as well as post-growth thermal treatment. Intrinsic defects are listed in TABLE 2.

### E1 Native Intrinsic Defects

The only native defect observed in LEC grown InP is the $P_{In}$ defect [27-29]. It has exclusively been observed by ODMR in p-type Zn doped InP. Its atomic configuration (substitutional, four equivalent P neighbours) has been confirmed by ODENDOR measurements [28]. Its concentration has been estimated to be in the $10^{15}$ $cm^{-3}$ range. The $P_{In}$ defect is a double donor with a 0/+ level at approximately $E_v$ + 0.8 eV [30]. Its second ionisation level, expected at $E_v$ + 0.5 eV, has not yet been found.

After thermal annealing under rather extreme conditions a different phosphorus anti-site related defect has been observed [27-30]. It has been attributed to a $P_{In}$ - $X_{In}$ complex.

### E2 Intrinsic Defects Produced by $e^-$ Irradiation

Irradiation of InP with electrons of near threshold energies leads to the formation of Frenkel pairs $V_{In}$ - $In_i$, $V_P$ - $P_i$, which transform further at distinct annealing stages [31]. Both vacancy and $P_{In}$ related defects have been observed in n-type and p-type material after room temperature irradiation [32-34].

TABLE 2 Intrinsic defects.

| Native defects | | | | | |
|---|---|---|---|---|---|
| Defect | g | A ($cm^{-1}$) | $A^{Lig}$ (MHz) | Material | Ref |
| $P_{In}$ | $2.006 \pm 0.005$ | $0.100 \pm 0.002$ | | Zn-LEC | [29] |
| $P_{In}$ | $2.000 \pm 0.003$ | $0.098 \pm 0.002$ | $A^P_{\|\|} = 367.3 \pm 0.2$<br>$A^P_{\perp} = 244.3 \pm 0.2$ | Zn-LEC | [28] |
| $P_{In}$ - X | $g_{\|\|} = 2.023 \pm 0.006$<br>$g_{\perp} = 2.009 \pm 0.006$<br>$D = 0.110 \pm 0.003$<br>$E = 0.038 \pm 0.001$ | $0.041 \pm 0.002$ | | | [30] |
| **e-irradiated** | | | | | |
| Defect | g | A ($cm^{-1}$) | ΔB (G) | Material | Ref |
| $P_{In}$ | $2.012 \pm 0.005$ | $0.104 \pm 0.002$ | | Zn-LEC | [34] |
| $P_{In}$ | $1.991 \pm 0.005$ | $0.099 \pm 0.002$ | | Sn-LEC | [34] |
| $P_{In}$ | $1.992 \pm 0.008$ | $0.092 \pm 0.005$ | 350 | Zn-LEC<br>Sn-LEC | [32] |
| $V_P$ | $2.05 \pm 0.04$ | | 1800 | Zn-LEC | [35] |
| $V_{In}$ | $1.986 \pm 0.005$ | | 500 | Zn-LEC | [36] |
| $Fe^{3+}$-X | $g_{\|\|} = 2.035 \pm 0.006$<br>$g_{\perp} = 8.05 \pm 0.1$ | | | Fe-LEC | [37] |
| $Fe^{3+}$-X | $g_{\|\|} = 2.0646 \pm 0.03$<br>$g_{\perp} = 8.058 \pm 0.03$ | | | Fe-LEC | [38] |
| **Neutron irradiated** | | | | | |
| Defect | g | A ($cm^{-1}$) | ΔB (G) | Material | Ref |
| $P_{In}$ | $1.995 \pm 0.003$ | $0.0947 \pm 0.001$ | 448 | undoped n-type LEC | [39] |
| $P_{In}$ | $2.004 \pm 0.003$ | $0.0920 \pm 0.001$ | 542 | undoped n-type LEC | [39] |
| $P_{In}$ | $2.007 \pm 0.003$ | $0.0933 \pm 0.001$ | 570 | Fe-LEC | [39] |
| $V_P$ | $2.28 \pm 0.06$ | | 2750 | | [39] |
| $Fe^{3+}$-$X^1$ | $g_{\|\|} = 2.32 \pm 0.06$<br>$g_{\perp} = 4.32 \pm 0.03$ | | | Fe-LEC | [39] |

Surprisingly, the irradiation leads to the formation of two slightly different $P_{In}$ anti-sites [34]. One of them is identical to the native one; the second one, observed in n-type InP, has a smaller central hyperfine interaction and a smaller Lande g-factor. In p-type material both of them are simultaneously seen in the EPR spectrum.

In p-type Zn doped InP two additional paramagnetic defects were observed; they have been attributed respectively to the $V_P$ [35] and the $V_{In}$ defect [36]. The $V_P$ defect is characterised by an extremely large linewidth of 1800 G due to hyperfine interaction with the four nearest In neighbours.

Electron irradiation of semi-insulating Fe doped material leads to a transformation of the substitutional $Fe_{In}^{3+}$ defect into a trigonal $Fe^{3+}$-X nearest neighbour complex [37-38]. The high introduction rate of this defect implies that X is an intrinsic defect.

## E3 Neutron and Ion Implantation Induced Defects

Irradiation with heavy particles with 0.1 MeV to 1 MeV energies introduces more complex defects. $P_{In}$ anti-site defects have been observed in neutron irradiated InP [39]; their EPR parameters are different from those observed as native defects or in e-irradiated material and change with the dose. In B-implanted InP [40] $P_{In}$ - X complexes with spin S = 1 have been seen. Unidentified complex defects tentatively attributed to Fe - X (X = $In_i$) have equally been reported [39]. Additionally, the same authors [39] have observed a spectrum of 2750 G line width, effective g-factor 2.28, which they relate to the $V_P$ defect.

## ACKNOWLEDGEMENT

We are very grateful to B. Clerjaud for fruitful discussions and critical reading of the manuscript.

## REFERENCES


[1] V.K. Bashenov [ *Phys. Status Solidi A (East Germany)* vol.10 (1972) p.9 ]

[2] U. Kaufmann, J. Schneider [ *Festkoerperprobleme (West Germany)* vol.20 (1980) p.87 ]

[3] U. Kaufmann, J. Schneider [ *Adv. Electron. Electron Phys. (UK)* vol.58 (1982) p.81 ]

[4] V.F. Masterov [ *Sov. Phys.-Semicond. (USA)* vol.18 no.1 (1984) p.1-13 ]

[5] M. Ennen, J. Schneider [ *Proc. 13th Int. Conf. on Defects in Semiconductors* 1984 (Metallurg. Soc. AIME, USA, 1986) p.115 ]

[6] B. Clerjaud [ *J. Phys. C (UK)* vol.18 no.19 (1985) p.3615 ]

[7] K.P. O'Donnell, K.M. Lee, G.B. Watkins [ *Bull. Am. Phys. Soc. (USA)* vol.28 (1983) p.330 ]

[8] Y. Dawei, B.C. Cavenett, M.S. Skolnick [ *J. Phys. C (UK)* vol.16 (1983) p.L647 ]

[9] B. Lambert, H.J. Von Bardeleben, B. Deveaud [ *J. Phys. C (UK)* vol.18 no.23 (1985) p.L707-10 ]

[10] B. Clerjaud, F. Gendron, M. Obloh, J. Schneider, W. Wilkening [ *Phys. Rev. B (USA)* vol.40 no.4 (1989) p.2042 ]

[11] B. Lambert, Y. Toudic, G. Grandpierre, M. Gauneau, B. Deveaud [ *Semicond. Sci. Technol. (UK)* vol.2 no.2 (1987) p.78-82 ]

[12] G.H. Stauss, J.J. Krebs, R.L. Henry [ *Phys. Rev. B (USA)* vol.16 (1977) p.974 ]

[13] N.K. Goswani, R.C. Newman, J.E. Whitehouse [ *Solid State Commun. (USA)* vol.36 (1980) p.897 ]

[14] V.F. Masterov, Yu.V. Malotsev, V.K. Sobolevskii [ *Sov. Phys.-Semicond. (USA)* vol.15 no.11 (1981) p.1235-7 ]

[15] W.H. Koschel, U. Kaufmann, S.G. Bishop [ *Solid State Commun. (USA)* vol.21 (1977) p.1069 ]

[16] B. Lambert, B. Deveaud, Y. Toudic, B. Clerjaud [ *Physica B & C (Netherlands)* vol.116 (1983) p.467 ]

[17] U. Kaufmann, J. Schneider [ *Solid State Commun. (USA)* vol.25 (1978) p.113 ]

[18] B. Lambert, R. Coquille, M. Gauneau, G. Grandpierre, G. Moisan [ *Semicond. Sci. Technol. (UK)* vol.5 (1990) p.616 ]

[19] Y. Toudic, B. Lambert, R. Coquille, G. Grandpierre, M. Gauneau [ *Semicond. Sci. Technol. (UK)* vol.3 no.5 (1988) p.464-8 ]

[20] V.F. Masterov, V.V. Romanov, B.E. Samorukov, K.F. Shtel'makh [ *Sov. Phys.-Semicond. (USA)* vol.17 no.5 (1983) p.596-7 ]

[21] V.F. Masterov, K.F. Shtel'makh, L.F. Zakharenkov [ *Sov. Phys.-Semicond. (USA)* vol.21 no.2 (1987) p.223 ]

[22] V.F. Masterov, V.V. Romanov, K.F. Shtel'makh [ *Sov. Phys.-Solid State (USA)* vol.25 no.5 (1983) p.824-6 ]

[23] L.F. Zakharenkov, S.I. Markov, V.F. Masterov, K.F. Shtel'makh [ *Sov. Phys.-Semicond. (USA)* vol.19 no.10 (1986) p.1132-3 ]

[24] R. Kallenbach, H.J. Reyher, J. Unruh, A. Winnacker, H. Ennen [ *Mater. Sci. Forum (Switzerland)* vol.10-12 (1986) p.681-6 ]

[25] G. Aszodi et al [ *Phys. Rev. B (USA)* vol.31 no.12 (1985) p.7767-71 ]

[26] W. Pötz, D.K. Ferry [ *Phys. Rev. B (USA)* vol.31 no.2 (1985) p.968-73 ]; C. Delerue [ unpublished ]
[27] T.A. Kennedy, N.D. Wilsey [ *J. Cryst. Growth (Netherlands)* vol.83 no.2 (1987) p.198-201 ]
[28] D.Y. Jeon, H.P. Gislason, J.F. Donegan, G.D. Watkins [ *Phys. Rev. B (USA)* vol.36 no.2 (1987) p.1324-7 ]
[29] L.H. Robins, P.C. Taylor, T.A. Kennedy [ *Phys. Rev. B (USA)* vol.38 no.18 (1988) p.13227-36 ]
[30] T.A. Kennedy, N.D. Wilsey, P.B. Klein, R.L. Henry [ *Mater. Sci. Forum (Switzerland)* vol.10-12 (1986) p.271-6 ]
[31] A. Sibille [ Thesis, Universite Paris 7, 1985 ]
[32] T.A. Kennedy, N.D. Wilsey [ *Appl. Phys. Lett. (USA)* vol.44 (1984) p.1089 ]
[33] M. Deiri, A. Kana-ah, B.C. Cavenett, T.A. Kennedy, N.D. Wilsey [ *J. Phys. C (UK)* vol.17 (1984) p.L793 ]
[34] A. Kana-ah, M. Deiri, B.C. Cavenett, N.D. Wilsey, T.A. Kennedy [ *J. Phys. C (UK)* vol.18 no.20 (1985) p.L619-24 ]
[35] H.J. Von Bardeleben [ *Solid State Commun. (USA)* vol.57 no.2 (1986) p.137-40 ]
[36] M. Deiri, A. Kana-ah, B.C. Cavenett, T.A. Kennedy, N.D. Wilsey [ *Semicond. Sci. Technol. (UK)* vol.3 no.7 (1988) p.706-14 ]
[37] T.A. Kennedy, N.D. Wilsey [ *Inst. Phys. Conf. Ser. (UK)* no.59 (1981) p.257 ]
[38] E.Yu. Brailovski, I.G. Megela, N.H. Pambuchchyan, V.U. Teslenko [ *Phys. Status Solidi A (Germany)* vol.72 (1982) p.K109 ]
[39] A. Goltzene, B. Meyer, C. Schwab [ *J. Appl. Phys. (USA)* vol.62 no.11 (1987) p.4406 ]
[40] T.A. Kennedy [ *Proc. Int. Conf. Science & Technology of Defect Control in Semiconductors* Yokohama, Japan, 1989, Ed. K.Sumino ]

# 6.4 Passivation of defects in InP by hydrogenation

R. Murray

This Datareview appeared in *Properties of Indium Phosphide* (INSPEC, IEE, London, UK, 1991) and is reproduced here for the reader's convenience.

## A INTRODUCTION

Hydrogen passivates (neutralises) the electrical activity of zinc, cadmium and manganese acceptors and some deep levels in InP. H may be deliberately introduced by proton implantation or exposure to an H (or deuterium) plasma. It may also appear as an inadvertent impurity following growth or subsequent processing. Neutralisation is thought to occur when the H atom bonds to an impurity or its nearest neighbour host lattice atom. Passivation effects are reversible and subsequent anneals in the temperature range 275 - 450°C restore the electrical activity of the impurities and defects. In contrast to GaAs or Si there is no convincing evidence for donor passivation in InP. In addition, the injection of H from a gas plasma is not so straightforward, as energetic H atoms react with P at the sample surface to produce phosphine leaving behind In droplets [1]. Consequently most studies involving the introduction of H from a plasma use InP layers capped with a thin layer of InGaAs [2], Au [3] or silicon nitride [4]. However, Omeljanovsky et al [3,5] did not observe any degradation of their uncapped InP samples following passivation treatments in a high frequency (2.45 GHz) H plasma, showing that the energy of the incident H atoms is an important parameter in passivation studies [6].

## B BULK InP

Clerjaud et al [7] have reported a number of sharp lines which they observed in the infrared spectra of as-grown bulk InP samples and these are listed below (TABLE 1). These lines are extremely narrow (less than 0.1 $cm^{-1}$ at 5 K) and remain sharp at higher temperatures, shifting in energy by only a few wavenumbers between 5 K and room temperature. Such behaviour is typical of absorption by an impurity local vibrational mode (LVM) and the relatively high frequencies suggest that a light mass is involved. It was therefore proposed that the lines arise from the stretching modes of P-H bonds, by comparing the frequencies of the absorption lines with those observed in phosphine. Some of the lines correlate with the presence of transition metal impurities while the lines at 2202.4 and 2315.6 $cm^{-1}$ have

TABLE 1 Energies of vibrational transitions (5 K) of H in LEC InP.

| Energy ($cm^{-1}$) | Assignment |
|---|---|
| 2202.4 | |
| 2250.8 | |
| 2254.7 | |
| 2254.8 | |
| 2272 | Mn |
| 2273.4 | |
| 2282.8 | |
| 2285.7 | V |
| 2287.7 | Zn |
| 2300 | Ti |
| 2315.6 | |

been observed in proton implanted material [14]. As most transition metals act as acceptors in III-V semiconductors, the appearance of these sharp LVM lines is probably the result of acceptor passivation. In fact the model proposed by Clerjaud et al is exactly the same as that put forward to explain the passivation of B in Si [8] or group II acceptors in GaAs [9]: an H atom bonds to the nearest neighbour to the impurity resulting in an electrically inactive complex. Different impurities having different ionic radii will then influence the vibrational frequencies of the bonded H atom. The H is thought to originate from the phosphorus starting material although the strength of the LVM lines increases after the LEC growth indicating that the $B_2O_3$ encapsulant may also be a source of H. Based on the strengths of these lines Chevallier et al [10] have estimated the concentration of H in bulk LEC InP to be as high as $10^{16}$ $cm^{-3}$.

## C IMPLANTED InP

Hydrogen may be deliberately introduced into the InP crystal by proton implantation. This technique is commonly used to produce high resistivity regions in GaAs/AlGaAs for device isolation and would have similar applications in the InP/InGaAsP system. A resistivity limit of about $10^3$ Ω cm has been achieved in bulk n-type InP but much higher values around $10^8$ Ω cm have been reached in p-type material [11]. Prolonged bombardment of p-type InP converts it to n-type and introduces deep states which lead to band tailing and a reduction in the size of the bandgap. Miyoshi et al [12] have utilised these two effects to obtain photodetection out to 1.52 μm. A number of new vibrational absorption lines (listed in TABLE 2) are observed in proton implanted InP. A comparison of the frequencies of these lines with those detected in phosphine indicates that they arise from the stretching vibrations of P-H bonds. The multiplicity of lines is perhaps due to the influence of neighbouring vacancies and interstitials, produced as a result of the lattice damage, or because the defects contain more than one H atom. Confirmation that these lines are due to the vibrations of H atoms comes from the observation of an analogous set of lines in deuteron implanted material. Assuming anharmonic terms in the vibrational potential to be negligible, the frequency ratio, r, of the H and D related lines is given by

$$r = \sqrt{\{2(M+1)/(M+2)\}} \qquad (1)$$

where M is the mass of a phosphorus atom. The value of r is in most cases less than the theoretical value of 1.393 using the above equation but there is reasonable agreement with the value of 1.371 measured for a $PH_3$ molecule. Note that three of these lines at 2197, 2212 and 2315 $cm^{-1}$ are close in energy to the absorption lines observed in as-grown bulk InP discussed in the previous section.

TABLE 2 Frequencies ($cm^{-1}$) of H and D lines in implanted InP.

| InP:H | InP:D | r | Ref |
|---|---|---|---|
| 2197 | 1600 | 1.373 | [13] |
| 2212 | 1619 | 1.366 | [13] |
| 2308 | 1659 | 1.391 | [13] |
| 2335 | 1678 | 1.391 | [13] |
| 2202 | 1604 | 1.372 | [14] |
| 2315 | 1683 | 1.375 | [14] |
| 2328 | 1698 | 1.371 | $PH_3$ mol |

## D INADVERTENT PASSIVATION OF InP

Many optoelectronic devices based on InP or InGaAs depend on the growth of good quality doped epitaxial layers by metalorganic chemical vapour deposition (MOCVD) or liquid phase epitaxy (LPE) on semi-insulating InP substrates. Antell et al [15] and Cole et al [16] have measured substantial reductions up to a factor of 20 in the hole concentrations of Zn and Cd-doped InP layers grown at 650°C by MOCVD. SIMS profiles showed that the reduction in p correlated with the presence of H when the crystals were cooled in an atmosphere of $PH_3/AsH_3/H_2$. It was proposed that pyrolysis of the $AsH_3$ released H at a temperature where stable P-H bonds can form within the semiconductor leading to passivation of the Zn acceptors. Support for this view came from the observation that [p] was reduced by no more than a factor of two in material cooled under $PH_3/H_2$ alone. Layers could also be protected if capped with an n-type InGaAs layer but a p-type InGaAs layer did not prevent passivation indicating that InGaAs is permeable to H. The diffusion of H in InGaAs is therefore quite different in n- and p-type material. Similar effects have already been observed in Si [17]. Rapid thermal annealing (RTA) of ion implanted InP is performed at rather higher temperatures and it might be expected that this would lead to greater pyrolysis of the $PH_3$ and subsequent injection of H. Haussler [18] has shown that injection of H does indeed occur following 850°C RTA in unimplanted samples but its presence resulted in only a small reduction in the hole concentration of samples implanted with Be and then given the RTA treatment.

Hydrogen can also be an inadvertent contaminant during reactive ion etching (RIE). Hayes et al [19] detected passivation of Zn acceptors in bulk InP samples subjected to a $CH_4/D_2$ RIE. SIMS profiles indicate that the D diffuses about 0.2 μm into the sample resulting in almost complete passivation of the Zn. A subsequent RTA of 1 min at 350°C restored the acceptor activity and redistributed the D to a depth of 1.5 μm. Rather different behaviour was found for n-type material where the D diffused to 0.7 μm but no passivation was detected. Indeed a small increase in n was measured following the RIE but is probably due to In enrichment near the surface.

## E PLASMA TREATED InP

There have been several [2-5] studies of the passivation of acceptors in bulk and epitaxial InP by exposing samples to a hydrogen or deuterium plasma. There is no strong evidence that donors can be passivated by such treatments [5,20]. In most cases a capping layer has been used to protect the sample surface during plasma treatments although one group [5] has used a GHz plasma which does not lead to surface damage. Of the passivated species Zn has received most attention and a model deduced from SIMS, electrical measurements and infrared spectroscopy has been deduced. This model is analogous to that proposed to explain the passivation of Zn acceptors in GaAs [21]. SIMS profiles show that the D concentration in Zn doped InP layers capped with a D permeable lattice-matched InGaAs layer correlates with the Zn doping level over the temperature range 150 - 300°C [2]. The resistivity was found to increase by more than six orders of magnitude while the hole mobility increased by a factor of 10 over unpassivated layers. These results are understood in terms of Zn-H pairing and evidence to support this view comes from LVM studies of passivated layers [22]. Following passivation a new vibrational absorption line is observed at 2287.7 $cm^{-1}$ and is attributed to the stretching mode of a bonded H atom. The observation of a line at 1664.5 $cm^{-1}$ in samples treated in a D plasma confirms the involvement of the H(D) atom. As the lines are very narrow this precludes direct bonding of the H or D atom to a Zn or In atom since these elements have a number of naturally occurring isotopes which would lead to a much broader feature displaying fine structure. Passivation is therefore thought to occur when an H atom breaks the bond between a Zn acceptor and its P neighbour. The H bonds to the P atom while the Zn impurity relaxes back towards its three remaining P neighbours. The Zn atom is then tri-coordinated and electrically neutral. A dramatic increase in the near band edge photoluminescence intensity has also been reported in passivated Zn doped InP [23]. The disappearance of two luminescence bands associated with deep levels suggests that H passivates both shallow acceptors and deep levels as in GaAs [20].

Mn is thought to be a deep acceptor in InP and electrical measurements have shown that it can be passivated by exposing samples to a microwave plasma [3,5]. LVM measurements have yet to be made on Mn doped InP but it is likely that the absorption line at 2272 $cm^{-1}$ observed by Clerjaud et al [7] results from the passivation of Mn acceptors. Photoluminescence measurements on passivated Mn doped InP show a large increase in the Mn related feature at 1.13 eV after hydrogenation. Omeljanovsky et al [5] attribute this to the passivation of deep levels which act as non-radiative recombination centres in the as-grown material.

## REFERENCES


[1] J.H. Thomas, G. Kaganowicz, J.W. Robinson [ *J. Electrochem. Soc. (USA)* vol.135 no.5 (1988) p.1201-6 ]

[2] J. Chevallier et al [ *Semicond. Sci. Technol. (UK)* vol.4 no.2 (1989) p.87-90 ]

[3] E.M. Omeljanovsky, A.V. Pakhomov, A.Y. Polyakov [ *Semicond. Sci. Technol. (UK)* vol.4 no.11 (1989) p.947-50 ]

[4] W.C. Dautremont-Smith, J. Lopata, S.J. Pearton, A. Koszi, M. Stavola, V. Swaminathan [ *J. Appl. Phys. (USA)* vol.66 no.5 (1989) p.1993-6 ]

[5] E.M. Omeljanovsky, A.V. Pakhomov, A.Y. Polyakov [ *J. Electron. Mater. (USA)* vol.18 no.6 (1989) p.659-70 ]

[6] G.S. Jackson, J. Beberman, M.S. Feng, K.C. Hsieh, N. Holonyak Jr., J. Verdeyen [ *J. Appl. Phys. (USA)* vol.64 no.10 pt.1 (1988) p.5175-8 ]

[7] B. Clerjaud, C. Cote, C. Naud [ *J. Cryst. Growth (Netherlands)* vol.83 no.2 (1987) p.190-3 ]

[8] J.I. Pankove, C.E. Magee, R.O. Wance [ *Appl. Phys. Lett. (USA)* vol.47 no.7 (1985) p.748-50 ]

[9] P.S. Nandhra et al [ *Semicond. Sci. Technol. (UK)* vol.3 no.4 (1988) p.356-60 ]

[10] J. Chevallier, B. Clerjaud, B. Pajot [ to be published ]

[11] J.P. Donnelly, C.E. Hurwitz [ *Solid-State Electron. (UK)* vol.20 no.8 (1977) p.727-30 ]

[12] T. Miyoshi, P.K. Tien, R.J. Martin, D.M. Tennant, A.M. Johnson, P.M. Downey [ *Appl. Phys. Lett. (USA)* vol.44 no.1 (1984) p.128-30 ]

[13] V. Riede, H. Neumann, H. Sobotta, C. Ascheron, R. Grotschel [ *Solid State Commun. (USA)* vol.65 no.9 (1988) p.1063 ]

[14] J. Tatarkiewicz, B. Clerjaud, D. Cote, F. Gendron, A.M. Hennel [ *Appl. Phys. Lett. (USA)* vol.53 no.5 (1988) p.382-4 ]

[15] G.R. Antell et al [ *Appl. Phys. Lett. (USA)* vol.53 no.9 (1988) p.758-60 ]

[16] S. Cole, J.S. Evans, M.J. Harlow, A.W. Nelson, S. Wong [ *J. Cryst. Growth (Netherlands)* vol.93 (1988) p.607-12 ]

[17] N.M. Johnson, C. Herring [ *Mater. Sci. Forum (Switzerland)* vol.38-41 pt.3 (1989) p.961-6 ]

[18] W. Haussler [ *Appl. Phys. Lett. (USA)* vol.54 no.25 (1989) p.2595-6 ]

[19] T.R. Hayes, W.C. Dautremont-Smith, H.S. Luftman, J.W. Lee [ *Appl. Phys. Lett. (USA)* vol.55 no.1 (1989) p.56-8 ]

[20] W.C. Dautremont-Smith [ *Mater. Res. Soc. Symp. Proc. (USA)* vol.104 (1988) p.313-24 ]

[21] N.M. Johnson, R.D. Burnham, R.A. Street, R.L. Thornton [ *Phys. Rev. B (USA)* vol.33 no.2 (1986) p.1102-5 ]

[22] B. Pajot, J. Chevallier, A. Jalil, B. Rose [ *Semicond. Sci. Technol. (UK)* vol.4 no.2 (1989) p.91-4 ]

[23] V. Swaminathan, J. Lopata, S.E.G. Slusky, W.C. Dautremont-Smith, S.J. Pearton [ *Electron. Lett. (UK)* vol.25 no.23 (1989) p.1584 ]

# 6.5 Deep states in InP associated with the 3d transition metals

A.M. Hennel (revised by A.R. Peaker)

June 1998

## A INTRODUCTION

Transition metals are important impurities in semiconductors and indium phosphide is no exception. The 3d transition metals are often present as inadvertent impurities transported to the semiconductor from stainless steel during growth or deposited on the semiconductor surface during cleaning and processing. Some of these metals are fast diffusants with moderate solubilities leading to precipitates and extended defects. In this Datareview, the electronic role of the metal in solution is considered although in most cases the exact siting is not known. Iron and titanium form a level at approximately midgap and are added as intentional impurities to produce semi-insulating materials.

The properties of the deep states are listed in order of the element whose adequate data are available, namely, titanium, vanadium, chromium, manganese, iron, cobalt, nickel and copper.

## B TITANIUM

Titanium can act as a midgap donor in InP and has been incorporated as an interstitial in order to produce semi-insulating material. Its level is located at about 0.5 eV below the bottom of the InP conduction band. The actual values obtained by different workers show some considerable scatter, as shown in TABLE 1.

TABLE 1.

| Measurement type and ref | Value (eV) |
|---|---|
| DLTS [1] | $E_c$ - 0.63 |
| Temperature dependent Hall effect [2] | $E_c$ - 0.62 |
| DLTS [3] | $E_c$ - 0.56 |
| Temperature dependent Hall effect [4] | $E_c$ - 0.53 |
| DLTS [5] | $E_c$ - 0.54 |
| Photoconductivity under hydrostatic pressure [6] | $E_c$ - 0.50 |

The hydrostatic pressure coefficient of this level is 87 ± 25 meV/GPa [7] relative to the InP conduction band.

From absorption [8] and EPR [9] measurements this level has been identified as the donor $Ti^{4+}/Ti^{3+}$ level, i.e. as the level between the neutral $Ti^{3+}(3d^1)$ charge state and the ionised $Ti^{4+}(3d^0)$ charge state. The degeneracy factor of the $Ti^{4+}/Ti^{3+}$ level is equal to $g(Ti^{4+}/g(Ti^{3+}) = g(^1A_1)/g(^2E) = 1/4$.

## C VANADIUM

A vanadium donor level has been observed at about 0.21 eV above the top of the InP valence band in DLTS measurements of p-type InP [10]. From absorption and DLOS measurements [11] this level has been identified as the donor $V^{4+}/V^{3+}$ level, i.e. as the level between the ionised $V^{4+}(3d^1)$ charge state and the neutral $V^{3+}(3d^2)$ charge state. The degeneracy factor of the $V^{4+}/V^{3+}$ level is equal to $g(V^{4+})/g(V^{3+}) = g(^2E)/g(^3A_2) = 4/3$.

## D CHROMIUM

The donor level of chromium is located at about 0.6 eV above the top of the InP valence band. This level has been observed using the temperature dependent Hall effect [12] at $E_v$ + 0.56 and in DLTS and DLOS [13] at $E_v$ + 0.61 eV. From the quantitative EPR measurements [4] this level has been identified as the $Cr^{4+}/Cr^{3+}$ level, i.e. as the level between the ionised $Cr^{4+}(3d^2)$ charge state and the neutral $Cr^{3+}(3d^3)$ charge state. The degeneracy factor of the $Cr^{4+}/Cr^{3+}$ donor level is equal to $g(Cr^{4+})/g(Cr^{3+}) = g(^3A_2)/g(^4T_1) = 1/4$.

A chromium acceptor level has also been reported at about 0.4 eV below the bottom of the InP conduction band. This level was observed using the temperature dependent hall effect [14] at $E_c$ - 0.39 and using DLTS and DLOS [15] at $E_c$ - 0.41 eV. From the absorption [16] and EPR [17] measurements this level has been identified as the acceptor $Cr^{3+}/Cr^{2+}$ level, i.e. as the level between the neutral $Cr^{3+}(3d^3)$ charge state and the ionised $Cr^{2+}(3d^4)$ charge state. The degeneracy factor of the $Cr^{3+}/Cr^{2+}$ level is equal to $g(Cr^{3+})/g(Cr^{2+}) = g(^4T_1)/g(^5T_2) = 4/5$.

## E MANGANESE

Manganese impurity acts as an acceptor in InP. Its level is located at about 0.22 eV above the top of the valence band. The manganese acceptor level has been observed in optical absorption photoionisation [18] at $E_v$ + 0.22 and in DLTS [19] at $E_v$ + 0.25 eV.

Contrary to the situation for some other transition metal impurities in InP, this level cannot be identified as the acceptor $Mn^{3+}/Mn^{2+}$ level, i.e. as the level between the neutral $Mn^{3+}(3d^4)$ charge state and the ionised $Mn^{2+}(3d^5)$ charge state. The observation of sharp lines in the photoionisation [18] and luminescence excitation spectra [20] of manganese allows us to interpret this centre as characterised by the effective-mass theory. In that case the neutral state of the manganese acceptor corresponds to the $Mn^{2+}(3d^5)$ charge state with a hole in a delocalised orbit.

## F IRON

Iron acts as a deep acceptor in InP and is used to obtain semi-insulating material. Its level is located about 0.78 eV above the top of the InP valence band. Measurements using the temperature dependent Hall effect [14] give an energy of $E_c$ - 0.65 eV, DLTS gives values between $E_c$ - 0.63 [21] and $E_c$ - 0.56 eV [22], and absorption photoionisation [23] gives $E_v$ + 0.785 eV. From the intracentre absorption and photoluminescence [24] and also EPR [17] measurements, this level has been identified as the acceptor $Fe^{3+}/Fe^{2+}$ level, i.e. as the level between the neutral $Fe^{3+}(3d^5)$ charge state and the ionised $Fe^{2+}(3d^6)$ charge state. The degeneracy factor of the $Fe^{3+}/Fe^{2+}$ level is equal to $g(Fe^{3+})/g(Fe^{2+}) = g(^6A_1)/g(^5T_2) = 2/5$.

## G COBALT

The cobalt acceptor level is located at about 0.3 eV above the top of the InP valence band. Measurements using the temperature dependent Hall effect and DLTS [25] give an energy of $E_v$ + 0.32 eV and ODLTS [26] gives $E_v$ + 0.24 eV.

From the photoluminescence, EPR and electrical and optical space-charge spectroscopy measurements [25,26], this level has been identified as the acceptor level between a neutral cobalt charge state and the ionised $Co^{2+}(3d^7)$ charge state. There are no experimental data concerning a neutral cobalt charge state.

## H NICKEL

The nickel acceptor level is located at about 0.48 eV above the top of the InP valence band and is observed in p-type InP. Two nickel levels have been reported in DLTS experiments [27]. The double acceptor level has been observed in n-type material at about 0.27 eV below the bottom of the conduction band [28].

In analogy with properties of the Ni impurity in GaAs and GaP crystals [29], these levels have been identified as the acceptor $Ni^{3+}/Ni^{2+}$ level and the double acceptor $Ni^{2+}/Ni^{+}$ level, i.e. as the levels between the neutral $Ni^{3+}(3d^7)$ charge state and the ionised $Ni^{2+}(3d^8)$ charge state, and the latter state and the double ionised $Ni^{+}(3d^9)$ charge state, respectively. The degeneracy factors of these levels are equal to $g(Ni^{3+})/g(Ni^{2+}) = g(^4A_2)/g(^3T_1) = 4/9$ and $g(Ni^{2+})/g(Ni^{+}) = g(^3T_1)/g(^2T_2) = 3/2$.

## I COPPER

Copper can act as a deep acceptor in InP. There are two levels tentatively attributed in the literature to copper or copper-related complexes: Cu(A) and Cu(B).

The copper (A) and copper (B) acceptor levels are located above the top of the InP valence band at about 0.3 eV and 0.6 eV respectively. These levels were observed in DLTS and photocapacitance measurements [30].

There is no clear identification of these two levels. Both of them probably do not correspond to isolated copper impurity, but to its complexes with other impurities or defects.

## REFERENCES

[1] C.D. Brandt et al [ *Appl. Phys. Lett. (USA)* vol.48 no.17 (1986) p.1162-4 ]
[2] G.W. Isler, B.S. Ahern [ *Appl. Phys. Lett. (USA)* vol.48 no.24 (1986) p.1656-7 ]
[3] G. Bremond et al [ *J. Phys. C (UK)* vol.19 no.24 (1986) p.4723-8 ]
[4] Y. Toudic, R. Coquille, M. Gauneau, G. Grandpierre, L. Le Marechal, B. Lambert [ *J. Cryst. Growth (Netherlands)* vol.83 (1987) p.184-9 ]
[5] D.G. Knight, C.J. Miner, W. Benyon [ *Proc. 5th Conf. on Semi-Insulating III-V Materials* Malmo, Sweden, 1998, Eds. G. Grossmann, L.A. Ledebo (Adam Hilger, Bristol, 1988) p.259-64 ]
[6] D. Wasik, M. Baj, A.M. Hennel [ *Proc. 19th Int. Conf. on Physics of Semiconductors* Warsaw, 1988, Ed. W. Zawadzki (Polish Acad. of Sci., 1989) p.1095-8 ]
[7] D.D. Nolte, W. Walukiewicz, E.E. Haller [ *Phys. Rev. Lett. (USA)* vol.59 no.4 (1987) p.501-4 ]
[8] C.D. Brandt, A.M. Hennel, T. Bryskiewicz, K.Y. Ko, L.M. Pawlowicz, H.C. Gatos [ *J. Appl. Phys. (USA)* vol.65 no.9 (1989) p.3459-69 ]
[9] B. Lambert, Y. Toudic, G. Grandpierre, M. Gaunneau, B. Deveaud [ *Semicond. Sci. Technol. (UK)* vol.2 no.2 (1987) p.78-82 ]
[10] B. Deveaud et al [ *J. Appl. Phys. (USA)* vol.59 no.9 (1986) p.3126-30 ]

[11] B. Clerjaud, D. Cote, C. Naud, G. Bremond, G. Guillot, A. Nouailhat [ *J. Cryst. Growth (Netherlands)* vol.83 (1987) p.194 ]

[12] Y. Toudic, B. Lambert, R. Coquille, G. Grandpierre, M. Gaunneau [ *Semicond. Sci. Technol. (UK)* vol.3 no.5 (1988) p.464-8 ]

[13] G. Bremond, G. Guillot, B. Lambert, Y. Toudic [ *Proc. 5th Conf. on Semi-Insulating III-V Materials* Malmo, Sweden, 1988, Eds G. Grossmann, L.A. Ledebo (Adam Hilger, Bristol, 1988) p.319-24 ]

[14] G.W. Iseler [ *Gallium Arsenide and Related Compounds* St. Louis, MO, USA, 24-27 Sept. 1978 (Institute of Physics, London, UK, 1979) p.144-53 ]

[15] G. Bremond, G. Guillot, A. Nouailhat, G. Picoli [ *J. Appl. Phys. (USA)* vol.59 no.6 (1986) p.2038-43 ]

[16] B. Clerjaud, C. Naud, G. Picoli, Y. Toudic [ *J. Phys. C (UK)* vol.17 no.35 (1984) p.6469-76 ]

[17] G.H. Stauss, J.J. Krebs, R.L. Henry [ *Phys. Rev. B (USA)* vol.16 no.2 (1977) p.974-7 ]

[18] B. Lambert, B. Clerjaud, C. Naud, B. Deveaud, G. Picoli, Y. Toudic [ *Proc. 13th Int. Conf. on Defects in Semiconductors* Coronado, CA, USA, 1984 (Metallurg. Soc. AIME, Warrendale, PA, USA, 1985) p.1141-7 ]

[19] K. Huang, B.W. Wessels [ *J. Appl. Phys. (USA)* vol.67 no.11 (1990) p.6882-5 ]

[20] B. Plot-Chan, B. Deveaud, A. Rupert, B. Lambert [ *J. Phys. C (UK)* vol.18 no.29 (1985) p.5651-7 ]

[21] G. Bremond, A. Nouailhat, G. Guillot, B. Cockayne [ *Electron. Lett. (UK)* vol.17 no.1 (1981) p.55-6 ]

[22] D.J. Nicholas, D. Allsopp, B. Hamilton, A.R. Peaker, S.J. Bass [ *J. Cryst. Growth (Netherlands)* vol.68 (1984) p.326 ]

[23] A. Juhl, A. Hoffmann, D. Bimberg, H.J. Schulz [ *Appl. Phys. Lett. (USA)* vol.50 no.18 (1987) p.1292-4 ]

[24] This volume

[25] M.S. Skolnick, R.G. Humphreys, P.R. Tapster, B. Cockayne, W.R. MacEwan [ *J. Phys. C (UK)* vol.16 no.36 (1983) p.7003-18 ]

[26] P. Rojo et al [ *J. Appl. Phys. (USA)* vol.55 no.2 (1984) p.395-401 ]

[27] K. Korona, K. Karpinska, A. Babinski, A.M. Hennel [ *Acta Phys. Pol. A (Poland)* vol.77 no.1 (1990) p.71 ]

[28] K. Korona, A.M. Hennel [ *Appl. Phys. Lett. (USA)* vol.55 no.11 (1989) p.1085-7 ]

[29] B. Clerjaud [ *J. Phys. C (UK)* vol.18 no.19 (1985) p.3615-61 ]

[30] N. Kullendorff, L. Jansson, L.-A. Ledebo [ *J. Appl. Phys. (USA)* vol.54 no.6 (1983) p.3203-12]

# 6.6 Defect energy levels in semi-insulating InP

J.P. Fillard (revised by A.R. Peaker)

June 1998

## A INTRODUCTION

InP is a wide gap (1.35 eV) semiconductor which, when grown as undoped LEC, tends to be n-type. Compensation using shallow acceptors is not easily achieved and so impurities forming deep levels are intentionally introduced to produce semi-insulating material. The situation is complex because the deep level species react with other impurities and intrinsic defects in the material. This results in resistivity striations and significant changes in resistivity with temperature cycling.

Significant improvements have been achieved during recent years in the fabrication of SI LEC materials. They rely on precise control of the growth process itself and the use of co-dopants in conjunction with the primary deep state species.

Semi-insulating substrate materials are often required when InP is used for MISFETs or JFETs in high speed ICs. Fortunately the commonly used defect centres in SI InP do not lead to severe charge retention effects which manifest as side or back gating perturbations, a feature often observed in GaAs [1]. In InP incident light [2] is required to charge the surface traps and induce a side gate effect on the implanted channel of an FET.

Investigation of the various defect levels can be performed by DLTS techniques [3-5] when the bulk resistance is low enough (high temperatures and n-type undoped materials: i.e. deep defects). However truly semi-insulating materials normally require optical excitation to change the initial charge state of the defect (PICTS, optical DLTS or DLOS) although some success has been achieved by the use of TSC.

## B DOPING SPECIES

Direct compensation is obtained by deep acceptors such as Fe, Cd, Zn and Cr which pin the Fermi level at the centre of the bandgap. Among these various atom species, Fe is preferred because it does not give rise to pronounced striations and also because $Fe^{3+}$ occupies the indium site in the lattice so providing a level close to the midgap ($E_c$ - 0.63 eV). Cr has also been considered for the same reasons ($E_c$ - 0.4 eV) and has been widely used in GaAs. However there is some debate about the iron related centres in p-type material. Although a dominant deep hole trap is revealed by both DLTS and Hall measurements in the vicinity of the valence band edge at $E_v$ + 0.2 eV, the spectroscopic analysis unambiguously invalidates previous speculations on the existence of a second energy level of the isolated iron impurity in the bandgap of InP, i.e. an $Fe^{4+}/Fe^{3+}$ donor level. From the axial concentration profile and a comparison with an LEC-grown p-type InP crystal doped with Zn only, it seems that the trap is not even iron-related, in contrast to tentative assignments often found in the literature. Native or Zn-related defects which depend on the particular growth conditions used are assumed to account for this level [6].

Deep levels close to the middle of the bandgap are also observed for Rh and Ru doping, making them interesting candidates for manufacturing semi-insulating InP. The transition-metal-related deep level concentrations achieved so far are somewhat above $10^{15}$ $cm^{-3}$. The electrically active concentration of these transition metals is strongly dependent on their reactivity with hydrogen. In the case of Rh doping the electrically active concentration can be enhanced by a factor of five using nitrogen carrier gas instead of hydrogen [7].

The level of doping is generally rather critical so it has been suggested that a more stable n/p balance can be achieved by a co-doping process; S, Sn, Se, Te, Hg, Ti were considered as donor atoms associated with Cd, Ge, Zn as acceptors [8-11]. This approach has proved to be successful when p-type layers are grown directly on SI material. In contact with p-type layers semi-insulating characteristics are thermally unstable because of pronounced inter-diffusion of Fe and p-type dopants. Co-doping of InP:Fe with Ti is shown to be a reproducible process for the preparation of thermally stable high-resistivity layers. Fe with Ti co-doping can compensate both excess shallow donors and excess shallow acceptors up to concentrations of $8 \times 10^{16}$ and $2 \times 10^{16}$ $cm^{-3}$, respectively. In contrast to InP:Fe, resistivities in excess of $10^7$ $\Omega$ cm are obtained in contact with both symmetric n- and p-type current injecting contacts. Moreover, co-doping of semi-insulating InP:Fe with Ti is found to suppress the inter-diffusion of Fe and p-type dopants. Therefore, the out-diffusion and accumulation of Fe in other regions of complex device structures can be significantly reduced [12].

## REFERENCES

[1] J. Bonnafé, J.P. Fillard, M. Castagné, T. Serviere, P.N.B. Favennec [ *Mater. Res. Soc. Symp. Proc. (USA)* vol.69 (1986) ]

[2] A. Roizes, J.P. David, J.F.F. Faucheux, T. Phampham [ *Mater. Sci. Monogr. (Netherlands)* vol.31 (1985) p.253 ]

[3] G. Brémont, G. Guillot, B. Lambert, Y. Toudic [ *Proc. 5th Conf. on Semi-Insulating III-V Materials* Malmo, Sweden, 1-3 June 1988, Eds. G. Grossmann, L. Lebedo (Adam Hilger, Bristol, UK, 1988) p.319-24 ]

[4] A. Chantre, G. Vincent, D. Bois [ *Phys. Rev. B (USA)* vol.23 no.10 (1981) p.5335-9 ]

[5] H. Lin, G. Sagne, G. Bastide, M. Rouzeyre [ *J. Appl. Phys. (USA)* vol.53 (1982) p.3317 ]

[6] T. Wolf et al [ *J. Appl. Phys. (USA)* vol.73 (1993) p.226-32 ]

[7] A. Dadgar et al [ *J. Cryst. Growth (Netherlands)* vol.170 (1997) p.173-6 ]

[8] Y. Toudic, B. Lambert, R. Coquille, G. Grandpierre, M. Gauneau [ *Semicond. Sci. Technol. (UK)* vol.3 no.5 (1988) p.464-8 ]

[9] Y. Toudic, R. Coquille, R. Gauneau, G. Grandpierre, L. Le Marechal, B. Lambert [ *J. Cryst. Growth (Netherlands)* vol.83 no.2 (1987) p.184-9 ]

[10] R. Fornari, J. Kumar, M. Curti, G. Zuccalli [ *J. Cryst. Growth (Netherlands)* vol.96 no.4 (1989) p.795-801 ]

[11] G. Brémond et al [ *Mater. Sci. Forum (Switzerland)* vol.10-12 pt.2 (1986) p.657-62 ]

[12] D. Bimberg et al [ *J. Cryst. Growth (Netherlands)* vol.145 (1994) p 455-61 ]

# 6.7 Defect energy levels in LPE InP

A.R. Peaker

June 1998

## A INTRODUCTION

The growth of InP from indium solution produces material with a very low concentration of unwanted impurities, essentially because the distribution coefficient for the most common metallic impurities is very small (typically $10^{-6}$). Consequently as-grown LPE InP material is essentially free from the well documented traps associated with 3d transition metals [1,2], a situation similar to the case of LPE GaAs and GaP [3,4]. As a result there are very few reports of deep states in InP grown from the liquid phase and some authors have stated that no traps can be detected in LPE InP [5]. Considering typical sensitivities for deep level measurements it appears to be possible in favourable circumstances, to grow LPE InP with deep state concentrations (both hole and electron traps) less than $10^{11}$ $cm^{-3}$.

## B ELECTRON TRAPS

Rhee and Bhattacharya [6] have detected an electron trap in n-type material grown by slow cooling of the indium melt from 680°C to room temperature (producing an epitaxial layer on (100) indium phosphide substrates). The trap has an activation energy of $E_a$ = 0.38 eV and an apparent (intercept) cross-section of 5 × $10^{15}$ $cm^{-3}$. These authors studied the effect of pre-baking the melt at different temperatures and found that although the background shallow doping concentration and the mobility varied with the pre-bake the trap concentration remained unchanged at NT = 8 × $10^{12}$ $cm^{-3}$. The same authors measured the capture cross-section directly and obtained a value of 2 × $10^{-17}$ $cm^2$ at 200 K increasing with increasing temperature (cross-section activation barrier 0.13 eV).

A deep state with almost identical electron emission characteristics has been reported in LPE InP by Nickel and Kuphal [7] ($E_a$ = 0.36, intercept cross-section 7 × $10^{-15}$ $cm^2$, NT = 2 × $10^{13}$ to $10^{14}$ $cm^{-3}$). These authors have also measured the true capture cross-section and attained a value of 1.4 × $10^{-17}$ invariant with temperature over the range 220 - 270 K. There seems to be no report of a trap with similar properties in VPE, MOVPE or LEC material or in InP doped with transition metals and so it seems probable that the 0.38 eV state is a native defect specific to LPE material.

An electron trap with an activation energy of 0.37 eV has also been observed [8] by MCTS in p-type InP with a concentration of 5 × $10^{15}$ $cm^{-3}$; however, its apparent intercept cross-section was 1.1 × $10^{-17}$ and so it may not be the same as the trap seen in n-type material and described above. Nickel and Kuphal [7] have observed a deeper electron trap ($E_a$ = 0.51 eV, apparent intercept cross-section $10^{-15}$ $cm^2$) at concentrations in the range 2 × $10^{13}$ to 8 × $10^{14}$ $cm^{-3}$, using DLTS. A similar state has also been observed in VPE material.

Forrest and Kim [9] have studied deep states in the vicinity of an isotype (n) heterojunction ($InP/In_{0.53}Ga_{0.47}As$). Using admittance spectroscopy (AS) they observe one defect in the InP with an activation energy of 0.2 eV and an apparent intercept cross-section of 7 × $10^{-16}$ $cm^2$. The concentration was very high ($10^{15}$ $cm^{-3}$) and increased towards the p-type InP substrate on which the LPE layer was grown.

Work has also been done [10] on solute grown bulk material which is produced under very similar conditions to LPE. The following states were observed.

TABLE 1.

| $E_a$ (eV) | Intercept (apparent) cross-section ($cm^2$) | Typical concentration ($cm^{-3}$) | Technique |
|---|---|---|---|
| 0.18 | $1.6 \times 10^{-19}$ | $\sim 10^{12}$ | DLTS |
| 0.24 | $4.4 \times 10^{-16}$ | $\sim 10^{13}$ | DLTS & AS |
| 0.24 | $1.7 \times 10^{-13}$ | $2 \times 10^{12}$ | DLTS |
| 0.76 | $3.2 \times 10^{-12}$ | $8 \times 10^{12}$ | DLTS |

## C HOLE TRAPS

Several investigators have searched specifically for hole traps and found none [7,8]. However, there are two reports, one by Rhee and Bhattacharya [6] who using DLTS have observed a low concentration of a hole trap with an activation energy of 0.22 eV and an apparent (intercept) cross-section of $1.5 \times 10^{-23}$ $cm^2$. The other observation is by Takeda et al [11] who report a state with an activation energy of 0.12 eV with an apparent (intercept) cross-section of $9 \times 10^{-18}$ $cm^2$. This was found to be present in quite high concentrations ($6 \times 10^{13}$ $cm^{-3}$) near an isotype heterojunction (p) of $InP/In_{0.52}Ga_{0.48}As$ (lattice matched).

## D PHOTOCAPACITANCE

Chiao and Antypas [12] have used photocapacitance techniques to examine deep states in LPE material and compared it with LEC and VPE grown layers. Two states are reported at 0.78 eV and 1.15 eV below the conduction band, both present in concentrations of the order of $10^{14}$ $cm^{-3}$. Both states were also observed in VPE and LEC material.

## REFERENCES


[1] B. Clerjaud [ in *Landolt-Bernstein New Series III/22b* (Springer-Verlag, Berlin, 1990) section 5.5.5 p.685 ]

[2] A.M. Hennel, A.R. Peaker [ Datareview in this book: *6.5 Deep states in InP associated with the 3d transition metals* ]

[3] A.R. Peaker [ in *Properties of Gallium Arsenide, 3rd Edition* (INSPEC, IEE, London, UK, 1997) p.358 ]

[4] R.F. Brunwin, B. Hamilton, J. Hodgkinson, A.R. Peaker, P.J. Dean [ *Solid-State Electron. (UK)* vol.24 no.3 (1981) p.249-56 ]

[5] O. Wada, A. Majerfeld, A.N.M.M. Choudhury [ *J. Appl. Phys. (USA)* vol.51 no.1 (1980) p.423-32 ]

[6] J.K. Rhee, P.K. Bhattacharya [ *J. Electrochem. Soc. (USA)* vol.130 no.3 (1983) p.700-3 ]

[7] H. Nickel, E. Kuphal [ *Phys. Status Solidi A (Germany)* vol.65 (1981) p.583-8 ]

[8] P.K. Bhattacharya, J.W. Ku, S.J.T. Owen, S.H. Chiao, R. Yeats [ *Electron. Lett. (UK)* vol.51 no.23 (1979) p.753-5 ]

[9] S.R. Forrest, O.K. Kim [ *J. Appl. Phys. (USA)* vol.53 no.8 (1982) p.5738-45 ]

[10] K. Tsubaki, E. Kubota, K. Sugiyama [ *Electron. Lett. (UK)* vol.15 no.17 (1979) p.513-5 ]

[11] Y. Takeda, S. Takigawa, M. Kasu, A. Sasaki [ *Inst. Phys. Conf. Ser. (UK)* no.79 (1986) p.301-6 ]

[12] S.H. Chiao, G.A. Antypas [ *J. Appl. Phys. (USA)* vol.49 no.1 (1978) p.466-8 ]

# 6.8 Defect energy levels in VPE and OMVPE InP

F.R. Bacher (revised by A.R. Peaker)

June 1998

## A INTRODUCTION

This Datareview summarises the traps found in InP grown by vapour-phase epitaxy (VPE), including organometallic vapour-phase epitaxy (OMVPE). Omitted by intent are references to irradiated or ion-implanted material. In these studies, the InP epilayers were grown on InP substrates. The data are derived from deep level transient spectroscopy (DLTS). (Photoluminescence spectra are summarised elsewhere in this volume.)

The most rigorous studies on VPE-grown material date back to the late 1970s. OMVPE is a newer technique, and studies on material grown by this method have added some new defect energy levels, but there is little reason to believe that these two growth techniques produce structurally different material. Most of the defects are unidentified or ambiguously identified, so no attempt is made here to explain them.

## B ELECTRON TRAPS

The electron traps found in undoped n-type InP are listed in TABLE 1. The energy levels are grouped together to form thirteen 'distinct' levels. This was done because of the typical ±20 meV uncertainty in the DLTS measurement, suggesting that the grouped levels may be the same or related, but this grouping should not be taken as fact - it was merely intended to make the information more ordered. The levels between 560 and 600 meV are generally ascribed to iron, and called E1. This is almost certainly the $Fe^{2+}$ to $Fe^{3+}$ transition. Other levels are observed to be dependent on In-rich (the 440 meV level) or P-rich (the 800 meV level, see [14]) conditions. A detailed study of the effect of growth conditions and the adduct purity in MOCVD growth has shown that pre-synthesised adducts can be a significant source of impurity related deep level defects [10]. The concentration of these deep states has been shown to relate directly to the photo-luminescent efficiency [11].

## C HOLE TRAPS

There is less information available for hole traps. Reference [15] contains a useful summary, although little specific to vapour-grown material. TABLE 2 shows the various levels observed, with those near 200, 290, and 400 meV being the most common. No definitive structural information is available but most are presumed to be an impurity/native defect complex (e.g. the 400 meV level probably involves a P vacancy/dopant interaction). It should be noted that 3d transition metals have been detected by chemical techniques in some vapour phase material, and deep states with emission characteristics similar to those reported in TABLE 2 are associated with transition metals [23].

TABLE 1 Electron traps found in InP by DLTS.

| $E_a$ (meV below conduction band) | Growth method | Ref |
|---|---|---|
| 120 | OMVPE | [10] |
| 130 | VPE | [4[a]] |
| 140 | VPE | [17] |
| 150 | VPE | [18] |
| 160 | OMVPE | [10] |
| 170 | VPE | [4[a]] |
| 220 | OMVPE | [10] |
| 230 | OMVPE | [13] |
| 240 | OMVPE | [10,21] |
| 270 | OMVPE | [10] |
| 290 | OMVPE | [13] |
| 300 | OMVPE | [10] |
| 350 | VPE | [4[a]] |
| 370 | VPE | [7[a]] |
| 380 | OMVPE | [13] |
| 400 | VPE | [20] |
| 420 | VPE | [20] |
| 430 | VPE | [4[a],9,18] |
| 440 | OMVPE | [21] |
| 450 | OMVPE | [3[a],11[a],10] |
| 510 | VPE | [4[a],17] |
| 520 | VPE | [20] |
| 530 | OMVPE | [12] |
| 550 | OMVPE | [18] |
| 560 | OMVPE | [10] |
| 580 | OMVPE | [3[a],11[a],20] |
| 600 | OMVPE and VPE | [7[a],8,12] |
| 610 | OMVPE | [13] |
| 620 | VPE | [4[a]] |
| 630 | OMVPE and VPE | [3[a],9,18] |
| 660 | OMVPE | [10] |
| 680 | VPE | [20] |
| 800 | OMVPE and VPE | [5,19[a],21] |

TABLE 2 Hole traps found in InP by DLTS.

| $E_a$ (meV above valence band) | Growth method | Doping | Ref |
|---|---|---|---|
| 530 | VPE | N | [16] |
| 500 | VPE | Zn | [6] |
| 440 | VPE | N | [16] |
| 430 | VPE | N | [16] |
| 410 | VPE | Zn | [6] |
| 400 | OMVPE | Mg | [1] |
| 290 | VPE | N, Zn | [6,16] |
| 270 | VPE | n-type | [2] |
| 250 | OMVPE | Mn | [22] |
| 220 | VPE | Zn | [6] |
| 200 | OMVPE | Mg | [1] |
| 190 | VPE | Zn | [4[a]] |
| 120 | VPE | Zn | [4[a]] |
| 90 | VPE | Zn | [6] |

For both tables:
[a]Details of $E_a$ calculation not given in reference and/or no apparent capture cross-section data.

**REFERENCES**


[1] F.R. Bacher, H. Cholan, W.B. Leigh [ *Mater. Res. Soc. Symp. Proc. (USA)* vol.90 (1987) p.5-12 ]

[2] A.N.M.M. Choudhury, P.N. Robson [ *Electron. Lett. (UK)* vol.15 (1979) p.247 ]

[3] N.D. Gerrard, D.J. Nicholas, J.O. Williams, A.C. Jones [ *Chemtronics (UK)* vol.3 no.1 (1988) p.17-30 ]

[4] B.H. Hu, B.L. Zhou, Z.X. Chen [ *J. Lumin. (Netherlands)* vol.40/41 (1988) p.371-2 ]

[5] M. Inuishi, B.W. Wessels [ *Thin Solid Films (Switzerland)* vol.103 no.1-2 (1983) p.141 ]

[6] M. Inuishi, B.W. Wessels [ *Electron. Lett. (UK)* vol.17 (1981) p.685 ]

[7] T.I. Kol'chenko, S.E. Moroz [ *Sov. Phys.-Semicond. (USA)* vol.23 no.3 (1989) p.340-1 ]

[8] A.T. Macrander, W.D. Johnston [ *J. Appl. Phys. (USA)* vol.54 no.2 (1983) p.806-13 ]

[9] A. Majerfeld, O. Wada, A.N.M.M. Choudhury [ *Appl. Phys. Lett. (USA)* vol.33 no.11 (1978) p.957-9 ]

[10] D.J. Nicholas, D. Allsopp, B. Hamilton, A.R. Peaker, S.J. Bass [ *J. Cryst. Growth (Netherlands)* vol.68 (1984) p.326 ]

[11] D.J. Nicholas et al [ *Inst. Phys. Conf. Ser. (UK)* no.91 (1988) p.295-8 ]

[12] M. Ogura, M. Mizuta, N. Hase, K. Kukimoto [ *Jpn. J. Appl. Phys. (Japan)* vol.22 no.4 (1983) p.658-62 ]

[13] M.A.A. Pudensi, K. Mohammed, J.L. Merz, D. Kasemset, K.L. Hess [ *J. Appl. Phys. (USA)* vol.57 no.8 pt.1 (1985) p.2788-92 ]

[14] L.H. Robins, P.C. Taylor, T.A. Kennedy [ *Phys. Rev. B (USA)* vol.38 no.18 (1988) p.13227-36 ]

[15] A. Singh, W.A. Anderson, Y.S. Lee, K. Jiao [ *Proc. SPIE - Int. Soc. Opt. Eng. (USA)* vol.1144 (1989) p.61-8 ]

[16] S.W. Sun, B.W. Wessels [ *J. Appl. Phys. (USA)* vol.57 no.10 (1985) p.4616-8 ]

[17] S.W. Sun, A.P. Constant, C.D. Adams, B.W. Wessels [ *J. Cryst. Growth (Netherlands)* vol.64 (1983) p.149 ]

[18] O. Wada, A. Majerfeld, A.N.M.M. Choudhury [ *J. Appl. Phys. (USA)* vol.51 no.1 (1980) p.423-32 ]

[19] B.W. Wessels, M. Inuishi [ *Proc. SPIE - Int. Soc. Opt. Eng. (USA)* vol.323 (1982) p.55 ]

[20] A.M. White, A.J. Grant, B. Day [ *Electron. Lett. (UK)* vol.14 (1978) p.409 ]

[21] N. Yamamoto, K. Uwai, K. Takahei [ *J. Appl. Phys. (USA)* vol.65 no.8 (1989) p.3072-5 ]

[22] K. Huang, B.W. Wessels [ *J. Appl. Phys. (USA)* vol.67 no.11 (1990) p.6882-5 ]

[23] A.M. Hennel, A.R. Peaker [ Datareview in this book: *6.5 Deep states in InP associated with the 3d transition metals* ]

# 6.9 Defect energy levels in MBE InP

F.R. Bacher

This Datareview appeared in *Properties of Indium Phosphide* (INSPEC, IEE, London, UK, 1991) and is reproduced here for the reader's convenience.

## A INTRODUCTION

Deep levels in InP grown by molecular beam epitaxy (MBE) have not been the subject of many studies. Data from three surveys of bulk electron traps are presented here; no hole trap studies are known to the author. Deep levels in MBE-grown InP are not unlike those found in vapour-grown methods, the latter methods being more fully characterised. The elemental sources typically used for MBE introduce more impurities (and defect levels) into InP epitaxial layers than do vapour sources [1].

## B ELECTRON TRAPS

The often-cited work of Asahi et al [2] gave three electron traps at -0.24, -0.32, and -0.44 eV. (All levels are relative to the conduction band. Arrhenius plots and capture cross-section data can be found in the references.) The level at -0.32 eV has not been reported in epitaxial InP grown by other techniques. The -0.44 eV level most likely corresponds to the -0.43 level found by Iliadis et al [3-5] and may relate to P vacancies. Other traps found were at -0.53 and -0.57 eV. The -0.57 level is probably the same one observed by Park and Stanley [6] at -0.59 eV. The presence of these levels in MBE-grown InP appears to be influenced by growth temperature as well as source impurities; the deeper levels (-0.53 to -0.59 eV) probably are a crystal defect (In vacancy or P interstitial).

## REFERENCES


[1] D.A. Andrews, S.T. Davey, C.G. Tuppen, B. Wakefield, G.J. Davies [ *Appl. Phys. Lett. (USA)* vol.52 no.10 (1988) p.816-8 ]

[2] H. Asahi, Y. Kawamura, M. Ikeda, H. Okamoto [ *J. Appl. Phys. (USA)* vol.52 no.4 (1981) p.2852-9 ]

[3] A.A. Iliadis, S.C. Laih, E.A. Martin, D.E. Ioannou [ *J. Appl. Phys. (USA)* vol.65 no.12 (1989) p.4805-8 ]

[4] A.A. Iliadis, S.C. Laih, E.A. Martin [ *Appl. Phys. Lett. (USA)* vol.54 no.15 (1989) p.1436 ]

[5] A. Iliadis, S.C. Laih, D.E. Ioannou, E.A. Martin [ *Inst. Phys. Conf. Ser. (UK)* no.91 (1988) p.219-22 ]

[6] R.M. Park, C.R. Stanley [ *Electron. Lett. (UK)* vol.17 (1981) p.669 ]

# 6.10 Defect energy levels in irradiated or implanted InP

A. Sibille and A.R. Peaker

June 1998

## A INTRODUCTION

Ion implantation is often used in the fabrication of InP devices, but unfortunately the process produces unwanted defects. Some of these can be removed (or modified) by annealing. Often the results of such processes are complex secondary defects which are detectable by sensitive techniques such as DLTS but do not exist in adequate concentrations to use methods which give true structural information. Deep level defects have also been observed as a result of plasma etching. These process induced defects arise from the bombardment of the surface of the indium phosphide with low energy ions.

Electron irradiation is a well known technique for creating, in a controllable way, intrinsic defects (vacancies or interstitials) in semiconductors, which can then react with impurities and other lattice defects. These defects provide a knowledge base on which interpretation of grown-in or process-induced defects depends. We review here the available data obtained by deep level transient spectroscopy in InP.

Sections B and C address the case of electron damage in n- and p-InP respectively and provide a compilation of irradiation induced levels. This compilation includes (a) the activation energy of the DLTS peak, (b) the various labels used in the publications, (c) an order of magnitude of the introduction rates (ratio of trap concentration to electron dose, expressed in $cm^{-1}$), (d) the typical formation or annealing temperature if any, and (e) a partial list of publications mentioning each deep level. Section D gives a brief overview of the present state of understanding of the formation and annealing reactions of these deep levels. Section E summarises the much less comprehensive literature on process induced defects resulting from ion implantation and proton damage.

## B ELECTRON IRRADIATION INDUCED ELECTRON TRAPS

The main electron traps produced by electron irradiation (typically 1 MeV) are presented in TABLE 1. The following points are worthy of note:

(a) Several deep levels have been detected by various authors, but the introduction rates are strongly sample dependent. This very clearly points to an involvement of residual impurities and/or native defects in the microscopic structure of these irradiation induced deep levels. Such an involvement has besides been directly demonstrated in the case of the dominant electron trap no.7 of TABLE 1 [10].

(b) The total introduction rate of electron traps is generally much smaller than that expected for the primary Frenkel pairs on the phosphorus and indium sub-lattices, which is of the order of 10 $cm^{-1}$.

TABLE 1 Characteristics of the main electron traps in n-InP after irradiation. $T_a$ is the temperature of an annealing event and $T_r$ of a reaction event without disappearance of the centre.

| No. | Activation energy (eV) | Trap label and reference of the corresponding publication | | | | | | | Introduction rate ($cm^{-1}$) | Reaction temperature |
|---|---|---|---|---|---|---|---|---|---|---|
| | | [1] | [2] | [3] | [4,5] | [6] | [7,8] | [9] | | |
| 1 | 0.15 | E1 | | | E4 or EA | IE1 | E4 | 2 or E1 | $10^{-5}$ - $10^{-2}$ | |
| 2 | 0.20 | E2 | E6 | | E6 | IE2 | | E2 | $10^{-5}$ - $10^{-3}$ | |
| 3 | 0.38 | E3 | E7 | | E7 | IE3 | E7 | E3 | $10^{-5}$ - $10^{-3}$ | |
| 4 | 0.37 | | E8 | 2 | | | | E4 | $10^{-2}$ - $10^{-1}$ | |
| 5 | 0.7 - 0.9 | | | | E8 | | E8 | | | $T_a \sim 100°C$ |
| 6 | 0.65 | | | 5 | E9 | | E9 | E5 | $10^{-4}$ - $10^{-1}$ | $T_r \sim 150°C$ |
| 7 | 0.75 | | E11 | 6 | E10 | IE7 | 0.79 eV trap | E6 | $10^{-3}$ - $10^{-1}$ | $T_a \sim 100°C$ |

## C ELECTRON IRRADIATION INDUCED HOLE TRAPS

The main hole traps produced by electron irradiation are presented in TABLE 2.

TABLE 2 Main electron irradiation induced hole traps, detected essentially in InP:Zn, but also as minority traps in n-InP [3]. $T_a$ is an annealing temperature, and $T_f$ is a formation temperature.

| No. | Activation energy (eV) | Trap label and reference of the corresponding publication | | | Introduction rate ($cm^{-1}$) | Reaction temperature |
|---|---|---|---|---|---|---|
| | | [11] | [12] | [13,14] | | |
| 1 | 0.19 | H2 | | HN2 | ~0.02 | $T_a \sim 100°C$ |
| 2 | 0.32 | H3 | | HN3 | ~0.3 | $T_a \sim 100°C$ |
| 3 | 0.37 | H4 | H2 | H4, HN4 | 0.5 - 2 | $T_a \sim 100°C$ |
| 4 | 0.53 | H5 | H3 | H5 | ~0.2 | $T_f \sim 100°C$ |
| 5 | 0.59 | H7 | H4 | | ~0.1 | |

From TABLE 2 and the many publications on the subject, it appears clearly that the main hole traps (levels 2 - 4) are quasi-systematic in p-InP and the total introduction rate is not too far from that of the primary Frenkel pairs, particularly if one considers only the phosphorus sublattice (see below).

## D DEFECT REACTIONS

### D1 Origin of the Irradiation Induced Deep Levels

As explained in sections B and C, n- and p-type InP behave quite differently under electron irradiation.

(a) In n-InP, the concentration of trap levels is strongly dependent on the purity of the material, and much smaller than that of primary Frenkel pairs. These pairs are thus unstable at 300 K (and even below) and diffuse rapidly. When impurity trapping of vacancies or interstitials is efficient, because of (i) a large impurity concentration or (ii) a large primary defect capture cross-section, a significant number of impurity related levels stable at 300 K are formed. The vacancies and interstitials which were not trapped disappear by recombination after long distance diffusion. This picture is consistent with small carrier removal rates.

(b) In p-InP, the usual dopants employed (Zn, Cd) appear to be very efficient sinks for vacancies or interstitials, possibly due to Coulomb interaction. Consequently almost all of the primary defects get complexed with shallow acceptors following diffusion [15]. The various levels observed correspond to different microscopic configurations of these complexes. In contrast to n-InP, p-InP thus has much larger carrier removal rates. Specific contributions to hole removal by defect levels H3, H4, and H5 were determined from combined deep-level transient spectroscopy and measured carrier concentrations. An additional contribution was attributed to one or more defects not observed by the present DLTS measurements. The high trapping rate observed for H5 suggests that this defect, if present in relatively high concentration, could be dominant in p-type InP [16].

It is obviously important to relate the various traps to the sub-lattice involved in the formation of the primary defects. From the few published studies, it is clear that most of the damage occurs on the phosphorus sub-lattice [17-19]. The reason for this is probably that Frenkel pairs on the indium sub-lattice are unstable and dominantly recombine even well below room temperature. The small fraction of pairs which spontaneously separate are subsequently trapped by impurities or native defects and may be at the origin of some of the traps produced by electrons of energy sufficient to eject the heavy indium atom [17-19]. However, further studies are necessary to clarify these issues.

### D2 Stability and Reactions of the Irradiation Induced Deep Centres

Since defects produced by irradiation are far from thermodynamic equilibrium we can expect them to exhibit irreversible changes by annealing or electronic stimulation. Apart from a few defects which disappear around 100 - 200 K, the main annealing step is at ~100°C and results in the quasi-disappearance of most of the levels [2,15]. The detailed reaction scheme is complex and several irreversible transformations of defects have been observed in this temperature range [15,20]. Also all the reactions are dramatically enhanced by minority carrier injection and may well occur well below 300 K [21]. Finally these reactions are sensitive to the electric field, which brings some information on the nature of the electronic states [22]. This work has been carried further in terms of predicting the radiation hardness of InP solar cells by measurements after 1 MeV electron irradiation and 3 MeV proton damage. The radiation response of the cells was characterised by a series of quantum efficiency (QE) and deep level transient spectroscopy measurements. From the QE measurements, the base minority carrier diffusion length was determined as a function of particle fluence, and a diffusion length damage coefficient calculated. From the C-V measurements, the radiation-induced carrier removal rate was determined. The DLTS data showed the electron and proton irradiations to produce essentially the same defect spectra [23].

## E ION IMPLANTATION AND PROTON DAMAGE

Ion implantation and plasma etching are widely used in the fabrication of InP devices. However in contrast to the case for silicon and gallium arsenide there are few papers presenting a systematic study of defects resulting from these processes. We present below data on point defects: work on extended defects resulting from ion implantation is dealt with in Datareview 6.1 in this book.

Si implantation is classically used to make an n-type channel MESFET in an SI substrate. In general the resulting damage is dominated by a specific level at 0.68 eV [24,25]. It is possible to anneal out this defect but a shallow trap at 0.18 eV tends to remain. For the case of p-type implants magnesium has been studied in some detail. Five deep levels located in the upper half of the bandgap were detected in the junctions by DLTS measurements, three of which (at 0.6, 0.45 and 0.425 eV) were found to appear due to rapid thermal annealing. The origin of the other two levels, at 0.31 and 0.285 eV, was ascribed to implantation damage. Admittance spectroscopy measurements showed the presence of three levels at 0.44, 0.415 and 0.30 eV, in agreement with those found by DLTS. The DLTS measurements showed that the concentration of deep levels decreased after longer annealing times, and that the concentration of deep levels due to the implantation increased after additional P or Si implantations [26,27].

The effects of oxygen ion implantation on defect levels and the electrical properties of undoped InP (n-type) and Sn-doped InP have been investigated as a function of post-implant annealing at temperatures of 300 and 400°C [28]. Results of samples implanted at room temperature show a drastic change in the defect structure, with the disappearance of peaks originally present in un-implanted control samples. DLTS studies have also been conducted on hot-implanted n-InP and on samples that have been subjected to rapid thermal annealing (RTA) at 750°C. DLTS results from RTA samples show that there is a certain degree of lattice recovery as shown by the return of electron trapping [29].

Proton-implantation effects on InP solar cells fabricated from LEC (liquid encapsulated Czochralski) grown single crystals have been studied by Yamaguchi et al using DLTS [30]. In the case of high-energy proton irradiation (above 7 MeV), the major defect centres in InP are thought to be due to phosphorus-related defects. In the case of low-energy proton irradiation (below 5 MeV), the origins of the major defect centres that act as recombination centres are speculated to be due to complex defects such as vacancy-impurity complexes and multi-vacancies.

Walters et al [23,31] have conducted a deep level transient spectroscopy study of proton irradiation induced defects in $n^+p$ InP mesa diodes grown by metal-organic chemical vapour deposition. In contrast to results reported for InP grown by other methods, 3 MeV proton irradiation produced a DLTS spectrum similar to the one produced by 1 MeV electron irradiation with the addition of two new peaks. Six majority carrier peaks, HP1 ($E_a$ = 0.15 eV), H2 ($E_a$ = 0.20 eV), H3 ($E_a$ = 0.30 eV), H4 ($E_a$ = 0.37 eV), H5 ($E_a$ = 0.54 eV) and H7 ($E_a$ = 0.16 eV), and three minority carrier peaks, EA ($E_a$ = 0.26 eV), EB ($E_a$ = 0.74 eV) and EC ($E_a$ = 0.16 eV), were detected.

## F CONCLUSION

The nature and the concentration of deep levels created by irradiation in InP are strongly dependent on the impurity content of the sample. This is particularly marked in (lightly doped) n-InP where the production rates are generally small. On the other hand shallow acceptors appear to have a large capture radius for vacancies or interstitials of the phosphorus sublattice and the deep level production rates are high. The situation with ion implantation is quite complicated and much work needs to be done before generalisations can be made.

## REFERENCES


[1] J. Suski, J.C. Bourgoin, H. Lim [ *J. Appl. Phys. (USA)* vol.54 no.5 (1983) p.2852-4 ]

[2] J. Suski, A. Sibille, J.C. Bourgoin [ *Solid State Commun. (USA)* vol.49 no.9 (1984) p.875 ]

[3] P.R. Tapster, P.J. Dean, M.S. Skolnick [ *J. Phys. C (UK)* vol.15 no.28 (1982) p.L1007-12 ]

[4] M. Levinson, J.L. Benton, H. Temkin, L.C. Kimerling [ *Appl. Phys. Lett. (USA)* vol.40 no.11 (1982) p.990-2 ]

[5] J.L. Benton, M. Levinson, A.T. Macrander, H. Temkin, L.C. Kimerling [ *Appl. Phys. Lett. (USA)* vol.45 no.5 (1984) p.566-8 ]

[6] S.W. Sun, A.P. Constant, C.D. Adams, B.W. Wessels [ *Proc. 2nd NATO Workshop on Materials Aspects of InP* Lancaster, 1983 ]

[7] K. Ando, M. Yamaguchi, C. Uemura [ *Jpn. J. Appl. Phys. 1 (Japan)* vol.23 no.10 (1984) p.1406-7 ]

[8] J. Koyama, J. Shirafuji, Y. Inuishi [ *Electron. Lett. (UK)* vol.19 no.16 (1983) p.609-11 ]

[9] T.I. Kolochenko, V.M. Lomako, S.E. Moroz [ *Sov. Phys.-Semicond. (USA)* vol.21 no.6 (1987) p.656-7 ]

[10] A. Sibille [ *Appl. Phys. Lett. (USA)* vol.48 no.9 (1986) p.593-5 ]

[11] A. Sibille, J.C. Bourgoin [ *Appl. Phys. Lett. (USA)* vol.41 no.10 (1982) p.956-8 ]; see also A. Sibille [ Thesis, Universite Paris VII, 1985 ]

[12] M. Levinson, H. Temkin, W.A. Bonner [ *J. Electron. Mater. (USA)* vol.12 no.2 (1983) p.423-32 ]

[13] K. Ando, M. Yamaguchi, C. Uemura [ *J. Appl. Phys. (USA)* vol.55 no.12 (1984) p.4444-6 ]

[14] M. Yamaguchi, K. Ando [ *J. Appl. Phys. (USA)* vol.60 no.3 (1986) p.935-40 ]

[15] A. Sibille, J. Suski, M. Gilleron [ *J. Appl. Phys. (USA)* vol.60 no.2 (1986) p.595 ]

[16] I. Weinberg, C.K. Swartz, P.J. Drevinsky [ *J. Appl. Phys. (USA)* vol.72 no.11 (1992) p.5509-11 ]

[17] A. Sibille, J. Suski, G. Le Roux [ *Phys. Rev. B (USA)* vol.30 no.2 (1984) p.1119-21 ]

[18] H.J. Von Bardeleben [ *Solid State Commun. (USA)* vol.57 no.2 (1986) p.137-40 ]

[19] B. Massarani, J.C. Bourgoin [ *Phys. Rev. B (USA)* vol.34 no.4 (1986) p.2470-4 ]

[20] T. Bretagnon, G. Bastide, M. Rouzeyre [ *Phys. Rev. B (USA)* vol.41 no.2 (1990) p.1028 ]
[21] A. Sibille [ *Phys. Rev. B (USA)* vol.35 no.8 (1987) p.3929-36 ]
[22] A. Sibille [ *Phys. Rev. Lett. (USA)* vol.56 no.5 (1986) p.476-9 ]
[23] R.J. Walters et al [ *J. Appl. Phys. (USA)* vol.82 no.5 (1997) p.2164-75 ]
[24] J.K. Rhee, P.K. Bhattacharya, S.J.T. Owen, R.Y. Koyama [ *Proc. Conf. on Semi-Insulating III-V Materials* Evian, France, 1982 (Shiva, Nantwich, England, 1982) ]
[25] J.K. Rhee, P.K. Bhattacharya [ *J. Appl. Phys. (USA)* vol.53 no.6 (1982) p.4247 ]
[26] J.M. Martin, S. Garcia, I. Martil, G. Gonzalezdiaz, E. Castan, S. Duenas [ *J. Appl. Phys. (USA)* vol.78 no.9 (1995) p.5325-30 ]
[27] L. Quintanilla et al [ *J. Appl. Phys. (USA)* vol.81 no.7 (1997) p.3143-50 ]
[28] C.C. Tin, P.A. Barnes, T.T. Bardin, J.G. Pronko [ *Nucl. Instrum. Methods Phys. Res. B (Netherlands)* vol.59 (1991) p.623-6 ]
[29] L. He, W.A. Anderson [ *J. Electron. Mater. (USA)* vol.21 no.10 (1992) p.937-45 ]
[30] M. Yamaguchi, K. Ando, H. Okazaki, T. Takamoto, M. Ohmori [ *Sol. Energy Mater. Sol. Cells (Netherlands)* vol.30 no.3 (1993) p.245-56 ]
[31] R.J. Walters, G.P. Summers [ *J. Appl. Phys. (USA)* vol.69 no.9 (1991) p.6488-94 ]

CHAPTER 7

# PROCESSING TECHNOLOGIES

# 7.1 Wet, photoelectrochemical and laser-assisted etching of InP

S. Adachi

July 1999

## A INTRODUCTION

Wet chemical etching processes have three general applications: (1) defect or damage revealing, (2) polishing and (3) pattern formation or device shaping. Certain chemical etchants provide the effect of producing surface features on the semiconductor which can be related to the defects in the crystal. This is essentially true of dislocations, which often form etch pits at their points of intersection with a surface, thus providing a simple way of estimating dislocation densities. Chemical etching is also used for removing the damage layer of material close to the surface, for producing a high-quality shiny flat surface, and for shaping semiconductor devices, as well as for characterising structural and compositional features (staining and p-n junction delineation). Many etching characteristics of InP have been studied by a number of groups. A useful description of the theoretical and practical aspects of etching is given by Sangwal [1] and Notten et al [2].

## B WET CHEMICAL ETCHING

### B1 Defect Etchants

Wet chemical etching is by far the most convenient way to reveal structural defects in semiconductors. There are many studies on etching characteristics of such crystallographic defects in InP [3-24]. TABLE 1 is the summary of these studies (etchant, crystal plane and etching condition).

Since chemical polishing did not usually give a mirror-like finish on the (111)In surface, most defect etch studies were performed on the (111)P surface. Thiel et al [8] used Nalcoag 1060 as the polishing etchant for {111}InP wafers. (This chemical resulted in an excellent surface finish on both (111)In and (111)P faces.) By etching with a modified AB solution (10 ml $H_2O$:40 mg $AgNO_3$:5 g $CrO_3$:8 ml HF), they then obtained triangular pits on the (111)In surface. Tuck and Baker [5] obtained hexagonal etch pits on the (111)In surface with 1% solution of bromine in methanol. On the (111)P surface, the etch pits have the morphology of a pyramidal (triangular) [5,7,11,15,20,24] or a conical (circular) shape [6,8,10,11].

On the (100) surface, the etch pits have a rectangular [5,7,15-17,20,21,24], oval [6,10,20] or square (or circular) shape [5-7,10,12,16,18,19,21,24]. The rectangular pits demonstrate the non-equivalence of $\langle 110 \rangle$ and $\langle \bar{1}10 \rangle$ type directions lying in a (100) plane in the III-V intermetallic compounds. This non-equivalence is known to arise from the $\langle 111 \rangle$ crystallographic polarity of the zincblende type crystal. The rectangular etch pits, therefore, provide a means for the unique determination of crystallographic directions between the $\langle 110 \rangle$ and $\langle \bar{1}10 \rangle$ directions [17,20,24].

Etched (110)InP surfaces revealed ridge features but some have hillocks and boat-shaped pits associated with them [6,10]. The boat-like etch pits were formed on the (221)InP surface [24]. A computer algorithm for automatic etch-pit density counting with an optical microscope has been presented [25].

TABLE 1 Summary of dislocation etchants for InP.

| Etchant | Plane | Comments | Ref |
|---|---|---|---|
| 0.4N $Fe^{3+}$:HCl | {111} | 25°C, 1.5 min | [3] |
| $20HNO_3$:10HCl:0.25Br | {111}, (100) | <5 s | [4] |
| 1HCl:$1HNO_3$ | {111}, (100) | 26°C | [5] |
| HCl | | | [5] |
| 0.4N $Fe^{3+}$:HCl | | | [5] |
| 1% BrMe | | | [5] |
| 1HCl:$2H_2SO_4$:$3HNO_3$ | | | [5] |
| $2H_3PO_4$:1HBr | (111)P, (100), (110) | 24°C, 2 min | [6] |
| HBr:HF | (111)P, (100) | | [7] |
| HBr:$CH_3COOH$ | | | [7] |
| 6HCl:$6H_2O$:$1HNO_3$ | {111}, (100), (110) | 60°C, 1.5 min | [8] |
| modified AB | | 75°C, 30 min | [8] |
| 40HCl:$80HNO_3$:1Br | | 25°C, 10 s | [8] |
| $2H_3PO_4$:1HBr | | 25°C, 2 min | [8] |
| $H_2SO_4$:$H_2O_2$:$H_2O$ | (111)P, (100) | RT | [9] |
| $2H_3PO_4$:1HBr | (111)P, (100), (110) | RT, 1 - 2 min | [10] |
| AB etchant | | 60°C, 30 - 90 min | [10] |
| AB etchant | (111)P, (100) | 60°C, 15 min | [11] |
| 1HBr:5HF | | RT, 2 min | [11] |
| $2H_3PO_4$:1HBr | (100) | ambient, up to 2 min | [12] |
| $6HNO_3$:$6H_2O$:1HCl | (100) | 60°C, 1.5 min | [12] |
| 40HCl:$80HNO_3$:1Br | (100) | 25°C, 10 s | [12] |
| modified AB | (100) | 75°C, 30 min | [12] |
| PEC1 | (100) | | [13] |
| PEC2 | (111)P, (100) | | [14] |
| $1HNO_3$:3HBr | (111)P, (100) | RT, 10 - 30 s | [15] |
| HBr:$CH_3COOH$:$K_2Cr_2O_7$ | (100) | RT | [16] |
| HBr:$H_3PO_4$:$K_2Cr_2O_7$ | | RT | [16] |
| $1H_2SO_4$:$1H_2O_2$:$1H_2O$ | (100) | 80 - 90°C | [17] |
| $1HNO_3$:3HBr | (100) | 10 s | [18] |
| $CrO_3$:HF:$H_2O$ | (100) | PE | [19] |
| 20HBr:$2H_2O_2$:20HCl:$20H_2O$ | (111)P, (100) | RT, 2 min | [20] |
| nHBr:$2H_2O_2$:20HCl:$20H_2O$ (n = 20, 40, 60 and 80) | (100) | RT, 2 min | [21] |
| 1HBr:$2H_3PO_4$ | (100) | $\alpha$ and $\beta$ dislocations | [22] |
| 0.3 - 10% BrMe | (100) | 25°C, 2 min | [23] |
| HBr:$K_2Cr_2O_7$:$H_2O$ | (221), (111)In, (111)P, (100) | RT | [24] |

modified AB (10 ml $H_2O$:40 mg $AgNO_3$:5 g $CrO_3$:8 ml HF), AB etchant (2 ml $H_2O$:8 mg $AgNO_3$: 1 g $CrO_3$:1 ml HF), PEC1 (photoelectrochemical etching in a 14M orthophosphoric acid), PEC2 (photoelectrochemical etching in a 1 mol/litre NaOH solution), 1% BrMe (1% bromine in methanol solution), RT (room temperature), PE (photoetching)

## B2 Polishing Etchants

Chemical polishing of semiconductors plays an essential role in semiconductor sciences and device technologies. Important factors determining the choice of an etchant are, generally, the etching rates for the materials in question and the degree of surface quality. Etchants for use in device shaping have also to satisfy factors such as the degree of undercutting, the solution chemical aggressiveness toward photoresist masks and the desired etching profile for the relevant purpose. TABLE 2 gives a summary of chemical polishing results for InP.

TABLE 2 Summary of InP chemical polishing results.

| Etchant | Plane | Etch rate (μm/min) | Comment | Ref |
|---|---|---|---|---|
| $1HCl:1HNO_3$ | (111)P > (100) > (111)In | | 26°C, reveals pits and hillocks | [5] |
| HCl | (111)P > (100) > (111)In | | 26°C, reveals pits and hillocks | [5] |
| 0.4N $Fe^{3+}$:HCl | (111)P = (100) | | 26°C, reveals pits | [5] |
| 1% BrMe | (100) > (111)P > (111)In | | 26°C, reveals pits | [5] |
| HBr:HF | (100) | CP | 27°C, reveals pits | [7] |
| $HBr:CH_3COOH$ | (100) | CP | 27°C, reveals pits | [7] |
| $HBr:H_3PO_4$ | (100) | CP | 27°C, reveals pits | [7] |
| $0.92H_2SO_4:100H_2O_2:5H_2O$ | (111)P | 0.06 | RT, reveals pits and scratches | [9] |
| $HBr:CH_3COOH:K_2Cr_2O_7$<br>$HBr:H_3PO_4:K_2Cr_2O_7$ | (100)<br>(100) | CP<br>CP | 20°C, no pits, no mask attack | [16]<br>[16] |
| $nHBr:2H_2O_2:20H_2O:20HCl$ (n = 20, 40, 60 and 80) | (100) | CP | RT, pit morphology changes with HBr concentration | [21] |
| HCl | (100) > (111)P > (111)In | | RT | [26] |
| $H_3PO_4$ | (111)P > (111)In > (100) | | 90°C | [26] |
| $3H_2SO_4:1H_2O_2:1H_2O$ | (100) > (111)P > (111)In | | RT | [26] |
| $3HCl:1HNO_3:5CH_3COOH$ | (111)P > (100) > (111)In | | RT | [26] |
| $1HCl:1HNO_3:2H_2O$ | (100) | 5 | RT | [26] |
| $1HCl:1HClO_4$ | (111)P | 4 | RT | [26] |
| $HCl:H_3PO_4$ | (111)P > (100) > (111)In | | 35°C, illum. effect, AE[(111)In]: 6.89 kcal/mol, AE[(111)P] : 4.96 kcal/mol | [27] |
| $HCl:CH_3COOH:H_2O_2$ | (100) | CP | 20°C, no pits | [28] |
| $K_2Cr_2O_7:H_2SO_4:HCl$ | (100) | CP | 25 - 60°C, no pits, no mask attack | [29] |
| $1HCl:1H_3PO_4$<br>$1C_3H_8O_3:2HCl:2HClO_4$ | (100) | | AE: 12.5 kcal/mol<br>AE: 14.1 kcal/mol | [30]<br>[30] |
| $1HBr:1CH_3COOH$<br>$1HBr:1H_3PO_4$ | (100)<br>(100) | 3<br>2 | 25°C<br>25°C | [31]<br>[31] |
| $1HCl:1H_3PO_4$ | (100) | 4.0 (bulk)<br>6.4 (LPE) | RT | [32] |
| $Br_2:H_2O:H_3PO_4$ | (100) | CP | RT, no mask attack, submicron fabrication | [33] |
| $HCl:H_3PO_4$ | (100) | CP | 20°C | [34] |
| BrMe<br>BrIs | (100)<br>(100) | CP<br>CP | -10°C<br>-10°C | [35]<br>[35] |
| $1H_2SO_4:1H_2O_2:10H_2O$ | (100)<br>(111)P | negligible<br>0.003 | RT<br>RT | [36] |
| $HCl:H_3PO_4$ | (100) | CP | 23°C, reproducible etching rate at 5% HCl | [37] |
| $Br_2:HBr:H_2O$ | (100) | CP | 22°C, submicron fabrication | [38] |
| 1% BrMe | (100) | | AE: 0.17 eV at high T, 0.29 eV at low T | [39] |
| $HCl:H_3PO_4:CH_3COOH$ | (100) | CP | selective etching of InP layer over InAlAs | [40] |
| citric acid:$H_2O_2$ | (100) | CP | selective etching of InGaAs and InAlAs layers over InP | [41] |
| $HCl:H_2O$ | (100) | CP | 5 - 35°C, AE: >40 kJ/mol | [42] |
| $CH_3CHOHCOOH:HIO_3:H_2O$ | (100) | CP | AE: 0.49 - 0.98 eV | [43] |
| $CH_3CHOHCOOH:HNO_3:HCl$ | (100) | CP | selective etching of InGaAs layer over InP | [43] |
| $CH_3CHOHCOOH:H_3PO_4:HCl$ | (100) | CP | AE: 0.36 - 0.66 eV | [43] |

BrMe (bromine in methanol solution), BrIs (bromine in isopropanol solution), CP (etching rate depends on etchant composition proportion), RT (room temperature), AE (activation energy)

Ion bombardment to amorphise the surface can greatly increase the etching rate for chemical etching. Inada et al [44] have shown that a highly enhanced etching rate of 0.08 micron/min in an HF solution is obtained from the Mg-ion bombarded amorphous InP layer.

### B3 Etching Profiles on (100)InP

Formation of crystallographically defined mesas and grooves by highly anisotropic etchants is desirable especially for optoelectronic and circuit applications. The intermetallic compound InP has two types of (111) surface, i.e. (111)In and (111)P, resulting from the crystallographic polarity of the <111> directions. The outermost atom layer in each surface consists of either In or P atoms which are triply bonded to the crystal lattice [3]. The In and P atoms have three and five bonding electrons per atom, respectively. The (111) surface In atoms have no spare electrons because they are all used up in hanging on to the crystal lattice. The (111) surface P atoms, on the other hand, have two electrons per atom available to take part in a chemical reaction. Thus, the (111)P planes are very reactive compared with the (111)In planes. In (100) planes, both In and P atoms are bonded by two bonds to the crystal. Neither kind will be preferred. Consequently, (100) planes may have medium chemical activity. Similarly, in (110) planes both In and P atoms are bonded by one bond to the crystal (and two bonds are parallel to their own (110) planes). Details of the electronic distribution in the crystal lattice may determine the chemical reactivity of (100) and (110) planes, $(100) > (110)$ or $(100) < (110)$, depending on the kind of etchant. As a result, it becomes clear that most etch profiles are determined by slowly-etched (111)In or near-(111)In crystallographic planes.

There have been some reports focused on preferential etching characteristics of InP [16,21,28-34,37-39,42,43,45-58]. All these studies were performed on (100) crystallographic planes. Although the profiles were mostly determined by (111)In planes (i.e. reverse mesa and ordinary mesa profiles by mask stripes parallel to the [110] and $[\bar{1}10]$ directions, respectively), some etchants produced ordinary mesa or nearly vertical walls by stripes parallel to the [110] direction [31].

In many practical device fabrications, the resultant etching profiles have been reported to be dependent on the surface treatment, mask quality and material-masking interactions [43,50,55,56,59]. For example, the presence of native oxide on the InP surface resulting from different surface treatments has been shown to affect the etching profile, and thus the desirable etching profile can be tailored by careful control of processing parameters [50,55]. It has also been demonstrated [59] that poor-quality etch mask can cause defective etching profiles, such as vertical sidewalls and extra mask undercutting in InP.

### B4 Etched-Surface Consideration

It is often assumed that after chemical polishing treatments the surface of a semiconductor is clean and devoid of contamination. Such an assumption is not necessarily true. Indeed, different forms of chemical processing produce residual surface contamination that at an atomistic level assumes radically different forms [60]. The investigations of the InP surface states after chemical treatments were carried out by means of Auger spectroscopy [60], spectroscopic ellipsometry [61,62], X-ray photoelectron spectroscopy (XPS) [63-67] and atomic force microscopy (AFM) [68]. Both X-ray diffraction (XRD) and XPS were also used, not only to investigate chemical composition, but also to characterise the structural features of the chemical oxide grown on InP in HCl and $HNO_3$ etches [69].

## C PHOTOCHEMICAL ETCHING

Lasers or other high intensity photon sources can produce enhanced etching rates by thermal heating, photochemical generation of reactants, photogeneration of electrons and holes, or some combination of these processes [70]. TABLE 3 summarises the photochemical wet etching results for InP.

TABLE 3 Summary of InP photochemical etching results.

| Etchant | Light source | Comments | Ref |
|---|---|---|---|
| $1HCl:2HNO_3:30H_2O$ | $Ar^+$ (514.5 nm) and He-Ne lasers | SI-InP | [71] |
| $1H_3PO_4:9H_2O$ | $Ar^+$ laser (514.5 nm) | SI-, n- and p-InP(100), thermally activated process | [72] |
| $FeCl_3:H_2O$ | He-Ne laser | n-InP(100), maskless groove formation | [73] |
| $1HCl:5HNO_3:21H_2O$ | He-Ne laser | n-InP(100), reveals surface defects | [74] |
| $1HCl:1HNO_3:20H_2O$ and $1HF:10H_2O$ | $Ar^+$ laser (514.5 nm) and incoherent light (350 - 360 nm) | n-InP, crystal face-dependent etching rate | [75] |
| $2HCl:5HNO_3:10H_2O$, $5HNO_3:10H_2O$, etc. | He-Ne laser | n-InP(100) | [76] |
| $1H_3PO_4:9H_2O$ | $Ar^+$ laser (488 nm) | n-InP(100), reaction-rate limited process with an activation energy ≥0.34 eV | [77] |
| $1HCl:1HNO_3:20H_2O$ | tungsten-halogen | stain etching | [78] |
| $FeCl_3:HCl:H_2O$ | tungsten-halogen | difference in etching profile between n- and p-InP | [79] |
| $I_2:KI:HCl:H_2O$ | tungsten-halogen | n- and p-InP, crystal face-dependent etching rate | [80] |

SI (semi-insulating)

Photochemical etching usually involves direct participation of photogenerated carriers in the etching reaction. In n-type semiconductors, the band bending at the electrolyte/sample interface causes majority carriers (electrons) to travel into the bulk and minority carriers (holes) to the surface. The illuminated regions, as a result, act as local anodes. Then, the photogenerated holes induce the formation of an oxide, which is immediately removed by the etching solution. However, in p-type material the band bends downward, repelling the holes from the surface. Therefore, the etching rate for p-InP is usually much slower than that for n-InP [75,77]. By intense photon flux, the process has been shown to be thermally activated because of localised heating of the InP sample [72]. This process promises similar etching properties for all types of InP, such as undoped, n- or p-type, and semi-insulating InP. In the absence of light, no etching takes place [72,74,75,77]. The laser etching rate can be controlled externally by the application of a secondary light source [75].

Fabrication of maskless photochemical gratings on InP has been demonstrated [71,74]. The resolution of an etching process, in the regime where local heating is unimportant, has been reported to be limited by the lateral spreading of the photogenerated carriers [71].

Photochemical etching has been proved to be very sensitive for the delineation of defects and inhomogeneities in substrates and epitaxial layers ([19]; see TABLE 1).

## D PHOTOELECTROCHEMICAL ETCHING

Photoelectrochemical etching requires ohmic contacts to the InP and an external electronic power supply. The etching rate is controlled by the applied current density while photons are used to create a plentiful supply of holes in n-InP so anodic oxidation and oxide dissolution can occur readily [70]. We summarise in TABLE 4 the photoelectrochemical etching results for InP obtained using a laser as the light source. TABLE 5 summarises the photoelectrochemical etching results using a tungsten-halogen lamp as the light source.

TABLE 4 Summary of InP photoelectrochemical etching results using a laser.

| Etchant | Laser | Comments | Ref |
|---|---|---|---|
| HF, HCl and HBr | He-Ne | n-InP, crystal face-dependent etching rate | [81] |
| HF:KOH:$H_2O$ | He-Ne, $Ar^+$ and He-Cd | n-InP, PV: 0.2 - 0.5 V, fabrication of a 0.5 μm period grating | [82] |
| 1HCl:20$H_2O$ | $Ar^+$ | fabrication of a deep/narrow groove | [83] |
| $H_2SO_4$:$H_2O_2$:$H_2O$, HCl:$H_2O$, etc. | He-Cd | n-InP, PV: 0.3 - 0.5 V, fabrication of a >500 $mm^{-1}$ grating | [84] |
| 1HCl:3$HNO_3$:X$H_2O$ | He-Cd | n-InP, using a thin-film cell | [85] |
| $H_2SO_4$:$H_2O$, HCl:$H_2O$, etc. | He-Ne | p-InP, AE: 0.12 eV | [86] |
| 1HF:20$H_2O$ and 1HCl:1$HNO_3$:20$H_2O$ | $Ar^+$ and $Ar^+$ (SHG) | n-InP, crystal face-dependent etching rate | [87] |
| $Br_2$ in HBr | He-Cd and He-Ne | p-InP, PV: -(0.1 - 1.4) V, using a two-component cell | [88] |
| $HNO_3$:$H_2O$ | He-Ne | p-InP, acid concentration-dependent etching rate | [89,90] |

PV (potential versus a saturated calomel reference electrode), AE (activation energy), SHG (second harmonic generation)

TABLE 5 Summary of InP photoelectrochemical etching results using a tungsten-halogen lamp.

| Etchant | Comments | Ref |
|---|---|---|
| unstated | PV: +(0.1 - 0.4) V, fabrication of integral lens on LEDs | [91] |
| 2M $H_2SO_4$ | n-InP, PV: +0.2 V, a mask patterned by a focused ion beam at low dose | [92] |
| 2M HF/0.5 M KOH and 2M $H_2SO_4$ | n-InP, PV: +(0.20 - 0.25) V, a mask patterned by ion implantation | [93] |
| 1M HCl | n-InP, formation of surface textures | [94,95] |
| HCl:$H_2O$ (pH = 1.0) | n-InP, uniform and clean etched surface, extremely high etched depth controllability | [96] |

PV (potential versus a saturated calomel reference electrode), LEDs (light emitting diodes)

The photoelectrochemical etching rate is dependent upon crystal face, temperature, electrode potential, photon flux and the cycle time of the potential step. The order in the reactivity of surface orientation is found to be the same as that in conventional chemical etches [81,87]. A major disadvantage of the photoelectrochemical etching is that many p-type semiconductors cannot be etched; the reason for this is the same as that for the case of photochemical etching (Section C).

Unique methods have been devised to photoelectrochemically etch p-type semiconductors. Kohl et al [86] etched p-InP by first reducing the surface to form an indium-rich layer followed by anodic stripping of the surface film. Van de Ven and Nabben [88] used a two-component cell in order to modify charge transfer processes by the reactions at different surfaces in different electrolytes. Quinlan

[89,90] reported p-InP etching in nitric acid electrolyte in the negative potential region. UV illumination is also found to etch not only n-type, but also p-type InP [87]. Ratios of the etching rates under illumination and in the dark of the order of $10^3$ and higher have been demonstrated [87,88].

A deep groove and an etched hole have been demonstrated on n- and p-InP by photoelectrochemical etching [83,86,87,89,90]. A diffracting grating has been fabricated on n-InP by using a holographic photoelectrochemical technique for distributed feedback lasers [82]. Photoelectrochemical etching has also been used to fabricate integral lenses on light-emitting diodes (LEDs) [91]. The resulting LEDs give an improvement in the light coupled into an optical fibre.

Photoelectrochemical etching has been proved to be very sensitive for the delineation of structural defects, such as mechanical damage and grown-in defects ([13,14]; see TABLE 1).

## E CONCLUSION

Etching is one of the most commonly employed techniques in semiconductor device and material science fields. A survey of the wet chemical, photochemical and photoelectrochemical etching of InP has been presented. Because of inexpensive and easy application, wet chemical etching has been most frequently used for many InP-based device applications. Also, it is widely used for revealing dislocations and inhomogeneities in substrates and epitaxial layers. Photochemical and photoelectrochemical etching have become important for the fabrication of various types of localised structures on InP and its related materials. One major disadvantage of these etching techniques is that p-type semiconductors cannot be selectively etched. However, unique methods have been devised to etch p-type materials. More work still needs to be done to obtain an improved understanding of the reaction chemistry.

## REFERENCES


[1] K. Sangwal [ *Etching of Crystals* (North-Holland, Amsterdam, 1987) ]

[2] P.H.L. Notten, J.E.A.M. van den Meerakker, J.J. Kelly [ *Etching of III-V Semiconductors* (Elsevier, Oxford, 1991) ]

[3] H.C. Gatos, M.C. Lavine [ *J. Electrochem. Soc. (USA)* vol.107 no.5 (1960) p.427-33 ]

[4] R.C. Clarke, D.S. Robertson, A.W. Vere [ *J. Mater. Sci. (UK)* vol.8 no.9 (1973) p.1349-54 ]

[5] B. Tuck, A.J. Baker [ *J. Mater. Sci. (UK)* vol.8 no.11 (1973) p.1559-66 ]

[6] A. Huber, N.T. Linh [ *J. Cryst. Growth (Netherlands)* vol.29 no.1 (1975) p.80-4 ]

[7] K. Akita, T. Kusunoki, S. Komiya, T. Kotani [ *J. Cryst. Growth (Netherlands)* vol.46 no.6 (1979) p.783-7 ]

[8] F.A. Thiel, R.L. Barns [ *J. Electrochem. Soc. (USA)* vol.126 no.7 (1979) p.1272-4 ]

[9] Y. Nishitani, T. Kotani [ *J. Electrochem. Soc. (USA)* vol.126 no.12 (1979) p.2269-71 ]

[10] G.T. Brown, B. Cockayne, W.R. MacEwan [ *J. Mater. Sci. (UK)* vol.15 no.10 (1980) p.2539-49 ]

[11] T. Kotani, S. Komiya, S. Nakai, Y. Yamaoka [ *J. Electrochem. Soc. (USA)* vol.127 no.10 (1980) p.2273-7 ]

[12] S. Mahajan, A.K. Chin [ *J. Cryst. Growth (Netherlands)* vol.54 no.1 (1981) p.138-49 ]

[13] C.R. Elliott, J.C. Regnault [ *J. Electrochem. Soc. (USA)* vol.128 no.1 (1981) p.113-6 ]

[14] A. Yamamoto, S. Tohno, C. Uemura [ *J. Electrochem. Soc. (USA)* vol.128 no.5 (1981) p.1095-100 ]

[15] S.N.G. Chu, C.M. Jodlauk, A.A. Ballman [ *J. Electrochem. Soc. (USA)* vol.129 no.2 (1982) p.352-4 ]

[16] S. Adachi [ *J. Electrochem. Soc. (USA)* vol.129 no.3 (1982) p.609-13 ]

[17] E.A. Caridi, T.Y. Chang [ *J. Electrochem. Soc. (USA)* vol.131 no.6 (1984) p.1440-1 ]

[18] J.A. Lourenco [ *J. Electrochem. Soc. (USA)* vol.131 no.8 (1984) p.1914-6 ]

[19] J.L. Weyher, L.J. Giling [ *J. Appl. Phys. (USA)* vol.58 no.1 (1985) p.219-22 ]

[20] D.T.C. Huo, J.D. Wynn, M.F. Yan, D.P. Wilt [ *J. Electrochem. Soc. (USA)* vol.136 no.6 (1989) p.1804-6 ]

[21] D.T.C. Huo, M.F. Yan, J.D. Wynn, D.P. Wilt [ *J. Electrochem. Soc. (USA)* vol.136 no.10 (1989) p.3094-7 ]

[22] A. Zozime, W. Schröter [ *Appl. Phys. Lett. (USA)* vol.57 no.13 (1990) p.1326-7 ]

[23] K.P. Quinlan [ *J. Electrochem. Soc. (USA)* vol.137 no.6 (1990) p.1991-5 ]

[24] J.L. Weyher, R. Fornari, T. Görög, J.J. Kelly, B. Erné [ *J. Cryst. Growth (Netherlands)* vol.141 (1994) p.57-67 ]

[25] E. Peiner, A. Schlachetzki [ *J. Electron. Mater. (USA)* vol.21 no.9 (1992) p.887-91 ]

[26] R. Becker [ *Solid-State Electron. (UK)* vol.16 no.11 (1973) p.1241-9 ]

[27] S.B. Phatak, G. Kelner [ *J. Electrochem. Soc. (USA)* vol.126 no.2 (1979) p.287-92 ]

[28] T. Kambayashi, C. Kitahara, K. Iga [ *Jpn. J. Appl. Phys. (Japan)* vol.19 no.1 (1980) p.79-85 ]

[29] S. Adachi, H. Kawaguchi, G. Iwane [ *J. Mater. Sci. (UK)* vol.16 no.9 (1981) p.2449-56 ]

[30] F. Fiedler, A. Schlachetzki, G. Klein [ *J. Mater. Sci. (UK)* vol.17 no.10 (1982) p.2911-8 ]

[31] S. Adachi, H. Kawaguchi [ *J. Electrochem. Soc. (USA)* vol.128 no.6 (1981) p.1342-9 ]

[32] K.L. Conway, A.G. Dentai, J.C. Campbell [ *J. Appl. Phys. (USA)* vol.53 no.3 (1982) p.1836-8 ]

[33] T. Saitoh, O. Mikami, H. Nakagome [ *Electron. Lett. (UK)* vol.18 no.10 (1982) p.408-9 ]

[34] P. Buchmann, A.J.N. Houghton [ *Electron. Lett. (UK)* vol.18 no.19 (1982) p.850-2 ]

[35] S. Aytac, A. Schlachetzki [ *Solid-State Electron. (UK)* vol.25 no.11 (1982) p.1135-9 ]

[36] G.A. Ferrante, J.P. Donnelly, C.A. Armiento [ *J. Electrochem. Soc. (USA)* vol.130 no.5 (1983) p.1222-4 ]

[37] S. Uekusa, K. Oigawa, M. Tacano [ *J. Electrochem. Soc. (USA)* vol.132 no.3 (1985) p.671-3 ]

[38] T. Matsuoka, H. Nagai [ *J. Electrochem. Soc. (USA)* vol.133 no.12 (1986) p.2485-91 ]

[39] D.T.C. Huo, M.F. Yan, J.D. Wynn, D.P. Wilt [ *J. Electrochem. Soc. (USA)* vol.136 no.6 (1989) p.1828-30 ]

[40] Y. He, B.W. Liang, N.C. Tien, C.W. Tu [ *J. Electrochem. Soc. (USA)* vol.139 no.7 (1992) p.2046-8 ]

[41] M. Tong, K. Nummila, A.A. Ketterson, I. Adesida, L. Aina, M. Mattingly [ *J. Electrochem. Soc. (USA)* vol.139 no.10 (1992) p.L91-3 ]

[42] S. das Neves, M.-A. De Paoli [ *J. Electrochem. Soc. (USA)* vol.140 no.9 (1993) p.2599-603 ]

[43] K. Ikossi-Anastasiou et al [ *J. Electrochem. Soc. (USA)* vol.142 no.10 (1995) p.3558-64 ]

[44] T. Inada, S. Taka, K. Kodama [ *J. Electrochem. Soc. (USA)* vol.131 no.6 (1984) p.1401-3 ]

[45] L.W. Stulz, L.A. Coldren [ *J. Electrochem. Soc. (USA)* vol.130 no.7 (1983) p.1628-30 ]

[46] L.A. Coldren, K. Furuya, B.I. Miller [ *J. Electrochem. Soc. (USA)* vol.130 no.9 (1983) p.1918-26 ]

[47] C.J. Keavney, H.I. Smith [ *J. Electrochem. Soc. (USA)* vol.131 no.2 (1984) p.452-3 ]

[48] L.E. Shea [ *GEC J. Res. (UK)* vol.3 no.1 (1985) p.55-7 ]

[49] L.N. Vozmilova, M.M. Berdichenko [ *Inorg. Mater. (USA)* vol.21 no.8 (1985) p.1110-2 ]

[50] D.T.C. Huo, J.D. Wynn, S.G. Napholtz, F.R. Lenzo, D.P. Wilt [ *J. Electrochem. Soc. (USA)* vol.134 no.11 (1987) p.2850-6 ]

[51] D.T.C. Huo, J.D. Wynn, S.G. Napholtz, D.P. Wilt [ *J. Electrochem. Soc. (USA)* vol.135 no.5 (1988) p.1231-4 ]

[52] D.T.C. Huo, J.D. Wynn, S.G. Napholtz, D.P. Wilt [ *J. Electrochem. Soc. (USA)* vol.135 no.9 (1988) p.2334-8 ]

[53] D.T.C. Huo, M.F. Yan, J.D. Wynn, D.P. Wilt [ *J. Electrochem. Soc. (USA)* vol.136 no.3 (1989) p.772-5 ]

[54] E. Inamura, Y. Miyamoto, S. Tamura, T. Takasugi, K. Furuya [ *Jpn. J. Appl. Phys. (Japan)* vol.28 no.10 (1989) p.2193-6 ]

[55] P.H.L. Notten [ *J. Electrochem. Soc. (USA)* vol.138 no.1 (1991) p.243-9 ]

[56] R. Klockenbrink, E. Peiner, H.-H. Wehmann, A. Schlachetzki [ *J. Electrochem. Soc. (USA)* vol.141 no.6 (1994) p.1594-9 ]

[57] M. Kappelt, D. Bimberg [ *J. Electrochem. Soc. (USA)* vol.143 no.10 (1996) p.3271-3 ]

[58] P. Bönsch, D. Wüllner, T. Schrimpf, A. Schlachetzki, R. Lacmann [ *J. Electrochem. Soc. (USA)* vol.145 no.4 (1998) p.1273-6 ]

[59] D.T. Huo, M.F. Yan, J.D. Wynn, D.P. Wilt [ *J. Electrochem. Soc. (USA)* vol.137 no.1 (1990) p.239-42 ]

[60] D.L. Kirk, C. Jones [ *J. Phys. D (UK)* vol.12 no.4 (1979) p.651-6 ]

[61] D.E. Aspnes, A.A. Studna [ *Appl. Phys. Lett. (USA)* vol.39 no.4 (1981) p.316-8 ]

[62] X. Liu, E.A. Irene, S. Hattangady, G. Fountain [ *J. Electrochem. Soc. (USA)* vol.137 no.7 (1985) p.2319-22 ]

[63] P.A. Bertrand [ *J. Vac. Sci. Technol. (USA)* vol.18 no.1 (1981) p.28-33 ]

[64] J.F. Wagner, D.L. Ellsworth, S.M. Gooduick, C.W. Wilurseu [ *J. Vac. Sci. Technol. (USA)* vol.19 no.3 (1981) p.513-8 ]

[65] G. Hollinger, E. Bergignat, J. Joseph, Y. Robach [ *J. Vac. Sci. Technol. A (USA)* vol.3 no.6 (1985) p.2082-8 ]

[66] J. Massies, F. Turco, J.-P. Contour [ *Jpn. J. Appl. Phys. (Japan)* vol.25 no.8 (1986) p.L6664-7 ]

[67] E. Kurth, A. Reif, V. Gottschalch, J. Finster, E. Butter [ *Cryst. Res. Technol. (Germany)* vol.23 no.1 (1988) p.117-26 ]

[68] Y. Morisawa, I. Kikuma, N. Takayama, M. Tekechi [ *Appl. Surf. Sci. (Netherlands)* vol.92 (1996) p.147-50 ]

[69] H.F. Hsieh, H.C. Shih [ *J. Appl. Phys. (USA)* vol.66 no.8 (1989) p.3542-7 ]

[70] C.I.H. Ashby [ in *Properties of Gallium Arsenide* 2nd Edition (INSPEC, IEE, London, UK, 1990) p.676-81 ]

[71] R.M. Osgood Jr., A. Sanchez-Rubio, D.J. Ehrlich, V. Daneu [ *Appl. Phys. Lett. (USA)* vol.40 no.5 (1982) p.391-3 ]

[72] J.E. Bjorkholm, A.A. Ballman [ *Appl. Phys. Lett. (USA)* vol.43 no.6 (1983) p.574-6 ]

[73] D. Moutonnet [ *Mater. Lett. (Netherlands)* vol.6 no.5,6 (1988) p.183-5 ]

[74] D. Moutonnet [ *Mater. Lett. (Netherlands)* vol.6 no.11,12 (1988) p.433-5 ]

[75] A.E. Willner, D.V. Podlesnik, H.H. Gilgen, R.M. Osgood Jr. [ *Appl. Phys. Lett. (USA)* vol.53 no.13 (1988) p.1198-200 ]

[76] V. Švorèík, V. Rybka, V. Myslík [ *Phys. Status Solidi A (Germany)* vol.106 (1988) p.K35-9 ]

[77] T.D. Lowes, D.T. Cassidy [ *J. Appl. Phys. (USA)* vol.68 no.2 (1990) p.814-9 ]

[78] J.B. Williamson, K.W. Carey [ *J. Electrochem. Soc. (USA)* vol.140 no.7 (1993) p.2125-8 ]

[79] I.E. Vermeir, W.P. Gomes, P. Van Daele [ *J. Electrochem. Soc. (USA)* vol.142 no.9 (1995) p.3226-32 ]

[80] I.E. Vermeir, W.P. Gomes [ *J. Electrochem. Soc. (USA)* vol.143 no.4 (1996) p.1319-25 ]

[81] P.A. Kohl, C. Wolowodiuk, F.W. Ostermayer Jr. [ *J. Electrochem. Soc. (USA)* vol.130 no.11 (1983) p.2288-93 ]

[82] R.M. Lum, A.M. Glass, F.W. Ostermayer Jr., P.A. Kohl, A.A. Ballman, R.A. Logan [ *J. Appl. Phys. (USA)* vol.57 no.1 (1985) p.39-44 ]

[83] J.E. Bowers, B.R. Hemenway, D.P. Wilt [ *Appl. Phys. Lett. (USA)* vol.46 no.5 (1985) p.453-5 ]

[84] R.M. Lum, F.W. Ostermayer Jr., P.A. Kohl, A.M. Glass, A.A. Ballman [ *Appl. Phys. Lett. (USA)* vol.47 no.3 (1985) p.269-71 ]

[85] H. Grebel, B. Iskandar, K.G. Sheppard [ *Appl. Phys. Lett. (USA)* vol.55 no.25 (1989) p.2655-7 ]

[86] P.A. Kohl, D.B. Harris, J. Winnick [ *J. Electrochem. Soc. (USA)* vol.138 no.2 (1991) p.608-14 ]

[87] M.N. Ruberto, X. Zhang, R. Scarmozzino, A.E. Willner, G.V. Podlesnik, R.M. Osgood Jr. [ *J. Electrochem. Soc. (USA)* vol.138 no.4 (1991) p.1174-85 ]

[88] J. van de Ven, H.J.P. Nabben [ *J. Electrochem. Soc. (USA)* vol.138 no.11 (1991) p.3401-6 ]

[89] K.P. Quinlan [ *J. Electrochem. Soc. (USA)* vol.143 no.9 (1996) p.L200-3 ]

[90] K.P. Quinlan [ *J. Electrochem. Soc. (USA)* vol.144 no.10 (1997) p.3469-73 ]

[91] F.W. Ostermayer Jr., P.A. Kohl, R.H. Burton [ *Appl. Phys. Lett. (USA)* vol.43 no.7 (1983) p.642-4 ]

[92] K.D. Cummings, L.R. Harriott, G.C. Chi, F.W. Ostermayer Jr. [ *Appl. Phys. Lett. (USA)* vol.48 no.10 (1986) p.659-61 ]

[93] G.C. Chi, F.W. Ostermayer Jr., K.D. Cummings, L.R. Harriott [ *J. Appl. Phys. (USA)* vol.60 no.11 (1986) p.4012-4 ]

[94] N.G. Gerreira, D. Soltz, F. Decker, L. Cescato [ *J. Electrochem. Soc. (USA)* vol.142 no.4 (1995) p.1348-52 ]

[95] D. Soltz, L. Cescato [ *J. Electrochem. Soc. (USA)* vol.143 no.9 (1996) p.2815-21 ]

[96] C. Kaneshiro, T. Sato, H. Hasegawa [ *Jpn. J. Appl. Phys. (Japan)* vol.38 no.2B (1999) p.1147-52 ]

# 7.2 Ion implantation in InP

B.J. Sealy

June 1998

## A INTRODUCTION

Ion implantation is widely used as a means of doping layers in silicon integrated circuit technology. However, the technique is not so easily applied to compound semiconductors which decompose during subsequent heat treatment. The process of ion implantation introduces damage to the lattice which for high doses of heavy ion species results in a highly disordered material which in the extreme can become amorphous. This damage can be removed by a post implantation heat treatment in which the damaged layer regrows by solid phase epitaxy onto the underlying single crystal substrate. The process is well controlled and reproducible for implants in silicon, but for compound semiconductors, annealing is difficult and often not reproducible. The problem is how to protect the surface from decomposing during the necessary annealing cycle. The most usual method employed for GaAs is to encapsulate the implanted surface with a thin layer of silicon nitride, which if done correctly allows samples to be processed to temperatures of around 1000°C without significant decomposition. Annealing temperatures employed by industry for processing GaAs devices and circuits tend to be rather less than this. There has been less interest in the doping of InP by ion implantation than in the doping of GaAs and, as a consequence, the problems of implanting and annealing InP are less well developed and understood. There are however many publications on the topic and it is the aim of this Datareview to summarise the data and provide a useful bibliography for those who wish to delve deeper into the literature.

## B ANNEALING AND ENCAPSULATION

There seems not to be a preferred way of preventing decomposition during the annealing of ion implanted InP, but despite this, good results have been obtained for donor ion species, particularly silicon ions. However, in contrast, it is difficult to activate acceptor ion implants. Although much lower temperatures of around 750°C are required to anneal implanted InP compared with GaAs, the problems of decomposition are exacerbated for InP because it tends to decompose very readily. Changes in surface composition seem to be accompanied by the production of an n-type conducting layer which improves the measured n-type activity values but tends to compensate p-type implants [1,2]. It is believed in cases where silicon containing encapsulants, such as silicon nitride, have been used, that the n-type surface layer is the result of the indiffusion of silicon. For example, room temperature implants of Xe and Hg ions both produce n-type layers when annealed above 700°C and 850°C respectively. However, Hg implants into heated substrates (200°C) produce p-type activity after annealing. It would appear that the presence of damage encourages silicon from the silicon nitride cap to diffuse into the surface regions. A useful overview of the high temperature treatment of InP and related materials is to be found in [3] which describes the use of various encapsulating layers, capless annealing and problems of controlling decomposition and impurity diffusion. A summary of encapsulants used to protect InP and a comparison of the use of silicon nitride and Ta-Si-N layers can be found in [4].

A particular problem with acceptor implants is the diffusion, both in and out of the implanted region. Good control of profiles is possible by performing dual implants of, for example, (Be + P) which seem to eliminate the diffusion of the acceptor atom [5] just as similar implants do in GaAs.

## C RESISTIVITY

The magnitude of the resistivity of implanted InP depends on the doping species and the nature of the substrate being implanted, that is, whether the substrate is doped or semi-insulating. Ion implantation is used to modify the electrical and/or optical properties by the creation of doped regions or regions of high resistivity used to isolate, electrically and/or optically, one device from another or one part of a circuit from another. Doped material is formed by implanting a donor or acceptor ion followed by a high temperature anneal to electrically activate the dopant. Conducting layers can be converted to semi-insulating by means of non-dopant implants of, for example, H, He or O which form trapping levels which can be removed by subsequent heat treatment and therefore do not form thermally stable layers. However, if heavy ions such as Fe, Co and Ti are used, layers are thermally stable because chemical deep levels are created which arise from the location of the impurities on indium sites.

### C1 High Resistivity Layers

Many authors have studied the removal of carriers in InP using ion implantation; however, the maximum resistivity achievable is lower than that for similar treatments of GaAs. Protons, for example, produce maximum resistivities of $10^3$ ohm cm and in excess of $10^8$ ohm cm for n-type and p-type InP respectively, providing that a multi-energy implant is performed [6]. It is thus more difficult to produce high resistivity layers in n-type InP than in p-type material. For an optimum dose, the high resistivity layers are stable up to about 400°C. Other ions such as B, O, N, Ar and Ne have also been studied [7-11]. In GaAs, these ions form highly compensated, highly resistive layers, but in InP the layers are not so resistive and also not thermally stable. The resistivity of n-type InP increases with increasing annealing temperature and goes through a maximum of about $10^6$ ohm/square at 200 - 400°C for ions such as B, O and H [10,11]. The maximum resistivity moves to higher temperatures with increasing ion mass and dose which is similar to results for GaAs. Annealing above 500°C removes this effect.

A number of workers [7,10-13] have resorted to the use of iron implants since iron is used as a dopant to produce semi-insulating Czochralski ingots. Using multi-energy implants of iron, it is possible to convert a surface layer of n-type InP to a high resistivity of $6 \times 10^7$ ohm cm after annealing at 725°C for fifteen minutes [12]. The surface layer is about 0.8 μm thick if the implant is carried out at room temperature and about 0.55 μm if performed at 200°C. Other papers [10,11] indicate that high resistivities ($>10^6$ ohm/square) can be achieved providing the iron concentration exceeds the donor concentration. More recently, iron implants have been shown to produce a resistivity of about $2 \times 10^7$ ohm cm in n-type InP with $n = 1.4 \times 10^{18}$ $cm^{-3}$ [14]. However, many authors believe that the iron redistributes during annealing and thus control of the process is not ideal, but this can be overcome by carrying out the implant into a substrate held at 200°C [15]. A second method of overcoming the problem of the mobility of iron is to use titanium ions instead [16,17]. The energy of the titanium ions was 400 keV [16] and in the range 1 - 5 MeV [17], the latter being used to produce buried layers in p-type InP following both room temperature and 200°C implants. Another transition metal that has been tried is cobalt [18] which was shown to produce resistivities of $4 \times 10^7$ ohm cm following a 200°C implant and an anneal up to at least 800°C [18,19]. A completely different approach is to use a single, 5 MeV implant of oxygen ions [20]. For such an implant, the ion range is made several times the thickness of the layer to be isolated which buries the end-of-range disorder deep within the substrate. This process has been applied to $p^+$ epitaxial layers of InP to produce sheet resistivities of about $5 \times 10^6$ ohm/square. It was found that multi-energy implants which produce the same level of damage give rise to higher resistivities but with increased process complexity.

n-Type layers can be produced by implanting ions such as Si, S, Ge, Se, Sn and Te to give, for high doses, sheet resistivities in the range 10 - 30 ohm/square (see TABLE 1, [13,21-27]).

TABLE 1 Sheet resistivity of donor implanted InP.

| Ion | Dose ($cm^{-2}$) | Energy (keV) | $T_i$ (°C) | $T_a$ (°C) | Resistivity (Ω/sq) | Ref |
|---|---|---|---|---|---|---|
| Si | $1.2 \times 10^{13}$ | 200 | RT | 900/10 s | 250 | [13] |
| | $10^{14}$ | 200 | RT | 900/10 s | 90 | [13] |
| | $10^{15}$ | 400 | 200 | 750/15 min | 15 | [21] |
| S | $10^{13}$ | 120 | RT | 800/20 min | 800 | [22] |
| | $10^{14}$ | 120 | RT | 800/20 min | 80 | [22] |
| | $10^{15}$ | 120 | RT | 800/20 min | 20 | [22] |
| | $3 \times 10^{14}$ | 1000 | 200 | 735/10 min | 30 | [23] |
| Se | $10^{13}$ | 200 | RT | 700/3 min | 1000 | [24] |
| | $10^{14}$ | 200 | RT | 700/3 min | 100 | [24] |
| | $5 \times 10^{14}$ | 200 | RT | 700/3 min | 55 | [24] |
| | $10^{15}$ | 200 | RT | 700/3 min | 50 | [24] |
| | $10^{15}$ | 400 | 200 | 750/15 min | 16 | [21] |
| Si + P | $10^{15}$ | 150 | RT | 750/60 s | 25 | [25] |
| | $10^{15}$ | 150 | RT | 750/100 s | 20 | [25] |
| Si | $10^{15}$ | 150 | RT | 750/60 s | 28 - 30 | [25] |
| | $8 \times 10^{14}$ | 3000 | RT | 875/10 s | 21 | [27] |
| | $2 \times 10^{15}$ | 3000 | RT | 875/10 s | 12 | [27] |
| | $8 \times 10^{14}$ | 3000 | 200 | 875/10 s | 8 | [27] |

$T_i$ is the implant temperature and $T_a$ the annealing temperature.

p-Type layers are more difficult to generate since acceptor implants into semi-insulating InP do not activate readily. Measured resistivities are very high, minimum values recorded being around 800 ohm/square (see TABLE 2, [13,21,28-34]).

TABLE 2 Sheet resistivity of acceptor implanted InP.

| Ion | Dose ($cm^{-2}$) | Energy (keV) | $T_a$ (°C) | Resistivity (Ω/sq) | Ref |
|---|---|---|---|---|---|
| Mg | $10^{14}$ | 40 | 750/30 min | ~6000 | [28] |
| | $10^{15}$ | 350 | 750/30 min | ~5000 | [28] |
| | $10^{14}$ | 150 | 750/15 min | ≈1500* | [21] |
| | $2 \times 10^{14}$ | 1000 | 800 | ≈780* | [22] |
| | $10^{14}$ | 400 | 750 | 2000 | [33] |
| Mg | $1 \times 10^{13}$ | 80 | 850/10 s | 10850 | [34] |
| | $5 \times 10^{13}$ | 80 | 850/10 s | 2905 | [34] |
| | $1 \times 10^{14}$ | 80 | 850/10 s | 2500 | [34] |
| | $5 \times 10^{14}$ | 80 | 850/10 s | 1670 | [34] |
| Mg/P | $1 \times 10^{14}$ | 80/120 | 850/10 s | 2156 | [34] |
| Mg/As | $1 \times 10^{14}$ | 80/140 | 850/10 s | 1954 | [34] |
| Hg | $10^{14}$ | 100 | 850/1 min | ≈1400* | [31] |
| Be | $5 \times 10^{13}$ | 60 | 850/5 s | 12000 | [13] |
| | $5 \times 10^{13}$ | 60 | 875/5 s | 7000 | [13] |
| | $4 \times 10^{14}$ | 60 | 850/2 s + 800/10 s | 810 | [13] |
| | $10^{14}$ | 300 | 640/30 min | 950 - 1260 | [29] |
| | $3 \times 10^{14}$ | 400 | 750 | ≈760* | [30] |

$T_a$ is the annealing temperature, * refers to values estimated from sheet hole concentration and sheet mobility values.

## D CARRIER CONCENTRATIONS

### D1 Acceptor Ion Implants

The group II elements Be, Mg, Zn, Cd and Hg all produce p-type layers following implantation and annealing in InP. However, a major problem is the high diffusivity of these atoms which produces broad profiles and hence deep junctions. Beryllium has been studied most and can produce hole concentrations of 1 - 3 × $10^{18}$ $cm^{-3}$ when annealed at 750°C for fifteen minutes or 800°C for ten seconds [35]. Dual implants of (Be + P) can reduce diffusion and produce an increased level of electrical activation together with a factor of three increase in hole concentration [36,37]. In addition, implanting Ar together with the Be also reduces indiffusion and produces sharp profiles [37,38].

Mg implants can also give rise to broad, diffused profiles [34] with peak hole concentrations of 1 - 3 × $10^{18}$ $cm^{-3}$ [34]. However, it has been reported that diffusion of Mg is mainly limited to the damaged region, resulting in very strong outdiffusion compared with indiffusion [39]. Similarly to the case of Be, implants of (Mg + P) suppress diffusion and produce sharper profiles than for single Mg implants [34].

Zinc has not received much attention [40,41]. However, it is possible to generate peak hole concentrations in the range 1 - 3 × $10^{18}$ $cm^{-3}$ but with significant diffusional broadening of the profile [41]. Co-implanting with P reduces diffusion and increases the peak hole concentration by up to a factor of two [41]. Little has been published on the implantation of cadmium which produces results similar to the other acceptor implants [42]. The heaviest acceptor ion, Hg, has been studied, and in summary very shallow layers can be produced, but with a very low degree of activation [43]. It has been demonstrated that Hg implants activate more easily if carried out at 200°C rather than at room temperature, but profiles are much broader than predicted [31].

### D2 Donor Ion Implants

As with gallium arsenide, silicon is the most widely studied of the donor impurities in InP. However, all the group IV elements C, Si, Ge and Sn are donors in InP [44]. Of the group VI elements Se and S have been studied and both produce good levels of activation. Peak electron concentrations of at least $10^{19}$ $cm^{-3}$ are readily achieved for Si, Ge, S and Se implants [42,45-51]. It has also been demonstrated that (Si + P) dual implants increase the activation, such that the peak electron concentration increases from 1 × $10^{19}$ to 5 × $10^{19}$ $cm^{-3}$ [25,52,53]. The implantation of carbon has received less attention, but results indicate that high donor activation is possible with electron concentrations of about 1 × $10^{19}$ $cm^{-3}$ which is increased to 3 × $10^{19}$ $cm^{-3}$ for (C + P) dual implants [44,54].

## E CARRIER MOBILITY

TABLE 3 summarises results for Si implants whilst TABLE 4 contains data for other ions. The majority of data are for anneals performed in the range 700°C to 800°C, although there are some reports of anneals up to 900°C. The tables contain information from samples annealed by rapid thermal means and in a furnace. By comparing the data, it is evident that hot implants at 200°C produce higher mobilities than do implants performed at room temperature. It is also evident that the method of protecting the surface during annealing is important and this varies from paper to paper. The use of phosphosilicate glass, silicon nitride and a phosphine ambient are popular and successful ways of limiting decomposition during annealing. The sheet mobility for silicon implants performed at room temperature varies from about 2000 $cm^2$/V s to 2500 $cm^2$/V s, to 1200 $cm^2$/V s to 650 $cm^2$/V s as the dose increases from $5 \times 10^{13}$ $cm^{-2}$ to $10^{14}$ $cm^{-2}$ to $10^{15}$ $cm^{-2}$ respectively. Raising the implant temperature to 200°C increases the mobility by approximately 40% for all doses.

TABLE 3 Sheet mobility of silicon implanted InP as a function of ion dose, energy, implant temperature and annealing conditions.

| Dose ($cm^{-2}$) | Energy (keV) | $T_i$ (°C) | $T_a$ (°C) | Cap | Mobility ($cm^2$/V s) | Ref |
|---|---|---|---|---|---|---|
| $4 \times 10^{14}$ | 200 | 200 | 790/10 s | PSG/$SiO_2$ | 1033 | [47] |
| $4 \times 10^{14}$ | 200 | RT | 800/10 s | $Si_3N_4$ | 470 | [47] |
| $4 \times 10^{14}$ | 200 | RT | 775/10 min | $PH_3$ | 730 | [47] |
| $5 \times 10^{12}$ | 50 | RT | 700/15 s | $Al_2O_3$ | 2600 | [55] |
| $5 \times 10^{14}$ | 50 | RT | 700/15 s | $Al_2O_3$ | 750 | [55] |
| $10^{13}$ | 180 | RT | 750/15 min | $SiO_2$/capless | 2706 | [56] |
| $3 \times 10^{14}$ | 180 | RT | 750/15 min | $SiO_2$/capless | 960 | [56] |
| $10^{13}$ | 180 | 200 | 850/5 s | $SiO_2$/capless | 3466 | [56] |
| $3 \times 10^{14}$ | 180 | 200 | 850/5 s | $SiO_2$/capless | 1212 | [56] |
| $2.5 \times 10^{14}$ | 200 | RT | 775/15 s | - | 1000 | [57] |
| $2.5 \times 10^{14}$ | 200 | 200 | 775/15 s | - | 1500 | [57] |
| $5 \times 10^{12}$ | 30 | RT | 825/10 s | prox | 3000 | [58] |
| $5 \times 10^{12}$ | 70 | RT | 825/10 s | prox | 2600 | [58] |
| $3 \times 10^{14}$ | 1000 | 200 | 700 | $PH_3$ | 2200 | [59] |
| $3 \times 10^{14}$ | 1000 | RT | 700 | $PH_3$ | 960 | [59] |
| $3 \times 10^{14}$ | 1000 | 200 | 750 | $PH_3$ | 2700 | [59] |
| $3 \times 10^{14}$ | 1000 | RT | 750 | $PH_3$ | 1030 | [59] |
| $3 \times 10^{14}$ | 1000 | 200 | 810 | $PH_3$ | 3800 | [59] |
| $3 \times 10^{14}$ | 1000 | RT | 810 | $PH_3$ | 3350 | [59] |
| $5 \times 10^{12}$ | 300 | RT | 750/30 min | PSG | 2900 | [29] |
| $6 \times 10^{13}$ | 300 | RT | 750/30 min | PSG | 1100 | [29] |
| $10^{14}$ | 300 | RT | 750/30 min | PSG | 1100 | [29] |
| $10^{13}$ | 40 | RT | 800/10 s | PSG | 2200 | [49] |
| $5 \times 10^{13}$ | 40 | RT | 800/10 s | PSG | 1400 | [49] |
| $10^{14}$ | 40 | RT | 800/10 s | PSG | 1000 | [49] |
| $5 \times 10^{14}$ | 40 | RT | 800/10 s | PSG | 800 | [49] |
| $10^{15}$ | 40 | RT | 800/10 s | PSG | 650 | [49] |
| $10^{15}$ | 350 | RT | 800/10 s | PSG | 1400 | [49] |
| $5 \times 10^{13}$ | 200 | RT | 900/5 s + 800/25 s | prox | 1590 | [60] |
| $10^{13}$ | 150 | RT | 900/5 s + 800/25 s | prox | 2260 | [60] |
| $1.2 \times 10^{13}$ | 200 | RT | 900/10 s | - | 2500 | [13] |
| $10^{14}$ | 200 | RT | 900/10 s | - | 1400 | [13] |
| $5 \times 10^{12}$ | 200 | RT | 730/15 min | - | 2500 | [13] |
| $10^{13}$ | 200 | RT | 730/15 min | - | 1800 | [13] |
| $5 \times 10^{13}$ | 200 | RT | 730/15 min | - | 1300 | [13] |
| $10^{14}$ | 200 | RT | 730/15 min | - | 900 | [13] |
| $3 \times 10^{14}$ | 200 | RT | 730/15 min | - | 900 | [13] |
| $10^{14}$ | 400 | 200 | 750/10 min | PSG | 1800 | [44] |

$T_i$ is the implant temperature, $T_a$ is the annealing temperature, PSG is phosphosilicate glass, prox refers to proximity annealing in which two samples are placed face to face without the use of an encapsulant.

TABLE 4 Sheet mobility for various donor ions in InP as a function of ion dose, energy, implant temperature and annealing conditions.

| Ion | Dose ($cm^{-2}$) | Energy (keV) | $T_i$ (°C) | $T_a$ (°C) | Cap | Mobility ($cm^2/V$ s) | Ref |
|---|---|---|---|---|---|---|---|
| Se | $10^{14}$ | 400 | RT | 750/15 min | PSG | 1470 | [21] |
| | $10^{14}$ | 400 | 200 | 750/15 min | PSG | 1810 | [21] |
| | $10^{15}$ | 400 | 200 | 750/15 min | PSG | 1300 | [21] |
| | $10^{14}$ | 400 | 180 | 750/10 min | PSG | 1800 | [61] |
| | $10^{15}$ | 400 | 180 | 750/10 min | PSG | 1200 | [61] |
| | $10^{13}$ | 200 | RT | 700/3 min | $Si_3N_4$ | 1400 | [62] |
| | $10^{13}$ | 200 | RT | 700/3 min | $Si_3N_4$ | 1500 | [24] |
| | $10^{14}$ | 200 | RT | 700/3 min | $Si_3N_4$ | 1200 | [24] |
| | $5 \times 10^{14}$ | 200 | RT | 700/3 min | $Si_3N_4$ | 750 | [24] |
| | $10^{15}$ | 200 | RT | 700/3 min | $Si_3N_4$ | 650 | [24] |
| S | $10^{13}$ | 120 | RT | 800/20 min | PSG + $Si_3N_4$ | 1600 | [22] |
| | $10^{14}$ | 120 | RT | 800/20 min | PSG + $Si_3N_4$ | 1600 | [22] |
| | $10^{15}$ | 120 | RT | 800/20 min | PSG + $Si_3N_4$ | 900 | [22] |
| | $3 \times 10^{14}$ | 1000 | 200 | 735/10 min | $Si_3N_4$ | 1500 | [23] |
| C | $10^{14}$ | 400 | 200 | 750/10 min | PSG | 3000 | [44] |
| Ge | $10^{14}$ | 400 | 200 | 750/10 min | PSG | 1500 | [44] |
| Sn | $10^{14}$ | 400 | 200 | 750/10 min | PSG | 1450 | [44] |

$T_i$ is the implant temperature, $T_a$ is the annealing temperature, PSG is phosphosilicate glass.

Of the acceptor impurities, Mg and Be have received most attention. The measured sheet mobilities lie in the range 40 - 110 $cm^2/V$ s and are generally independent of ion dose and species. The optimum annealing conditions are in the range 700 to 900°C. TABLE 5 summarises the published data. Dual implants of (Be + P) used to limit the diffusion of the acceptor species and to increase the peak hole concentration cause a decrease in mobility compared with single ion implants [37]. However, similar work carried out for (Mg + P) implants shows that the mobility is similar for both single and dual implants [34]. Similar data have also been published for 1 MeV implants of Be in InP which show that mobilities decrease with increasing dose, for example, from 100 $cm^2/V$ s to 55 $cm^2/V$ s for doses of $2 \times 10^{13}$ $cm^{-2}$ and $6 \times 10^{14}$ $cm^{-2}$ respectively [63].

TABLE 5 Sheet hole mobility as a function of ion and dose following an anneal at 750°C.

| Ion | Dose ($cm^{-2}$) | Energy (keV) | Mobility ($cm^2/V$ s) | Ref |
|---|---|---|---|---|
| Mg | $10^{13}$ | 40 | 100 | [28] |
| | $10^{14}$ | 40 | 100 | [28] |
| | $10^{15}$ | 40 | 100 | [28] |
| | $10^{15}$ | 350 | 80 | [28] |
| | $10^{14}$ | 400 | 80 | [33] |
| | $10^{14}$ | 150 | 83 | [21] |
| | $10^{13}$ | 80 | 108* | [34] |
| | $10^{14}$ | 80 | 83* | [34] |
| Be | $3 \times 10^{13}$ | 400 | 90 | [30] |
| | $10^{14}$ | 400 | 110 | [30] |
| | $3 \times 10^{14}$ | 400 | 110 | [30] |
| | $10^{14}$ | 50 | 81 | [21] |
| Cd | $10^{14}$ | 400 | 90 | [21] |
| Hg | $10^{14}$ | 100 | 100 | [31] |

* Anneal at 850°C.

## F ION RANGES

When an energetic ion penetrates a solid target it will eventually come to rest at a depth characteristic of the ion energy and the mass and atomic number of both ion and target atoms. The ions lose their energy by excitation and ionisation of electrons and by elastic collisions with nuclei. These two processes are called electronic and nuclear stopping respectively. As an ion enters a target it is slowed down initially by electronic stopping, but as its velocity decreases, nuclear stopping becomes predominant until the ion comes to rest. The contribution of electronic stopping to lattice damage is very small, but nuclear stopping can cause severe damage to crystalline targets and, in the extreme, the surface layer may become amorphous.

Because the energy loss process is random in amorphous material, there is a distribution of ion ranges with depth from the surface which is approximately Gaussian and therefore characterised by a projected range, $R_p$, and a standard deviation, $\Delta R_p$.

In practice, experimentally determined atomic/electrical profiles are not Gaussian in shape, but often have a tail so that the profile is skewed. In order to model this distortion of the profile it is necessary to introduce two additional parameters representing the skewness, $\gamma_1$, and kurtosis, $\beta_2$, [64].

Favennec [65] has found experimentally using SIMS that the projected range, $R_p$, of ions in InP is a linear function of ion energy for a significant range of energies:

$$R_p\ (\mu m) = 3.1 \times 10^{-2}\ \frac{E\ (keV)}{M} \pm 10\% \qquad (1)$$

where M is the ion mass given by $9 \leq M < 40$. For larger masses:

$$R_p = 3.5 \times 10^{-2}\ \frac{E\ (keV)}{M} \pm 10\% \qquad (2)$$

for $40 < M \leq 140$. He also finds that for each ion the standard deviation of the range, $\Delta R_p$, is identical for GaAs and InP. TABLE 6 lists values of the constant K in the formula:

$$R_p = K\ (\text{target,ion})\ E \qquad (3)$$

and indicates the ion energy up to which this relationship is approximately true.

Apart from the above data, there are few detailed measurements of ion ranges reported in the literature. In general, where SIMS profiles have been performed they are said to agree approximately with LSS theory [66] (see [29] for an example). However, Favennec [65] concludes that experimental results do not agree with LSS predictions for many ion species. There has been a detailed study of Be [63] and Si [67] range statistics in InP in the energy range 0.4 to 3.0 MeV for Be and 10 to 20 MeV for Si. In both cases all four range parameters were determined experimentally as a function of ion energy using SIMS and compared with values calculated using the TRIM code [68,69] (see next section). The agreement between experiment and theory is very good for the projected range, but TRIM tends to overestimate the standard deviation, which for the case of Be amounts to about a 15% difference.

TABLE 6 Variation of K with ion mass for which the relationship $R_p$ = KE is approximately true [65].

| Ion | Mass | K ($10^{-4}$ μm/keV) | Valid up to energy E (keV) |
|---|---|---|---|
| Be | 9 | 40 | 100 |
| B | 11 | 28 | 200 |
| C | 12 | 26 | 200 |
| O | 16 | 20 | 400 |
| Mg | 24 | 13.4 | 500 |
| Si | 28 | 11.2 | 600 |
| S | 32 | 9.7 | 700 |
| Fe | 56 | 6.2 | 1000 |
| Zn | 64 | 5.7 | 1000 |
| Se | 80 | 4.7 | 1000 |
| Cd | 114 | 3.3 | 1000 |
| Te | 130 | 2.9 | 1000 |

## G RANGE PROFILE STATISTICS

The data presented in TABLE 7 have been obtained by use of the Monte Carlo program TRIM97.06 and the Projected Range Algorithm (PRAL) [68,69]. The calculations refer to amorphous targets. The terms mean, sigma, skew and kurt refer to: (i) the projected range, $R_p$; (ii) the standard deviation of the range, $\Delta R_p$; (iii) the skewness $\gamma_1$; and (iv) the kurtosis $\beta_2$.

Mean is the statistical first moment of the distribution and is characteristic of the range of the ions.

Sigma, the standard deviation, is the statistical second moment of the distribution about the mean and characterises the spread of the distribution.

Skew is the statistical third moment of the distribution about the mean and indicates the proportion of spread on either side of the mean. A skewness of zero indicates a symmetric distribution.

Kurtosis is the statistical fourth moment of the distribution about the mean and indicates the importance of the tail of the distribution. A value of 3.0 corresponds to a Gaussian distribution. The large values for the very high energy, light ion distributions indicate that the distributions cannot be described by simple Gaussian distributions. The more usual way to describe these is by a step function up to the mean plus a Gaussian distribution.

## H FURTHER READING AND DEVICES

There are many reviews published on the topic of ion implantation in GaAs and some of these mention other compounds such as InP [3,50,70-74]. There are, however, only two of these papers aimed only at InP and related compounds [3,50]. The reader is referred to these papers and the others [70-74] for additional information on implantation and annealing of InP. Both Si and Be implants have been used to fabricate JFETs and heterojunction FETs in InP [75-82].

TABLE 7 Range profile statistics for various ions in InP (in Angstroms).

Ion: Silicon (Z = 14; M = 28)
Target: InP (density 4.79 g/cc)

| | 50 keV | | | | 100 keV | | | |
|---|---|---|---|---|---|---|---|---|
| | mean | sigma | skew | kurt | mean | sigma | skew | kurt |
| TRIM | 626 | 311 | 0.28 | 2.70 | 1191 | 550 | 0.122 | 2.53 |
| PRAL | 554 | 375 | | | 1100 | 658 | | |

| | 200 keV | | | | 500 keV | | | |
|---|---|---|---|---|---|---|---|---|
| | mean | sigma | skew | kurt | mean | sigma | skew | kurt |
| TRIM | 2326 | 968 | -0.10 | 2.45 | 5730 | 1889 | -0.46 | 2.80 |
| PRAL | 2244 | 1144 | | | 5642 | 2175 | | |

| | 1000 keV | | | | 2000 keV | | | |
|---|---|---|---|---|---|---|---|---|
| | mean | sigma | skew | kurt | mean | sigma | skew | kurt |
| TRIM | 10346 | 2732 | -0.79 | 3.65 | 16432 | 3379 | -1.075 | 4.59 |
| PRAL | 10286 | 3059 | | | 16735 | 3751 | | |

Ion: Selenium (Z = 34; M = 78)
Target: InP (density 4.79 g/cc)

| | 50 keV | | | | 100 keV | | | |
|---|---|---|---|---|---|---|---|---|
| | mean | sigma | skew | kurt | mean | sigma | skew | kurt |
| TRIM | 293 | 142 | 0.47 | 3.03 | 502 | 233 | 0.39 | 2.90 |
| PRAL | 270 | 139 | | | 478 | 230 | | |

| | 200 keV | | | | 500 keV | | | |
|---|---|---|---|---|---|---|---|---|
| | mean | sigma | skew | kurt | mean | sigma | skew | kurt |
| TRIM | 919 | 406 | 0.23 | 2.70 | 2190 | 843 | 0.02 | 2.69 |
| PRAL | 896 | 399 | | | 2239 | 876 | | |

| | 1000 keV | | | | 2000 keV | | | |
|---|---|---|---|---|---|---|---|---|
| | mean | sigma | skew | kurt | mean | sigma | skew | kurt |
| TRIM | 4297 | 1477 | -0.25 | 2.72 | 8517 | 2365 | -0.55 | 3.29 |
| PRAL | 4709 | 1602 | | | 9892 | 2797 | | |

Ion: Boron (Z = 5; M = 11)
Target: InP (density 4.79 g/cc)

| | 50 keV | | | | 100 keV | | | |
|---|---|---|---|---|---|---|---|---|
| | mean | sigma | skew | kurt | mean | sigma | skew | kurt |
| TRIM | 1405 | 585 | -0.12 | 2.42 | 2620 | 935 | -0.44 | 2.71 |
| PRAL | 1383 | 879 | | | 2693 | 1342 | | |

| | 200 keV | | | | 500 keV | | | |
|---|---|---|---|---|---|---|---|---|
| | mean | sigma | skew | kurt | mean | sigma | skew | kurt |
| TRIM | 4709 | 1414 | -0.69 | 3.25 | 9415 | 1935 | -1.09 | 4.56 |
| PRAL | 5030 | 1891 | | | 10285 | 2573 | | |

TABLE 7 continued

| | 1000 keV | | | | 2000 keV | | | |
|---|---|---|---|---|---|---|---|---|
| | mean | sigma | skew | kurt | mean | sigma | skew | kurt |
| TRIM | 15024 | 2364 | -1.66 | 7.89 | 24412 | 2704 | -2.16 | 12.22 |
| PRAL | 16531 | 2990 | | | 26118 | 3370 | | |

Ion: Iron (Z = 26; M = 56)
Target: InP (density 4.79 g/cc)

| | 50 keV | | | | 100 keV | | | |
|---|---|---|---|---|---|---|---|---|
| | mean | sigma | skew | kurt | mean | sigma | skew | kurt |
| TRIM | 350 | 174 | 0.41 | 2.93 | 629 | 304 | 0.30 | 2.86 |
| PRAL | 326 | 187 | | | 602 | 321 | | |

| | 200 keV | | | | 500 keV | | | |
|---|---|---|---|---|---|---|---|---|
| | mean | sigma | skew | kurt | mean | sigma | skew | kurt |
| TRIM | 1209 | 527 | 0.07 | 2.55 | 2876 | 1073 | -0.13 | 2.50 |
| PRAL | 1178 | 573 | | | 3054 | 1251 | | |

| | 1000 keV | | | | 2000 keV | | | |
|---|---|---|---|---|---|---|---|---|
| | mean | sigma | skew | kurt | mean | sigma | skew | kurt |
| TRIM | 5644 | 1825 | -0.42 | 2.86 | 10812 | 2786 | -0.82 | 3.65 |
| PRAL | 6380 | 2182 | | | 12634 | 3445 | | |

Ion: Nitrogen (Z = 7, M = 14)
Target: InP (density 4.79 g/cc)

| | 50 keV | | | | 100 keV | | | |
|---|---|---|---|---|---|---|---|---|
| | mean | sigma | skew | kurt | mean | sigma | skew | kurt |
| TRIM | 1052 | 467 | -0.01 | 2.42 | 1964 | 735 | -0.31 | 2.56 |
| PRAL | 968 | 624 | | | 1884 | 985 | | |

| | 200 keV | | | | 500 keV | | | |
|---|---|---|---|---|---|---|---|---|
| | mean | sigma | skew | kurt | mean | sigma | skew | kurt |
| TRIM | 3587 | 1162 | -0.53 | 2.91 | 7507 | 1746 | -0.94 | 3.95 |
| PRAL | 3573 | 1449 | | | 7599 | 2099 | | |

| | 1000 keV | | | | 2000 keV | | | |
|---|---|---|---|---|---|---|---|---|
| | mean | sigma | skew | kurt | mean | sigma | skew | kurt |
| TRIM | 12124 | 2118 | -1.39 | 6.40 | 18943 | 2451 | -1.86 | 9.54 |
| PRAL | 12505 | 2521 | | | 19695 | 2875 | | |

Ion: Magnesium (Z = 12; M = 24)
Target: InP (density 4.79 g/cc)

| | 50 keV | | | | 100 keV | | | |
|---|---|---|---|---|---|---|---|---|
| | mean | sigma | skew | kurt | mean | sigma | skew | kurt |
| TRIM | 750 | 388 | 0.40 | 2.88 | 1476 | 692 | 0.18 | 2.46 |
| PRAL | 659 | 464 | | | 1338 | 826 | | |

TABLE 7 continued

| | 200 keV | | | | 500 keV | | | |
|---|---|---|---|---|---|---|---|---|
| | mean | sigma | skew | kurt | mean | sigma | skew | kurt |
| TRIM | 2948 | 1209 | -0.07 | 2.45 | 7127 | 2348 | -0.44 | 2.73 |
| PRAL | 2772 | 1441 | | | 6980 | 2701 | | |

| | 1000 keV | | | | 2000 keV | | | |
|---|---|---|---|---|---|---|---|---|
| | mean | sigma | skew | kurt | mean | sigma | skew | kurt |
| TRIM | 12395 | 3313 | -0.81 | 3.62 | 19023 | 3903 | -1.07 | 4.55 |
| PRAL | 12336 | 3657 | | | 19387 | 4330 | | |

Ion: Zinc (Z = 30; M = 65)
Target: InP (density 4.79 g/cc)

| | 50 keV | | | | 100 keV | | | |
|---|---|---|---|---|---|---|---|---|
| | mean | sigma | skew | kurt | mean | sigma | skew | kurt |
| TRIM | 312 | 157 | 0.44 | 2.81 | 554 | 263 | 0.33 | 2.76 |
| PRAL | 293 | 163 | | | 532 | 275 | | |

| | 200 keV | | | | 500 keV | | | |
|---|---|---|---|---|---|---|---|---|
| | mean | sigma | skew | kurt | mean | sigma | skew | kurt |
| TRIM | 1038 | 461 | 0.19 | 2.66 | 2457 | 942 | -0.08 | 2.63 |
| PRAL | 1020 | 485 | | | 2605 | 1063 | | |

| | 1000 keV | | | | 2000 keV | | | |
|---|---|---|---|---|---|---|---|---|
| | mean | sigma | skew | kurt | mean | sigma | skew | kurt |
| TRIM | 4957 | 1601 | -0.41 | 2.78 | 9545 | 2579 | -0.71 | 3.47 |
| PRAL | 5469 | 1897 | | | 11237 | 3166 | | |

Ion: Beryllium (Z = 4; M = 9)
Target: InP (density 4.79 g/cc)

| | 50 keV | | | | 100 keV | | | |
|---|---|---|---|---|---|---|---|---|
| | mean | sigma | skew | kurt | mean | sigma | skew | kurt |
| TRIM | 1830 | 800 | -0.09 | 2.47 | 3427 | 1221 | -0.37 | 2.69 |
| PRAL | 1696 | 1027 | | | 3256 | 1511 | | |

| | 200 keV | | | | 500 keV | | | |
|---|---|---|---|---|---|---|---|---|
| | mean | sigma | skew | kurt | mean | sigma | skew | kurt |
| TRIM | 6198 | 1797 | -0.72 | 3.44 | 12143 | 2382 | -1.11 | 4.95 |
| PRAL | 5940 | 2045 | | | 11902 | 2681 | | |

| | 1000 keV | | | | 2000 keV | | | |
|---|---|---|---|---|---|---|---|---|
| | mean | sigma | skew | kurt | mean | sigma | skew | kurt |
| TRIM | 18914 | 2830 | -1.56 | 7.15 | 30631 | 3180 | -2.31 | 14.43 |
| PRAL | 19161 | 3087 | | | 31158 | 3524 | | |

TABLE 7 continued

Ion: Hydrogen (Z = 1; M = 2)
Target: InP (density 4.79 g/cc)

| | 50 keV | | | | 100 keV | | | |
|---|---|---|---|---|---|---|---|---|
| | mean | sigma | skew | kurt | mean | sigma | skew | kurt |
| TRIM | 4491 | 1142 | -0.77 | 3.87 | 7878 | 1502 | -1.29 | 5.70 |
| PRAL | 4380 | 1338 | | | 8050 | 1700 | | |

| | 200 keV | | | | 500 keV | | | |
|---|---|---|---|---|---|---|---|---|
| | mean | sigma | skew | kurt | mean | sigma | skew | kurt |
| TRIM | 15630 | 2000 | -2.05 | 10.38 | 48226 | 3577 | -4.05 | 35.97 |
| PRAL | 15868 | 2250 | | | 48230 | 4328 | | |

| | 1000 keV | | | | 2000 keV | | | |
|---|---|---|---|---|---|---|---|---|
| | mean | sigma | skew | kurt | mean | sigma | skew | kurt |
| TRIM | 128331 | 7995 | -6.45 | 69.83 | 369123 | 17032 | -5.94 | 66.44 |
| PRAL | 128322 | 9646 | | | 366110 | 25523 | | |

## I CONCLUSION

Although the use of ion implantation to process InP has received much attention over many years, there are still problems that remain, particularly the need to develop a reliable and reproducible encapsulant to protect the surface during annealing. There is also a general lack of understanding of the implantation process at the atomic level. There are obvious device applications of ion implantation, some of which are mentioned in Section H. However, the future applications of ion implantation to InP lie almost certainly with multi-layer heterostructures used for high speed transistors, lasers and LEDs for which the use of ion beams is beneficial especially for creating high resistivity layers and promoting interdiffusion and hence local changes in composition.

## ACKNOWLEDGEMENT

The author would like to thank Dr R.P. Webb (University of Surrey) for performing the range statistics calculations presented in TABLE 7, Section G.

## REFERENCES


[1] W. Haussler, J. Muller, R. Trommer [ *Inst. Phys. Conf. Ser. (UK)* no.91 (1987) p.493-6 ]

[2] J.H. Wilkie, B.J. Sealy [ *Thin Solid Films (Switzerland)* vol.162 (1988) p.49-57 ]

[3] S.J. Pearton, A. Katz [ *Mater. Sci. Eng. B (Switzerland)* vol.18 (1993) p.153-68 ]

[4] J.S. Reid, R.P. Ruiz, E. Kolawa, J.S. Chen, J. Madok, M.-A. Nicolet [ *Mater. Res. Soc. Symp. Proc. (USA)* vol.260 (1992) p.555-60 ]

[5] K.W. Wang, C.L. Chang, S.M. Zima [ *Electron. Lett. (UK)* vol.23 no.20 (1987) p.1040-1 ]

[6] J.P. Donnelly, C.E. Hurwitz [ *Solid-State Electron. (UK)* vol.20 no.8 (1977) p.727-30 ]

[7] P.N. Favennec, J. L'Haridon, M. Salvi, M. Gauneau [ *Rev. Phys. Appl. (France)* vol.19 no.3 (1984) p.191-5 ]

[8] Y. Kamiya, K. Shinomura, T. Itoh [ *J. Electrochem. Soc. (USA)* vol.133 no.4 (1986) p.780-4 ]

[9] F. Xiong, T.A. Tombrello, T.R. Chen, H. Wang, Y.H. Zhuang, A. Yariv [ *Nucl. Instrum. Methods Phys. Res. B (Netherlands)* vol.39 no.1-4 (1989) p.487-91 ]

[10] C.R. Abernathy, S.J. Pearton, M.B. Panish, R.A. Hamm, L.M. Lunardi [ *Inst. Phys. Conf. Ser. (UK)* no.96 (1989) p.359-64 ]

[11] S.J. Pearton, C.R. Abernathy, M.B. Panish, R.A. Hamm, L.M. Lunardi [ *J. Appl. Phys. (USA)* vol.66 no.2 (1989) p.656-62 ]

[12] J.P. Donnelly, C.E. Hurwitz [ *Solid-State Electron. (UK)* vol.21 no.2 (1978) p.475-8 ]

[13] E.V.K. Rao, M.P. Keating, P.E. Thompson [ *J. Electron. Mater. (USA)* vol.17 no.4 (1988) p.315-20 ]

[14] A. Carnera et al [ *J. Appl. Phys. (USA)* vol.76 no.9 (1994) p.5085-94 ]

[15] J. Vellanki, R.K. Nadella, M.V. Rao, O.W. Holland, D.S. Simons, P.H. Chi [ *J. Appl. Phys. (USA)* vol.73 no.3 (1993) p.1126-32 ]

[16] H. Ullrich, A. Knecht, D. Bimberg, H. Kraütle, W. Schlaak [ *J. Appl. Phys. (USA)* vol.70 no.5 (1991) p.2604-9 ]

[17] J.M. Martin, R.K. Nadella, J. Vellanki, M.V. Rao, O.W. Holland [ *J. Appl. Phys. (USA)* vol.73 no.11 (1993) p.7238-43 ]

[18] M.V. Rao et al [ *J. Electron. Mater. (USA)* vol.21 no.9 (1992) p.923-8 ]

[19] F. Vidimari, M. Caldironi, A. Di Paola, R. Chen, S. Pellegrino [ *Electron. Lett. (UK)* vol.27 no.10 (1991) p.817-8 ]

[20] M.C. Ridgway, C. Jagadish, R.G. Elliman, N. Hauser [ *Appl. Phys. Lett. (USA)* vol.60 no.24 (1992) p.3010-2 ]

[21] J.P. Donnelly, C.E. Hurwitz [ *Appl. Phys. Lett. (USA)* vol.31 no.7 (1977) p.418-20 ]

[22] J. Kasahara, J.F. Gibbons, T.J. Magee, J. Peng [ *J. Appl. Phys. (USA)* vol.51 no.8 (1980) p.4119-24 ]

[23] D.E. Davies, J.P. Lorenzo, T.G. Ryan [ *Solid-State Electron. (UK)* vol.21 no.7 (1978) p.981-5 ]

[24] S.S. Gill, B.J. Sealy [ *J. Appl. Phys. (USA)* vol.56 no.4 (1984) p.1189-94 ]

[25] A. Dodabalapur, B.G. Streetman [ *J. Electron. Mater. (USA)* vol.18 no.1 (1989) p.65-8 ]

[26] M.C. Ridgway, P. Kringhoj, C.M. Johnson [ *Nucl. Instrum. Methods Phys. Res. B (Netherlands)* vol.96 (1995) p.311-4 ]

[27] M.V. Rao, R.K. Nadella, O.W. Holland [ *J. Appl. Phys. (USA)* vol.71 no.1 (1992) p.126-32 ]

[28] J. Kraütle [ *Proc. Int. Symp. on Applications of Ion Beams in Materials Science* (Hosei University Press, Tokyo, Japan, 1988) p.209-14 ]

[29] K.V. Vaidyanathan, C.L. Anderson, H.L. Dunlap, D.E. Holmes [ *Nucl. Instrum. Methods Phys. Res. (Netherlands)* vol.182/183 pt.2 (1981) p.631-6 ]

[30] J.P. Donnelly, C.A. Armiento [ *Appl. Phys. Lett. (USA)* vol.34 no.1 (1979) p.96-8 ]

[31] J.H. Wilkie, B.J. Sealy [ *Electron. Lett. (UK)* vol.22 no.24 (1986) p.1308-9 ]

[32] W.H. van Berlo, G. Landgren [ *J. Appl. Phys. (USA)* vol.66 no.7 (1989) p.3117-20 ]

[33] P.J. Topham [ *Proc. Conf. on Semi-insulating Materials* Kah-nee-ta, OR, USA, 1984, Eds. D.C. Look, J.S. Blakemore (Shiva Publishing, Nantwich, UK, 1984) p.49-52 ]

[34] J.M. Martin, S. Garcia, F. Calle, I. Martil, G. Gonzalez-Diaz [ *J. Electon. Mater. (USA)* vol.24 no.1 (1995) p.59-67 ]

[35] B. Molnar, G. Kelner, G.L. Ramseyer, G.H. Morrison, S.C. Shatas [ *Mater. Res. Soc. Symp. Proc. (USA)* vol.27 (1984) p.329-34 ]

[36] K.W. Wang [ *Appl. Phys. Lett. (USA)* vol.51 no.25 (1987) p.2127-9 ]

[37] C.O. Jeong, S.J. Kim, B.D. Choe [ *J. Electron. Mater. (USA)* vol.21 no.8 (1992) p.825-9 ]

[38] M.V. Rao, R.K. Nadella [ *J. Appl. Phys. (USA)* vol.67 no.4 (1990) p.1761-6 ]

[39] W.H. van Berlo, M. Ghaffari, G. Landgren [ *J. Electron. Mater. (USA)* vol.21 no.4 (1992) p.431-5 ]

[40] T. Inada, S. Taka, Y. Yamamoto [ *J. Appl. Phys. (USA)* vol.23 no.9 (1980) p.943-8 ]

[41] J.D. Woodhouse, M.C. Gaidis, J.P. Donnelly [ *Solid-State Electron. (UK)* vol.33 no.8 (1990) p.1089-96 ]

[42] J.P. Donnelly, C.E. Hurwitz [ *Solid-State Electron. (UK)* vol.23 no.9 (1980) p.943-8 ]

[43] P.N. Favennec, H. L'Haridon, J.M. Roquais, M. Salvi [ *Appl. Phys. Lett. (USA)* vol.48 no.2 (1986) p.154-6 ]

[44] J.P. Donnelly, G.A. Ferrante [ *Solid-State Electron. (UK)* vol.23 no.11 (1980) p.1151-4 ]

[45] T. Inada, S. Taka, Y. Yamamoto [ *J. Appl. Phys. (USA)* vol.52 no.7 (1981) p.4863-5 ]

[46] S. Gill, B.J. Sealy [ *J. Cryst. Growth (Netherlands)* vol.64 (1983) p.174-80 ]

[47] J.P. Lorenzo, D.E. Davies, K.J. Soda, T.G. Ryan, P.J. McNally [ *Mater. Res. Soc. Symp. Proc. (USA)* vol.13 (1983) p.683-8 ]

[48] G. Bahir, J.L. Merz, J.R. Abelson, T.W. Sigmon [ *J. Electrochem. Soc. (USA)* vol.134 no.9 (1987) p.2187-93 ]

[49] H. Krautle [ *J. Appl. Phys. (USA)* vol.63 no.9 (1988) p.4418-21 ]

[50] B.L. Sharma [ *Solid State Technol. (USA)* vol.32 no.11 (1989) p.113-7 ]

[51] P. Kringhoj, V.V. Gribkovski, A. Nylandsted-Larsen [ *Appl. Phys. Lett. (USA)* vol.57 no.15 (1990) p.1514-6 ]

[52] S. Honglie, Y. Genqing, Z. Zuyao, Z. Shichang [ *Semicond. Sci. Technol. (UK)* vol.4 no.11 (1989) p.951-4 ]

[53] H. Shen, G. Yang, Z. Zhou, S. Zou [ *Appl. Phys. Lett. (USA)* vol.56 no.5 (1990) p.463-5 ]

[54] A. Katz, S.J. Pearton [ *J. Vac. Sci. Technol. B (Netherlands)* vol.8 no.6 (1990) p.1285-90 ]

[55] D.L. Lyle, D.A. Collins, C.R. Zeisse [ *IEEE Electron Device Lett. (USA)* vol.EDL-4 no.7 (1983) p.683-8 ]

[56] G. Bahir, J.L. Merz, J.R. Abelson, T.W. Sigmon [ *Mater. Res. Soc. Symp. Proc. (USA)* vol.45 (1985) p.297-302 ]

[57] B. Tell, K.F. Brown-Goebeler, C.L. Cheng [ *Appl. Phys. Lett. (USA)* vol.52 no.4 (1988) p.299-301 ]

[58] G.P. Kothiyol, K.S. Seo [ *J. Phys. D (UK)* vol.21 no.10 (1988) p.1504-7 ]

[59] D.E. Davies, J.J. Comer, J.P. Lorenzo, T.G. Ryan [ *Appl. Phys. Lett. (USA)* vol.35 no.2 (1979) p.172-94 ]

[60] M.V. Rao, P.E. Thompson [ *Appl. Phys. Lett. (USA)* vol.50 no.20 (1987) p.1444-6 ]

[61] J.D. Woodhouse, J.P. Donnelly, P.M. Nitishiri, E.B. Owens, J.L. Ryan [ *Solid-State Electron. (UK)* vol.27 no.7 (1984) p.677-86 ]

[62] S.S. Gill, B.J. Sealy, P.J. Topham, K.G. Stephens [ *Electron. Lett. (UK)* vol.17 no.17 (1981) p.623-4 ]

[63] R.K. Nadella, M.V. Rao, D.S. Simons, P.H. Chi [ *J. Appl. Phys. (USA)* vol.70 no.6 (1991) p.2973-8 ]

[64] H. Ryssel, I. Ruge [ *Ion Implantation* (Wiley, New York, 1986) ]

[65] P.N. Favennec, M. Gauneau, M. Salvi [ *Diffus. Defect Data B (Switzerland)* vol.1/2 (1988) p.377-416 ]

[66] J. Lindhard, M. Scharff, H. Schiott [ *Mat.-Fys. Medd. K. Dan. Vidensk. Selsk. (Denmark)* (1963) ]

[67] R.K. Nadella, M.V. Rao, D.S. Simons, P.H. Chi, H.B. Dietrich [ *J. Appl. Phys. (USA)* vol.70 no.11 (1991) p.7188-90 ]

[68] J.P. Biersack, L.G. Haggmark [ *Nucl. Instrum. Methods Phys. Res. (Netherlands)* vol.174 (1980) p.257-69 ]

[69] J.P. Biersack [ *Nucl. Instrum. Methods Phys. Res. (Netherlands)* vol.182/183 pt.1 (1981) p.199-206 ]

[70] B.L. Sharma [ *J. Inst. Electron. Telecommun. Eng. (India)* vol.36 no.3/4 (1990) p.149-55 ]

[71] S.J. Pearton, W.S. Hobson, C.R. Abernathy [ *Mater. Res. Soc. Symp. Proc. (USA)* vol.147 (1989) p.261-72 ]

[72] S.J. Pearton [ *Mater. Sci. Rep. (Netherlands)* vol.4 no.8 (1990) p.315-67 ]

[73] S.J. Pearton [ *Nucl. Instrum. Methods Phys. Res. B (Netherlands)* vol.59/60 (1991) p.970-7 ]

[74] S.J. Pearton [ *Int. J. Mod. Phys. B (Singapore)* vol.7 no.28 (1993) p.4687-761 ]

[75] J.B. Boos, H.B. Dietrich, T.H. Weng, K.J. Sleger, S.C. Binari, R.L. Henry [ *IEEE Electron Device Lett. (USA)* vol.EDL-3 no.9 (1982) p.256-8 ]

[76] J.B. Boos et al [ *IEEE Electron Device Lett. (USA)* vol.EDL-5 no.7 (1984) p.273-6 ]

[77] J.D. Woodhouse, J.P. Donnelly [ *IEEE Electron Device Lett. (USA)* vol.EDL-7 no.6 (1986) p.387-9 ]

[78] S.J. Kim, K.W. Wang, G.P. Vella-Coleiro, J.W. Lutz, Y. Ota, G. Guth [ *IEEE Electron Device Lett. (USA)* vol.EDL-8 no.11 (1987) p.518-20 ]

[79] S.J. Kim, G. Guth, G.P. Vello-Coleiro [ *IEEE Electron Device Lett. (USA)* vol.EDL-9 no.6 (1988) p.306-8 ]

[80] S.J. Kim, J. Jeong, G.P. Vello-Coleiro, P.R. Smith [ *IEEE Electron Device Lett. (USA)* vol.EDL-11 no.1 (1990) p.57-8 ]

[81] W. Haussler, D. Romer [ *J. Appl. Phys. (USA)* vol.67 no.7 (1990) p.3400-8 ]

[82] J.W. Watts, Ch. Lauterback, D. Romer, J. Muller, G. Ebbinghaus [ *Electron. Lett. (UK)* vol.29 no.18 (1993) p.1599-600 ]

# 7.3 Ohmic contacts to InP and related materials

A. Katz

This Datareview appeared in *Properties of Indium Phosphide* (INSPEC, IEE, London, UK, 1991) and is reproduced here for the reader's convenience.

## A INTRODUCTION

Ohmic contacts to InP and related materials are essential in the manufacturing technology of the following devices:

(a) GaAs-based electronic devices, in which an epitaxial thin graded heterojunction layer of InGaAs (starting at GaAs lattice matched structure and becoming InAs on top) is applied to reduce the effective metal/GaAs Schottky barrier height [1-8].

(b) InP-based electronic devices, such as field effect transistors (FET) [9,10], junction field effect transistors (JFET) [11,12], high electron mobility transistors (HEMT) and heterojunction bipolar transistors (HBT) [13,14].

(c) InP-based photonic devices, such as long wavelength laser diodes [15,16], light emitting diodes (LED) [17,18], and photodetectors.

(d) Solar cells [19].

The main concerns while selecting the metallisation scheme for the ohmic contact and designing the process are the following:

(a) Lowest possible contact resistance.

(b) Thermally stable microstructure over a wide temperature range.

(c) Stable microstructure through the device operating conditions.

(d) Eliminating excessive stress in the various contact layers.

(e) Being compatible with the interconnection metallisation.

(f) Fabrication through a robust process that fits as an integrated step into the overall device manufacturing scheme.

There is a fundamental difference in the nature of ohmic contacts to n-type and p-type InP. The energy distribution of the surface states density ($N_{ss}$) in InP is parabolic, with its minimum positioned near the conduction band edge and increasing toward the valence band [20]. As a consequence, the surface of n-type InP is only slightly depleted, leading to an easy realisation of low resistance ohmic contacts with low Schottky barrier heights to n-type InP, while it is much harder to identify a metallisation scheme which will perform with good ohmic characteristics to p-type InP [21].

## B Au BASED CONTACTS

### B1 p-Type Materials

A variety of Au alloys have been suggested as possible candidates to form adequate ohmic contacts to p-type InP and related materials. AuZn [22-29], evaporated by e-gun and alloyed at 400 to 450°C, yields a specific contact resistance of $2 \times 10^{-4}$ - $8 \times 10^{-5}$ $\Omega$ $cm^2$ when deposited onto InP (Zn doped $5 \times 10^{18}$ $cm^{-3}$), and $1 \times 10^{-5}$ $\Omega$ $cm^2$ on InGaAs (Zn doped $5 \times 10^{18}$ $cm^{-3}$). AuBe [30-34], usually alloyed with 1 - 3 wt% Be, is also e-gun deposited and is alloyed at 420 - 430°C and yields a low specific contact resistance of $8 \times 10^{-6}$ $\Omega$ $cm^2$ on $5 \times 10^{18}$ $cm^{-3}$ Zn doped InP. In addition to those two metal schemes there are reports on using Au-Mg [35], Au-Cd [36], Au-Ni [37], Au-Cr and some ternary alloys such as Au-Be-Cr [38] and Au-Zi-Ni [39] as ohmic contacts to p-type InP. All the Au-based contacts, which require an alloying thermal cycle in order to drive the dopants into the metal/semiconductor interfacial mixing layer, suffer some general deficiencies: (1) formation of a wide interfacial reaction layer, typically 2 - 3 times the thickness of the initial deposited metal layer [41]; (2) non-uniform depth of interaction and spatial composition, leading to non-uniform current density across the contact [31,41]; (3) lack of long-term metallurgical stability [42-44]. It was shown, however, that a better electrical behaviour, a much shallower interfacial reacted layer, and more stable structure may be achieved by alloying the Au-based contacts by means of rapid thermal processing [45].

### B2 n-Type Materials

The standard contacts to n-type InP and related materials are AuGe (12 wt% Ge) [46-49], AuGeNi [2,50-54], and AuSn [31,13,55,56]. These contacts are characterised by relatively low specific contact resistance in the range of $2 \times 10^{-6}$ - $2 \times 10^{-7}$ $\Omega$ $cm^2$, and have low eutectic melting temperatures (280 - 361°C), and thus a relatively poor thermal stability. Therefore, thick, uneven, and spiky interfacial reactions occur in the contacts while alloying in the temperature range of 300 - 420°C by conventional furnace as well as by rapid thermal processor, with a slight advantage, however, to the latter. In addition these contacts are very sensitive to any changes in the alloying conditions.

## C NEAR-NOBLE TRANSITION METAL (Co, Ni, Ti, Pd, Pt) BASED CONTACTS

In order to overcome the deficiencies involved in the processing and performance of the Au-based alloyed contacts to both p- and n-type materials and in particular their unstable nature, various alternatives have been suggested as metallisation schemes for ohmic contacts to p-type and n-type InP and related materials. Those contacts mainly consist of using the near-noble transition metals as the metallisation scheme for the ohmic contacts. Due to the more stable thermodynamic properties, and less reactive nature of these metals, they provide a better contact microstructure, which is less reactive and more stable through heat treatments up to 450°C. The ohmic nature of the contacts is obtained both by heavily doping the semiconductor layer under the metals and by the limited formation of narrow bandgap intermetallic phases in the metal/semiconductor interface through the sintering process. Most commonly used are the Ni and Ni-P [57,58], and Ti and Ti/Pt [59-67] structures. By depositing the metal scheme, either by e-gun evaporation or by sputtering, onto highly doped p- and n-type InP-based material (typically Zn and S doped, respectively, to the levels of $1 \times 10^{18}$ - $8 \times 10^{18}$ $cm^{-3}$), and optimising the process conditions, specific contact resistance values as low as $1 \times 10^{-7}$ - $3 \times 10^{-8}$ $\Omega$ $cm^2$ may be achieved.

## D REFRACTORY METAL (W, W-ALLOYS) BASED CONTACTS

For high speed devices operated in high temperature conditions, the thermal stability offered by the transition metals may not be sufficient. These contacts cannot be exposed to temperatures higher than 450°C through the contact processing and many suffer some severe degradation while operated at high temperature and high current density. While designing self-aligned processes for GaAs- and InP-based device fabrication sequences the metal of choice has to serve as a mark for selective protection of the semiconductor under it during various processes such as ion implantation, mesa wet and dry etching and blocking layer regrowth around the mesa, in addition to providing ohmic characteristics to the contact. Various tungsten alloys such as $W_xSi_y$, $WSi_z$, WN, WSiN, WTiN and WPtAg were used as self-aligned metallisation schemes for InP and related material layers, adjacent to either GaAs [68-76] or InP [77,78] substrates. Those contacts exhibited an inert behaviour through heating cycles up to 550 - 700°C, preserved an almost abrupt interface with the semiconductor and had a typical specific contact resistance of $5 \times 10^{-6}$ - $1 \times 10^{-7}$ $\Omega$ $cm^2$.

## E CONCLUSION

The major conclusions may be stated as follows:

(1) Ohmic contacts to InP and related materials can be fabricated by using Au and Au-alloys, near-noble transition metal or refractory metal schemes. The thermal stability of the contact depends solely on the nature of the chosen metal and improves when higher melting point metals are used.

(2) Careful design of the doping of the semiconductor layer under the metal, the choice of metal or alloy and the process parameters are crucial in achieving the required contact and device electrical properties.

(3) Use of refractory metals as the contact metallisation of choice enables implementation of advanced self-aligned processes to the device manufacturing sequence. Those contacts can be stable up to 700°C and under conditions of various processes such as ion implantation, mesa-etching and semiconductor regrowth.

## REFERENCES

[1] A.G. Milnes, P.L. Feucht [ *Heterojunctions and Metal-Semiconductor Junctions* (Academic Press, New York, 1972) ]

[2] I. Mehdi, U.K. Reddy, J. Oh, J.R. East, G.I. Haddad [ *J. Appl. Phys. (USA)* vol.65 (1989) p.867-73 ]

[3] C.K. Peng, T. Won, J. Chen, C. Litton, H. Morkoc [ *IEEE Trans. Electron Devices (USA)* vol.35 (1988) p.2445-6 ]

[4] J.M. Woodal, N. Braslau, J.L. Freeouf [ *Physics of Thin Films: Advances in Research and Development* (Academic Press, San Diego, CA, 1987) vol.13 p.199-225 ]

[5] C.K. Peng, J. Chen, H. Morkoc [ *Appl. Phys. Lett. (USA)* vol.53 (1988) p.1738-40 ]

[6] T. Nittono, H. Ito, O. Nakajima, T. Ishibashi [ *Jpn. J. Appl. Phys. (Japan)* vol.27 (1988) p.1718-24 ]

[7] C.K. Peng, G. Ji, N.S. Kumar, H. Morkoc [ *Appl. Phys. Lett. (USA)* vol.53 (1988) p.900-1 ]

[8] C.K. Peng, J. Chen, J. Chyi, H. Morkoc [ *J. Appl. Phys. (USA)* vol.64 (1988) p.429-32 ]

[9] T.C. Eschrich, R.D. Carroll, R.N. Sacks, W.J. Tanski [ *IEEE Trans. Electron Devices (USA)* vol.36 (1989) p.1213-5 ]

[10] J.A. DelAlamo, T. Mizutani [ *Solid-State Electron. (UK)* vol.31 (1988) p.1635-9 ]

[11] P.E. Hallali, P. Blanconnier, L. Bricard, J.C. Renaud [ *J. Phys. Colloq. (France)* vol.49 (1988) p.453-6 ]

[12] J.B. Boos, W. Kruppa [ *J. Vac. Sci. Technol. B (USA)* vol.7 (1989) p.502-4 ]

[13] H. Schumacher, J.R. Hayes, R. Bhat, M. Koza [ *Int. Electron Devices Meet. Tech. Dig. (USA)* (1987) p.852-3 ]

[14] A. Yoshida, H. Tamura, T. Fujii, S. Hasuc [ *Extended Abstracts of International Superconductivity Electronic Conference - ISEC (USA)* (1987) p.368-71 ]

[15] A. Katz et al [ *J. Appl. Phys. (USA)* vol.65 (1989) p.4319-23 ]

[16] A. Katz [ *China Mater. Res. Soc. Proc. (UK)* vol.4 (1990) p.7-19 ]

[17] H. Temkin, R.A. Logan, R.K. Karlicek Jr., K.E. Strege, J.P. Blaha, P.N. Gabla [ *Appl. Phys. Lett. (USA)* vol.53 (1988) p.1156-8 ]

[18] R. Kaumans, N. Grote, H.G. Bart, F. Tidorra [ *Inst. Phys. Conf. Ser. (UK)* no.91 (1987) p.501-6 ]

[19] D.L. Meier, D.R. Schroeder [ *IEEE Trans. Electron Devices (USA)* vol.31 (1984) p.647-51 ]

[20] D. Fritzche [ *Inst. Phys. Conf. Ser. (UK)* no.52 (1980) p.258-68 ]

[21] E. Kuphal [ *Solid-State Electron. (UK)* vol.24 (1981) p.69-78 ]

[22] J.B. Boos, W. Kruppa [ *Solid-State Electron. (UK)* vol.31 (1988) p.127-31 ]

[23] L. Wang, Q. Zhang [ *Chin. Phys. (USA)* vol.7 (1987) p.1139-43 ]

[24] F. Schulte, H. Beneking [ *Solid-State Electron. (UK)* vol.30 (1987) p.1039-42 ]

[25] O. Oparaku, L.L. Dargan, N.M. Pearsall, R. Hill [ *Semicond. Sci. Technol. (UK)* vol.5 no.1 (1990) p.65-8 ]

[26] P. Auvray, A. Guivarc'h, H. L'Haridon, J.P. Bercier, P. Henol [ *Thin Solid Films (Switzerland)* vol.127 (1985) p.39-43 ]

[27] E. Kaminska, A. Piotrowska, A. Barcz, J. Adamczewska, A. Turos [ *Solid-State Electron. (UK)* vol.29 (1986) p.279-83 ]

[28] A. Piotrowska, E. Kaminska, Z. Barcz, J. Adamczewska, A. Turos [ *Thin Solid Films (Switzerland)* vol.130 (1985) p.231-6 ]

[29] W.X. Chen, S.C. Hsueh, P.K.L. Yu, S.S. Lau [ *IEEE Electron. Device Lett. (USA)* vol.7 (1986) p.471-6 ]

[30] H. Temkin, R.J. McCoy, V.G. Karamidas, W.A. Bonner [ *Appl. Phys. Lett. (USA)* vol.36 (1980) p.444-6 ]

[31] V.G. Keramidas, H. Temkin, S. Mahajan [ *Inst. Phys. Conf. Ser. (UK)* no.56 (1981) p.293-6 ]

[32] B.P. Segner et al [ *J. Appl. Phys. (USA)* vol.64 (1988) p.3718-21 ]

[33] D.R. Myers et al [ *J. Electron. Mater. (USA)* vol.18 (1989) p.465-72 ]

[34] N.K. Dutta, T. Cella, A.B. Piccirilli, R.L. Brown [ *Appl. Phys. Lett. (USA)* vol.49 (1986) p.1227-9 ]

[35] L.P. Erickson, A. Waseem, G.Y. Robinson [ *Thin Solid Films (Switzerland)* vol.64 (1979) p.421-5 ]

[36] J.J. Kelly, J.M.G. Rikken, J.W.M. Jocobs, A. Valster [ *J. Vac. Sci. Technol. B (USA)* vol.6 (1988) p.48-52 ]

[37] D.G. Ivey, R. Bruce, G.R. Piercy [ *J. Electron. Mater. (USA)* vol.17 (1988) p.373-7 ]

[38] T.C. Hasenberg, E. Garmir [ *J. Appl. Phys. (USA)* vol.61 (1987) p.808-9 ]

[39] V. Malina, K. Vogel, J. Zelinka [ *Semicond. Sci. Technol. (UK)* vol.3 (1988) p.1015-21 ]

[40] C.C. Chang, T.T. Sheng, R.J. McCoy, S. Nakahara, V.G. Karamidas, F. Ermanis [ *J. Appl. Phys. (USA)* vol.50 (1979) p.7032-7 ]

[41] J.M. Vandenberg, H. Temkin, R.A. Hamm, M.A. Digiuseppe [ *J. Appl. Phys. (USA)* vol.53 (1982) p.7385-9 ]

[42] A.K. Chin et al [ *IEEE Trans. Electron Devices (USA)* vol.30 (1983) p.304-9 ]

[43] A.K. Chin, C.L. Zipfel, M. Geva, I. Camlibel, P. Skeath, B.H. Chin [ *Appl. Phys. Lett. (USA)* vol.54 (1984) p.37-9 ]

[44] S.N. Fatemi, V.G. Weizer [ *Mater. Res. Soc. Symp. Proc. (USA)* vol.181 (1990) p.417-22 ]

[45] A. Katz, P.M. Thomas, S.N.G. Chu, J.W. Lee, W.C. Dautremont-Smith [ *J. Appl. Phys. (USA)* vol.66 (1989) p.2056-60 ]

[46] H. Morkoc, T.J. Drummond, C.M. Stancha [ *IEEE Trans. Electron Devices (USA)* vol.28 (1981) p.1-4 ]

[47] D.V. Morgan, J. Frey, W.J. Devlin [ *J. Electrochem. Soc. (USA)* vol.127 (1980) p.1202-5 ]

[48] J. Dunn, G.B. Stringfellow [ *J. Electron. Mater. (USA)* vol.19 (1990) p.1-3 ]

[49] S.C. Binari, J.B. Boos [ *Electron. Lett. (UK)* vol.25 (1989) p.1207-9 ]

[50] D.A. Anderson, R.J. Graham, J.W. Steeds [ *Semicond. Sci. Technol. (UK)* vol.3 (1988) p.63-76 ]

[51] G. Bahir, J.L. Merz, J.R. Abelson, T.W. Sigmon [ *J. Electron. Mater. (USA)* vol.16 (1987) p.257-62 ]

[52] K.P. Pande, E. Martin, D. Gutierrez, O. Aina [ *Solid-State Electron. (UK)* vol.30 (1987) p.253-9 ]

[53] M.F.J. O'Keefe, R.E. Miles, M.J. Howes [ *Proc. SPIE - Int. Soc. Opt. Eng. (USA)* vol.1144 (1989) p.361-7 ]

[54] P.J. Mellor, J. Herniman [ *Proc. SPIE - Int. Soc. Opt. Eng. (USA)* vol.1144 (1989) p.347-53 ]

[55] P.A. Barnes, R.S. Williams [ *Solid-State Electron. (UK)* vol.24 (1981) p.907-10 ]

[56] I. Camlibel, A.K. Chin, F. Ermanis, M.A. Digiuseppe, J.A. Lourenco, W.A. Bonner [ *J. Electrochem. Soc. (USA)* vol.129 (1982) p.2585-90 ]

[57] A. Appelbaum, M. Robins, R. Schrey [ *IEEE Trans. Electron Devices (USA)* vol.34 (1987) p.1026-30 ]

[58] A. Appelbaum, L.C. Feldman, L.A. Koszi, P.M. Thomas, P.A. Barnes [ *AT&T Bell Labs. Internal Technical Report (USA)* no.20 (1985) ]

[59] A. Katz et al [ *Appl. Phys. Lett. (USA)* vol.54 (1989) p.2306-8 ]

[60] A. Katz, P.M. Thomas, S.N.G. Chu, W.C. Dautremont-Smith, R.G. Sobers, S.G. Napholtz [ *J. Appl. Phys. (USA)* vol.67 (1990) p.884-9 ]

[61] A. Katz, W.C. Dautremont-Smith [ *J. Appl. Phys. (USA)* vol.67 (1990) p.6237-46 ]

[62] S.N.G. Chu et al [ *J. Appl. Phys. (USA)* vol.67 (1990) p.3754-60 ]

[63] A. Katz, S.N.G. Chu, P.M. Thomas, W.C. Dautremont-Smith [ *Proc. SPIE - Int. Soc. Opt. Eng. (USA)* vol.1144 (1989) p.321-31 ]

[64] A. Katz, B.E. Weir, W.C. Dautremont-Smith [ *J. Appl. Phys. (USA)* vol.68 (1990) p.1123-8 ]

[65] A. Katz, S.N.G. Chu, W.C. Dautremont-Smith, M. Soler, B.E. Weir, P.M. Thomas [ *Proc. SPIE - Int. Soc. Opt. Eng. (USA)* vol.1189 (1989) p.142-9 ]

[66] R. Kaumanns, N. Grote, H.G. Bach, F. Fidorra [ *Proc. 14th Int. Symp. on Gallium Arsenide and Related Compounds* 1987, USA, p.501-4 ]

[67] W.C. Dautremont-Smith, P.A. Barnes, J.W. Stayt Jr. [ *J. Vac. Sci. Technol. B (USA)* vol.2 (1984) p.620-5 ]

[68] T. Oshnishi, N. Yokomaya, H. Onodera, S. Suzuki, A. Shibatomi [ *Appl. Phys. Lett. (USA)* vol.43 (1983) p.600-2 ]

[69] K. Ishii, T. Oshima, T. Futatsugi, T. Fujii, N. Yokoyama, A. Shibatomi [ *International Electron Devices Meeting (IEDM) Proc. (USA)* vol.11 (1986) p.274-8 ]

[70] A. Lahav, C.S. Wu, F.A. Biocchi [ *J. Vac. Sci. Technol. B (USA)* vol.6 (1988) p.1785-92 ]

[71] R.P. Gupta, W.S. Khokle, J. Wuerfl, H.L. Hartnagel [ *J. Electrochem. Soc. (USA)* vol.137 (1990) p.631-5 ]

[72] N. Uchitomi, M. Nagaoka, K. Shimada, T. Mizoguchi, N. Toyoda [ *J. Vac. Sci. Technol. B (USA)* vol.4 (1986) p.1392-6 ]

[73] K.M. Yu, J.M. Jaklevic, E.E. Haller, S.K. Cheung, S.K. Kwok [ *J. Appl. Phys. (USA)* vol.64 (1988) p.1284-9 ]

[74] K. Asai, H. Sugahara, Y. Matuoka, M. Tokumitsu [ *J. Vac. Sci. Technol. B (USA)* vol.6 (1988) p.1526-9 ]

[75] A.E. Geissberger, R.A. Sadler, M.L. Balzman, J.W. Crites [ *J. Vac. Sci. Technol. B (USA)* vol.5 (1987) p.1701-6 ]

[76] A. Lahav, F. Ren, R.F. Kopf [ *Appl. Phys. Lett. (USA)* vol.54 (1989) p.1963-5 ]

[77] A. Katz et al [ *Appl. Phys. Lett. (USA)* vol.55 (1989) p.2220-2 ]

[78] A. Katz, S.J. Pearton, M. Geva [ *J. Appl. Phys. (USA)* vol.68 (1990) p.3110-3 ]

# 7.4 Diffusion of impurities in InP

I. Harrison

February 1999

## A DIFFUSION OF Au, Cr, Cu AND Fe

The diffusion profiles for Cr, Cu, Au and Fe in InP are very similar and resemble those found for Cr and Fe into GaAs [1,2]. The profiles can be broken down into two parts [1]. In the high concentration part, which extends from the surface, the concentration of the dopant falls very rapidly with distance, whilst in the low concentration part which extends much further into the crystal, there is little variation with depth. The drop from the high surface concentration to the bulk concentration occurs approximately within the first 40 μm for all the dopants. The actual distance depends on dopant, diffusion temperature and time. Since the profiles in this region could not be described by a complementary error function, no diffusion coefficients have been determined for any of the dopants. It is generally accepted that when a dopant has a very high diffusion coefficient it most likely diffuses via the interstitial mechanism and so the diffusion process for these atoms into InP involves the interstitial species.

### A1 Cr Diffusion

The surface concentration of Cr obtained after an 850°C 60 minute diffusion was approximately $3 \times 10^{20}$ $cm^{-3}$ and the concentration in the bulk was approximately $5 \times 10^{15}$ $cm^{-3}$. C-V measurements, however, indicate that the Cr is not electrically active [3]. This lack of electrical activity contrasts with the growth of Cr doped InP crystals. In these crystals, the Cr gives rise to an acceptor level 1.0 eV above the valence band. It also gives rise to a photoluminescence line at 1.458 μm (0.85 eV). This photoluminescence line is not seen in Cr diffused samples [4].

As noted by Brozel et al [1] it seems unlikely that the very large surface concentration is entirely due to interstitial Cr. The simple incorporation onto the In sub-lattice has however been ruled out because Cr diffused samples show no electrical or optical activity [3]. This large concentration could however be accounted for by either complex formation or precipitation. Further work is required to identify the exact nature of the Cr.

Outdiffusion of Cr doped InP has been studied by Oberstar et al [5]. They annealed, for various temperatures, samples which were capped with $Si_3N_4$ and cut from a Cr doped LEC wafer. They measured the outdiffusion with SIMS and observed a large increase in the Cr concentration at the $Si_3N_4$/InP interface followed by a region that was depleted of Cr. For a 500°C anneal the depletion was within the first 0.1 μm, whereas for an 800°C anneal the depletion region extended from 0.1 μm to 0.5 μm. Moreover they also observed accumulation of Cr in the damaged areas of He implanted InP.

### A2 Fe Diffusion

After a 60 minute 900°C diffusion, the surface concentration was approximately $10^{19}$ $cm^{-3}$ and in the bulk of the crystal it was approximately $3 \times 10^{15}$ $cm^{-3}$. As was the case with Cr, the Fe is not electrically active [3], nor does it show the 3.54 μm (0.35 eV) photoluminescence line [4]. The actual location of the Fe in the crystal lattice is still open to debate. Since it is not electrically active, it is unlikely to be on a substitutional site. Other possible explanations are precipitates or complexes.

Holmes et al [6] grew, by LPE, a nominally undoped epilayer of InP on an Fe doped InP wafer and studied the outdiffusion of Fe from the substrate to the epilayer. They found that the resulting Fe profile in the epilayer could be described by a complementary error function. As noted by Tuck [7] this fit is rather fortuitous since the indiffusion mechanism for Fe is not described by Fick's law and secondly there is a moving boundary which will complicate the problem. When a similar experiment was performed using VPE growth, the Fe concentration in the epilayer was as high as in the substrate. Moreover, there was an accumulation of Fe at the epilayer/substrate interface [8].

Outdiffusion experiments with Fe doped, LEC grown, InP samples showed that there was a build-up of Fe at the interface of the InP and the $Si_3N_4$ mask followed by a region depleted of Fe. The depth of the depleted region varied with temperature: at 600°C it was 0.1 μm while at 800°C it was 1 μm. In a similar way to Cr, Fe also accumulated in the damaged regions of the crystal caused by He implantation [5].

Subsequently to the above work, Young and Fontijn [9] have studied the growth of an Fe doped layer on top of a Zn doped layer. They found that there was considerable diffusion between these two epilayers. Moreover, the Fe signal, as measured by SIMS, was greater in the Zn doped layer than in the layer that was originally doped with Fe. They observed, however, very little Fe or Zn diffusion between the Fe doped substrate and the Zn doped epilayer. The relationship between this study and the earlier diffusion is unclear since the effect of the growth on the diffusion of Fe is unknown.

**A3 Au Diffusion**

Rembeza [2] fitted a complementary error function to the low concentration part of the Au profile and from this derived the following expression for the diffusion coefficient:

$$D = 1.32 \times 10^{-4} e^{-\frac{0.48e}{kT}}$$

The solubility of Au is very small in InP ($10^{15}$ $cm^{-3}$ at 800°C) and this has been confirmed by Parguel et al [10]. The diffusion coefficient they obtained was, however, less than that predicted from the above equation. One should, however, note that Parguel et al studied the diffusion of Au in implanted InP. Since implantation causes damage to the crystal, the diffusion of Au is unlikely to be given by the above equation.

**A4 Cu Diffusion**

There is a slight difference between Cu diffusion profiles and those obtained for Cr, Au and Fe. There is a large dip between the two sections. Similar dips have been observed in Ag diffused InP, albeit much smaller, and have been assigned to the outdiffusion of the dopant during the cooling period. The low concentration profile has been fitted with a complementary error function and an expression for the diffusion coefficient obtained [11]:

$$D = 3.8 \times 10^{-3} e^{-\frac{0.69e}{kT}}$$

The solubility of Cu varies from $4.5 \times 10^{15}$ $cm^{-3}$ to $4.2 \times 10^{16}$ $cm^{-3}$ over the temperature range 600 - 900°C and is slightly retrograde with the maximum concentration of $6 \times 10^{16}$ $cm^{-3}$ at 800°C. One should note that the whole of the work described above used plated Cu as the diffusion source and the stress at the Cu/InP interface could affect the solubility. In Ag diffused samples, there is a considerable difference between the measured solubility between plated and unplated diffused samples. Unlike Cr, Au and Fe, diffused Cu is both electrically and optically active [4].

## B DIFFUSION OF Ag

Ag diffusion into InP has been studied by using the radioactive tracer method [12-14]. All the experiments show the same general shape of diffusion profile, which cannot be fitted to any known solution of Fick's law. All the experiments show that Ag is a fast diffuser. They differ, however, in their observation on the effect of a P over pressure. Arseni and Boltaks [12] found that the solubility of Ag increases with P vapour pressure whilst Tuck and Jay [13] found the opposite. Tuck and Jay attributed the difference between the two observations to the different diffusion techniques employed. Arseni and Boltaks plated Ag on to the surface of the wafer before performing the diffusion anneal, whereas Tuck and Jay used a vapour source. It was proposed that the Ag plate would isolate the crystal from the P vapour. The study by Chaoui and Tuck [14] confirmed the earlier study of Tuck and Jay but also showed that the interaction occurring at the surface is not a simple process. In diffusion experiments, it is frequently assumed that the surface and the vapour quickly come into equilibrium and so over the timescale of the diffusion process the surface concentration can be assumed to be constant. This is not the case for the diffusion of Ag into InP. The surface concentration changes slowly with diffusion time. Chaoui and Tuck showed that it took several hours for the surface to reach equilibrium. Although the ternary phase diagram for the Ag-In-P system does not exist at present, the binary diagrams indicate that there are a variety of compounds that could form.

The shape of the Ag diffusion profile is reminiscent of those obtained for Cr, Cu, Au and Fe, namely it consists of two parts. Close to the surface there is a high concentration part that falls away very rapidly (within the first 40 μm) and a low concentration profile that extends deep into the crystal. Arseni and Boltaks fitted a complementary error function to the profile that extends deep into the crystal and obtained the following expression for the diffusion coefficient:

$$D = 3.6 \times 10^{-4} e^{-\frac{0.59e}{kT}}$$

One should, however, give a word of caution: the diffusion profiles only varied very slightly with depth and so there are large inherent errors and the above equation should only be used as a first estimate.

Since Ag diffuses very quickly, it has been proposed that it diffuses interstitially [12-14]. From the solubility studies, Chaoui and Tuck have proposed that the Ag interstitial is positively charged. Furthermore the variation of solubility with P overpressure suggests that the Ag becomes substitutional by combining with a P vacancy, or more likely it combines with a single indium vacancy and two P vacancies forming a neutral complex ($V_PAg_{In}V_P$). This kind of complex is thought to exist in Cd and Zn diffused samples [15,16].

## C DIFFUSION OF Zn INTO InP

Initial experiments, which used radioactive tracers to obtain the diffusion profile, suggest that the diffusion mechanism for Zn is similar to that for Zn into GaAs [17]. Subsequent experiments show, however, that there is a large discrepancy between the concentration of atomic Zn and the hole concentration [18,19]. Tuck and Zahari [19] showed that for homogeneously diffused InP, the hole concentration was relatively constant at approximately $6 \times 10^{18}$ $cm^{-3}$ even though the atom concentration varied from $10^{19}$ $cm^{-3}$ to $4 \times 10^{20}$ $cm^{-3}$. This difference cannot be explained by the simple substitutional-interstitial (S-I) mechanism that had been proposed for the diffusion of Zn into GaAs. A more extensive set of Zn diffusion profiles was obtained by Tuck and Hooper [15] who suggested a modification to the S-I mechanism. In the S-I mechanism, the Zn moves through the crystal as a positively charged interstitial. It becomes substitutional by combining with an In vacancy forming a singularly charged acceptor. Tuck and Hooper suggested that this charged acceptor could combine with two P vacancies to form a neutral complex. It is this complex that causes the difference between the hole and atomic Zn concentrations. This mechanism also has the advantage that it accounts for the decrease in Zn solubility when P is added to the diffusion source. The mechanism predicts that the diffusion coefficient for Zn will be independent of Zn concentration at high concentrations and this was confirmed by Tuck and Zahari [20]. Further evidence for this mechanism comes from positron lifetime measurements that show the existence of vacancy complexes in Zn doped InP [21].

Some of the atomic Zn profiles obtained by Tuck and Hooper could be described by a complementary error function suggesting that the diffusion process could be described by a single temperature dependent diffusion coefficient. To test this hypothesis, they performed a chemical diffusion and an isoelectronic diffusion using the same weights of Zn and P and hence the same surface conditions and found that the profiles were different. The diffusion coefficient of the former diffusion profile was an order of magnitude less. This implies that a single diffusion coefficient cannot be used to describe the diffusion of Zn into InP. This observed reduction in the diffusion coefficient and the anomalous profiles are probably indicative of a breakdown in the equilibrium of a defect species, most likely In vacancies. A similar suggestion has been made for Zn diffusion in GaAs [22].

### C1 Surface Features

Tuck and Hooper [48] have shown that surface features are present on some Zn diffused samples. The exact features critically depended on the external conditions. No features were observed on samples diffused with small amounts of InP or large amounts of excess P in the ampoules. For samples showing surface features, two different types were observed. For diffusion times in the region of 4 to 5 hours the surface deposits took the form of hemispheres whose size ranged from a few microns to 0.25 mm. These features were found to be In rich. However, on samples diffused for only 20 minutes, the defects were rectangular in shape and were aligned to the crystal surface. These features were Zn and P rich. Tuck and Hooper explained their results with the aid of phase diagrams and assigned the rectangular features to crystalline $Zn_3P_2$.

### C2 Junction Depths

In the light of the difficulty outlined above, it is unwise to quote diffusion coefficients. It has however been shown by several authors [23,25] that at small junction depths and low temperatures the junction depth scales as the square root of time. The junction depths for a variety of diffusion sources and temperatures are listed in TABLE 1. By appropriate scaling of these distances and times, an estimate of either the diffusion time to reach a given depth or the diffusion depth given a diffusion time can be found. It should be stressed that the values obtained should be used as an indicator and experimental work performed to verify the result.

The diffusion lengths given in TABLE 1 are for the high concentration regime. The Zn diffusion profile also exhibits a second diffusion tail. This tail means that the junction formed by the diffusion of Zn into a lightly doped n-type InP will no longer be abrupt but graded. This should increase the breakdown voltage of the PN junction and it has been confirmed experimentally [28]. The depths of the low concentration diffusion tail are given in TABLE 2.

TABLE 1 Junction depths for Zn diffused InP.

| Depth (μm) | Time (min) | Temperature (°C) | Doping ($cm^{-3}$) | Source/comments | Ref |
|---|---|---|---|---|---|
| 0.6 | 0.5 | 520 | SI | $Zn_3P_2$ film on the surface | [26] |
| 0.8 | 0.5 | 540 | SI | $Zn_3P_2$ film on the surface | [26] |
| ≈1 | 0.25 | 600 | SI | $Zn_3P_2$ film on the surface | [26] |
| ≈1 | 60 | 500 | $8 \times 10^{18}$ | $Zn_3P_2$ powder | [25] |
| <2 | 60 | 600 | $5 \times 10^{15}$ | $Zn_3P_2$ powder | [28] |
| 2.5 | 60 | 550 | $8 \times 10^{18}$ | $Zn_3P_2$ powder | [25] |
| 4 | 60 | 500 | $2 \times 10^{16}$ | $Zn_3P_2$ powder | [25] |
| 7.5 | 60 | 550 | $2 \times 10^{16}$ | $Zn_3P_2$ powder | [25] |
| 10 | 60 | 600 | $5 \times 10^{15}$ | $Zn_3P_2$ powder | [27] |
| 10 | 60 | 600 | $4 \times 10^{15}$ | $Zn_3P_2$ powder | [28] |
| 15 | 60 | 700 | $5 \times 10^{15}$ | $Zn_3P_2$ powder | [27] |
| 15 | 600 | 520 | $4 \times 10^{15}$ | $Zn_3P_2$ powder | [28] |
| 17 | 6840 | 450 | $4 \times 10^{15}$ | $Zn_3P_2$ powder | [28] |

TABLE 2 Junction depths for the low concentration diffusion tail in lightly doped InP wafers.

| Depth (μm) | Time (min) | Temperature (°C) | Doping ($cm^{-3}$) | Ref |
|---|---|---|---|---|
| 5 | 60 | 600 | $5 \times 10^{15}$ | [27] |
| 3 | 60 | 600 | $5 \times 10^{15}$ | [27] |
| 2.5 | 60 | 600 | $5 \times 10^{15}$ | [27] |
| 22 | 60 | 600 | $4 \times 10^{15}$ | [28] |

## C3 Diffusion during MOVPE

The diffusion of Zn during MOVPE growth has been studied by Blaauw et al [29] who grew NPNPN structures and observed the outdiffusion of the Zn from the p-type layers as a function of Zn concentration and Si concentration of the n-type spacer layers. In the regions of higher Zn concentration there was more Zn diffusion. The variation of the amount of Zn diffusion with Si doping concentration was one of the most interesting observations of this work. As the concentration of the Si in the spacer layers increased the diffusion length of the Zn also increased. However, when the Si concentration was greater than the Zn concentration the diffusion length of the Zn suddenly decreased. This decrease continued at higher Si concentrations albeit more slowly. However, those regions in which the Zn did diffuse, had a higher Zn concentration than those regions that were originally doped with Zn. Obviously there is another diffusion driving force that is making this 'up hill' diffusion possible.

Veuhoff et al [30] performed a similar experiment and confirmed that the diffusion of Zn increased with Zn diffusion level. They seem to suggest there is a doping level ($2 \times 10^{18}$ $cm^{-3}$) at which the Zn begins to diffuse. Blaauw et al also see significant diffusion above this level when using an undoped spacer layer. The two sets of experiments seem to disagree with each other on the effect of Si. However, both groups observe Zn build up ('up-hill diffusion') at the edges of the Zn diffused regions. This has also been observed by Lyubomirsky et al [31] who also explained the phenomenon in terms of the electric field present between the layers.

Zn doped InP is often used for the confinement layer in buried heterostructure laser diodes. This layer is normally a regrowth layer on top of a patterned active layer. If the active layers are grown by chemical beam epitaxy then the p-type dopant is often Be. During the regrowth, by MOCVD, of the Zn doped confinement layer, there can be significant co-diffusion of the Be and the Zn. This has been explained by Sugiura et al [52] in terms of the kick out mechanism They further showed that the co-diffusion of the Be and Zn is minimised, if the Be concentration is below $5 \times 10^{17}$ $cm^{-3}$. They also demonstrated that the performance of laser diodes improved when the diffusion of the Zn and Be is minimised.

Another interesting feature of MOVPE is the use of flowing dimethylzinc (DMZ) as a diffusion source [23]. Zn concentrations and hole concentrations in the region of $10^{18}$ can be obtained. This is, however, below the level obtainable with closed ampoule diffusions. An unusual aspect of this technique is the reduction of the diffusion coefficient with increasing temperature. This has since been confirmed [24]. Wada et al [23] attributed this to the evaporation from the surface sample of free atomic Zn. (In closed ampoule experiments there is a saturated Zn overpressure inside the ampoule, whereas in the open tube experiments, the DMZ decomposes on the surface of the sample liberating free Zn that can evaporate.) The junction depths for a variety of conditions are given in TABLE 3.

TABLE 3 Junction depths for the InP wafers diffused using DMZ.

| Depth (μm) | Time (min) | Temperature (°C) | Doping ($cm^{-3}$) | Source/comments | Ref |
|---|---|---|---|---|---|
| 0.4 | 60 | 500 | $5 \times 10^{16}$ | 50 sccm DMZ | [23] |
| 0.8 | 60 | 500 | $5 \times 10^{16}$ | 100 sccm DMZ | [23] |
| 1.7 | 60 | 500 | $5 \times 10^{16}$ | 200 sccm DMZ | [23] |
| 0.8 | 60 | 550 | $5 \times 10^{16}$ | 200 sccm DMZ | [23] |
| 0.4 | 60 | 570 | $5 \times 10^{16}$ | 200 sccm DMZ | [23] |
| 1.6 | 7.5 | 510 | $10^{16}$ | $P_{PH3}$ = 800 Pa, $P_{DEZ}$ = 3.23 Pa | [24] |
| 1.2 | 7.5 | 550 | $10^{16}$ | $P_{PH3}$ = 800 Pa, $P_{DEZ}$ = 3.23 Pa | [24] |

## D DIFFUSION OF Cd INTO InP

The diffusion of Cd into InP has been used in the manufacture of optical devices and has been the subject of many papers [16,18,28,32-37]. They have all been concerned directly or indirectly with the electrical activity profile caused by the diffusion of the Cd. As shown by Hooper et al [37] for Zn into InP there can be a large difference between the electrical activity and the atomic concentration. The early work by Kundukhov et al [18] provides some evidence for this to be the case in Cd diffused InP. In their experiments, they diffused Cd into selenium or tellurium doped substrates that had an electron concentration in the range $10^{16}$ $cm^{-3}$ to $10^{19}$ $cm^{-3}$. They then measured the junction depth by chemical staining. All the samples had an electrical activity in the range 3 - 5 × $10^{18}$ $cm^{-3}$, even the sample that was originally doped n-type to $10^{19}$ $cm^{-3}$.

The aspect of Cd diffused InP that is useful to the optical device engineer is the deep tail at low Cd concentrations that can be used to manufacture $p^+p^-n$ junctions [32,33]. The initial doping of the InP substrate needs to be less than $10^{16}$ $cm^{-3}$ before this tail can be observed.

Ohtsuka et al [16] have proposed that Cd and Zn diffuse in a similar way (see Section C). By assuming mass action is valid they showed that the diffusion of Cd was concentration dependent. However, they do not give a value for the concentration at which the diffusion coefficients given were measured. Therefore, it is difficult to use these values. Of more use are the junction depths formed by the diffusion of Cd into n-InP. The values given in TABLE 4 refer to the steep diffusion front that one would observe if the diffusion had been performed into a wafer that was originally doped to $10^{16}$ $cm^{-3}$. If the original doping was less than this, the junction depth given in the table refers to the $p^+p^-$ junction. The $p^-n$ junction will be slightly deeper.

TABLE 4 Junction depths of Cd diffused InP samples for a variety of temperatures and times.

| Temperature (°C) | Time (hr) | Source | Depth (μm) | Ref |
|---|---|---|---|---|
| 500 | 100 | $Cd_3P_2$ | 13 | [28] |
| 580 | 10 | $Cd_3P_2$ | 12 | [28] |
| 650 | 0.5 | $Cd_3P_2$ | 8 | [16] |
| 650 | 1 | $Cd_3P_2$ | 10 | [16] |
| 650 | 2 | $Cd_3P_2$ | 16 | [16] |
| 670 | 0.25 | $CdP_2$ | 4 | [35] |
| 680 | 1 | $Cd_3P_2$ | 12 | [28] |

## E DIFFUSION OF S INTO InP

Although S is frequently used as an n-type dopant in InP, there is very little diffusion data. Initially Chin and co-workers [38,39] studied the outdiffusion of S from an LEC grown crystal. The majority of this work was performed by cathodoluminescence. This is a technique that measures the electrical activity rather than the impurity concentration. To confirm that the decrease in carrier concentration was caused by the diffusion of S, they performed SIMS. In their work, they defined the diffusion front ($X_j$) to be where the S concentration began to decrease and they found that $X_j$ was inversely proportional to the P overpressure. They were able to obtain a diffusion coefficient for the S outdiffusion process by considering the temperature dependence of the $X_j$. The value of the diffusion coefficient obtained was very large. If this value were correct then one would have expected significant redistribution of S during device fabrication or growth. This is not the case. However, the above results could be explained by indiffusion of impurities.

$Ga_2S_3$ has been used as a source for S indiffusion experiments [40]. In this series of experiments the $Ga_2S_3$ was placed on the InP surface and capped with $SiO_2$ to prevent P outdiffusion. The diffusion coefficient was assessed by measuring the sheet conductance and assuming the profile is a complementary error function. They oberved that the diffusion coefficient could be described by the following equation:

$$D = 3.6\times10^{-4} e^{-\frac{1.94e}{kT}}$$

However, there was a delay time before the start of the diffusion process. This delay time not only depended on the diffusion temperature but also on the thickness of the $Ga_2S_3$ diffusion source. Parat and Ghandhi [40] explained this observation by suggesting that the S was generated at the $SiO_2/Ga_2S_3$ interface and so had to diffuse through the $Ga_2S_3$ film. The delay times taken from the graphs presented by Parat and Ghandhi [40] are given in TABLE 5.

TABLE 5 Delay times for the diffusion of S.

| Temperature (°C) | $Ga_2S_3$ thickness (Å) | Delay time (min) |
|---|---|---|
| 585 | 200 | 160 |
| 625 | 200 | 13 |
| 625 | 560 | 71 |
| 625 | 770 | 115 |
| 670 | 200 | 9 |
| 708 | 200 | 1 |

## F IMPURITY INDUCED DISORDERING IN InGaAsP/InP

Impurity disordering is a very important technique for the fabrication of integrated optoelectronic components. The range of devices and the techniques have been discussed in a review [49] and a special issue [50]. This report will concentrate on the physics behind the diffusion processes. A more comprehensive discussion on the intermixing of the GaInAsP alloys can be found in [51].

Thermal annealing of lattice matched InGaAs/InP quantum wells has been performed by Temkin et al [41], who used photoluminescence and transmission electron microscopy to investigate the intermixing. They observed that the interfaces of the wells stayed very sharp even though the photoluminescence results indicated that there was considerable interdiffusion. If the wells were gradually to disorder like those in GaAs/AlAs then there would be an increase in the strain. It was proposed that the wells, to avoid an increase in the strain, disorder in such a way as to stay lattice matched with barriers.

The first report of disordering of the GaInAsP/InP system lattice matched to InP by a dopant was by Razeghi et al [42] who diffused Zn into $In_{0.75}Ga_{0.25}P_{0.5}As_{0.5}$/InP quantum wells for a variety of temperatures and times. The interdiffusion process was monitored by observing the shift in the PL peak. They found that in the InGaAsP/InP system, as in the AlGaAs/GaAs system, the diffusion of Zn enhances the diffusion of the group III atoms. They also observed evidence for the intermixing of the group V elements at 700°C. Since the enhancement of the interdiffusion of the group III atoms moves the absorption edge to lower energies Zn impurity induced disordering would not be appropriate for the manufacture of transverse junction lasers or waveguides [42,43].

To examine the enhancement of the interdiffusion of the group III and group V elements, Park et al [44] annealed an $In_{0.72}Ga_{0.28}P_{0.39}As_{0.61}$/InP heterostructure. Their results confirm the previous results. However, in their discussion they attempted to relate the diffusion mechanism for Zn in InP to the intermixing process. The first diffusion mechanism they discussed was that of Yamada et al [27]. They noted that if Zn diffused in InP and related compounds by Zn-$V_V$ pairs then one would expect a significant enhancement of the group V diffusion and this is not seen in experiments. They then dismissed the model proposed by Tuck and Hooper [15] since they assumed that the source of $V_P$ was the surface.

The situation is slightly more complex since not only does intermixing occur but also the formation of $Zn_3P_2$ and $Zn_3As_2$ superlattices in the region where the InGaAs and InP layers were [45-47]. The conversion of these layers depended on the Zn pressure in the ampoule. Although not mentioned in the original text these observations fit well with the earlier work of Tuck and Hooper described above.

Si intermixes InGaAs/InP when the silicon concentration is in the region of $5 \times 10^{18}$ $cm^{-3}$ However at higher concentrations no intermixing of the layers was observed. No explanation as to the cause of this has been given [47].

## REFERENCES


[1] M.R. Brozel, E.J. Foulkes, B. Tuck [ *Electron. Lett. (UK)* vol.17 (1982) p.532-3 ]

[2] S.I. Rembeza [ *Sov. Phys.-Semicond. (USA)* vol.3 (1969) p.519-20 ]

[3] M.R. Brozel, E.J. Foulkes, B. Tuck [ *Phys. Status Solidi A (Germany)* vol.72 (1982) p.K159-63 ]

[4] M.S. Skolnick, E.J. Foulkes, B. Tuck [ *J. Appl. Phys. (USA)* vol.55 (1984) p.2951-61 ]

[5] J.D. Oberstar, B.G. Streetman, J.E. Baker, P. Williams [ *J. Electrochem. Soc. (USA)* vol.128 (1981) p.1814-7 ]

[6] D.E. Holmes, R.G. Wilson, P.W. Yu [ *J. Appl. Phys. (USA)* vol.52 (1981) p.3396-9 ]

[7] B. Tuck [ *Atomic Diffusion in III-V Semiconductors* (Adam Hilger, 1988) ]

[8] J. Chevrier, M. Armand, A.M. Huber, N.T. Linh [ *J. Electron. Mater. (USA)* vol.9 (1980) p.745 ]

[9] E.W.A. Young, G.M. Fontijn [ *Appl. Phys. Lett. (USA)* vol.56 (1990) p.146-7 ]

[10] V. Parguel et al [ *J. Appl. Phys. (USA)* vol.62 (1987) p.824-7 ]

[11] K.A. Arseni [ *Sov. Phys.-Solid State (USA)* vol.10 (1969) p.2263-4 ]

[12] K.A. Arseni, B.I. Boltaks [ *Sov. Phys.-Solid State (USA)* vol.10 (1969) p.2190-4 ]

[13] B. Tuck, P.R. Jay [ *J. Phys. D (UK)* vol.11 (1978) p.1413-20 ]

[14] R. Chaoui, B. Tuck [ *J. Phys. D (UK)* vol.16 (1983) p.1721-35 ]

[15] B. Tuck, A. Hooper [ *J. Phys. D (UK)* vol.8 (1975) p.1806-21 ]

[16] K. Ohtsuka, T. Matsui, H. Ogata [ *Jpn. J. Appl. Phys. (Japan)* vol.27 (1988) p.253-9 ]

[17] L.L. Chang, H.C. Casey Jr. [ *Solid-State Electron. (UK)* vol.7 (1964) p.481-5 ]

[18] R.M. Kundukhov, S.G. Metreveli, N.V. Siukaev [ *Sov. Phys.-Semicond. (USA)* vol.1 (1967) p.765-9 ]

[19] B. Tuck, M.D. Zahari [ *J. Phys. D (UK)* vol.10 (1977) p.2473-9 ]

[20] B. Tuck M.D. Zahari [ *Inst. Phys. Conf. Ser. (UK)* no.33a (1977) p.177 ]

[21] G. Dlubek, O. Brummer, F. Plazaola, P. Hautojarvi, N. Naukkarinen [ *Appl. Phys. Lett. (USA)* vol.46 (1985) p.1136-8 ]

[22] M.D. Zahari, B. Tuck [ *J. Phys. D (UK)* vol.18 (1985) p.1585-95 ]

[23] M. Wada, M. Seko, K. Sakakibara, Y. Sekiguchi [ *Jpn. J. Appl. Phys. 2 (Japan)* vol.28 (1989) p.L1700-3 ]

[24] J. Wisser, M. Glade, H.J. Schmidt, K. Heime [ *J. Appl. Phys. (USA)* vol.71 (1992) p.3234-7 ]

[25] G.J. van Gurp, P.R. Boudewijn, M.N.C. Kempeners, D.L.A. Tjaden [ *J. Appl. Phys. (USA)* vol.61 (1987) p.1846-55 ]

[26] K.W. Wang, S.M. Parker, C.L. Cheng, J. Long [ *J. Appl. Phys. (USA)* vol.63 (1988) p.2104-9 ]

[27] M. Yamada, P.K. Tien, R.J. Martin, R.E. Nahory, A.A. Ballman [ *Appl. Phys. Lett. (USA)* vol.43 (1983) p.594-6 ]

[28] H. Ando, N. Susa, H. Kanbe [ *Jpn. J. Appl. Phys. 2 (Japan)* vol.20 (1981) p.L197-200 ]

[29] C. Blaauw, F.R. Shepherd, D. Egar [ *J. Appl. Phys. (USA)* vol.66 (1989) p.605-10 ]

[30] E. Vuehoff, H. Baumeister, J. Rieger, M. Gorgel, R. Treichler [ *J. Electron. Mater. (USA)* vol.20 (1991) p1037-41 ]

[31] I. Lyubomirsky, V. Lyahovitskaya, D. Cahen [ *Appl. Phys. Lett. (USA)* vol.70 (1997) p.613-5 ]

[32] A.K. Chin, B.V. Dutt, H. Temkin, W.A. Bonner, D.D. Roccasecca [ *Appl. Phys. Lett. (USA)* vol.36 (1980) p.924-6 ]

[33] W. Kuebart, O. Hildebrand, H.W. Marten, N. Arnold [ *Inst. Phys. Conf. Ser. (UK)* no.65 (1983) p.597-604 ]

[34] K. Ohtsuka, T. Nishino, Y. Hmakawa [ *Jpn. J. Appl. Phys. 1 (Japan)* vol.21 (1982) p.1170-3 ]

[35] Y. Takanashi, Y. Horikoshi [ *Jpn. J. Appl. Phys. 1 (Japan)* vol.19 (1980) p.687-91 ]

[36] R.K. Tien, B.I. Miller [ *Appl. Phys. Lett. (USA)* vol.34 (1979) p.701-4 ]

[37] A. Hooper, B. Tuck, A.J. Baker [ *Solid-State Electron. (UK)* vol.17 (1974) p.531 ]

[38] A.K. Chin, I. Camlibel, T.T. Sheng, W.A. Bonner [ *Appl. Phys. Lett. (USA)* vol.43 (1983) p.495-7 ]

[39] B.V. Dutt, A.K. Chin, I. Camlibel, W.A. Bonner [ *J. Appl. Phys. (USA)* vol.56 (1984) p.1630-5 ]

[40] K.K. Parat, S.K. Ghandhi [ *Solid-State Electron. (UK)* vol.31 (1988) p.1053-6 ]

[41] H. Temkin, S.N.G. Chu, M.B. Panish, R.A. Logan [ *Appl. Phys. Lett. (USA)* vol.50 (1987) p.956-8 ]

[42] M. Razeghi, O. Acher, F. Launay [ *Semicond. Sci. Technol. (UK)* vol.2 (1987) p.793-6 ]

[43] I.J. Pape, P. Li Kam Wa, J.P.R. David, P.A. Claxton, P.N. Robson, D. Sykes [ *Electron. Lett. (UK)* vol.24 (1988) p.910-1 ]

[44] H.H. Park, B.K. Kang, E.S. Nam, Y.T. Lee, J.H. Kim, D.O. Kwon [ *Appl. Phys. Lett. (USA)* vol.55 (1989) p.1768-70 ]

[45] G.J. van Gurp, W.M. van de Wijgert, G.M. Fontijn, P.J.A. Thijs [ *J. Appl. Phys. (USA)* vol.67 (1990) p.2919-26 ]
[46] D.M. Hwang et al [ *Appl. Phys. Lett. (USA)* vol.54 (1989) p.1160-2 ]
[47] S.A. Schwarz et al [ *Appl. Phys. Lett. (USA)* vol.53 (1988) p.1051-3 ]
[48] B. Tuck, A. Hooper [ *J. Mater. Sci. (UK)* vol.11 (1976) p.491-9 ]
[49] J.H. Marsh [ *Semicond. Sci. Technol. (UK)* vol.8 (1993) p.1136 ]
[50] E.H. Li, E.S. Koteles, J.H. Marsh (Eds.) [ *IEEE J. Sel. Top. Quantum Electron. (USA)* vol.4 no.4 (1998) ]
[51] R.M. Cohen [ *J. Appl. Phys. (USA)* vol.73 (1993) p. 4903-15 ]
[52] H. Sugiura, S. Kondo, M. Mitsuhara, S. Matsumoto, M. Itoh [ *Appl. Phys. Lett. (USA)* vol.70 (1997) p.2846-8 ]

# 7.5 In-situ etching (ISE) of InP and InGaAsP alloys

D. Bertone

December 1998

## A INTRODUCTION

Modern optical and optoelectronic systems require, at the same time, high reliability, high performance and low cost of the optical components. The challenge for mass production of optoelectronic devices is to develop new technological processes in order to reduce the number of steps and to improve the process yield. In particular, for the realisation of integrated photonic devices, the etching and the following (eventually selective) regrowth represent the most crucial technological steps. The realisation of buried active stripes for high speed lasers (i.e. SI-BH semi insulating - buried heterostructure lasers), butt-coupling between active and passive waveguides for tunable lasers, and etching of Fabry-Perot laser facets, directly performed on the wafer, for low cost application, are only a few possible examples. Generally the etching step is performed by using the RIE (reactive ion etching) technique. This is based on a plasma effect, which can damage the quality of the interfaces. A damaged interface can promote interdiffusion of the doping species and can affect the process yield and the device characteristics. A process in which a simple dry chemical selective etching (without plasma effect) immediately followed (without air exposure) by a selective regrowth step burying the active region of the device (if the regrowth is required) is highly desirable.

Although many works have been published on the GaAs-based alloys leading to device realisation [1,2], up to now only sporadic etching experiments have been done on InP [3,4]. In this work we report on the realisation of vertical and oblique etched walls suitable for different device applications, by using ISE, in-situ etching.

Selective etching has been obtained on InP and InP-based laser structures containing a stack of InGaAsP/InGaAsP MQWs, by using different chlorinated compounds.

## B EXPERIMENTAL

The etching experiments were performed in a home-made MOVPE reactor at a constant pressure of 76 torr. To investigate how the reaction atmosphere influences the etching behaviour, either pure argon or a mixture of argon + hydrogen was used as carrier gas. Four different chlorinated compounds, 2-chloropropane ($C_3H_7Cl$), dichloromethane ($CH_2Cl_2$), chloroform ($CHCl_3$) and carbon-tetrachloride ($CCl_4$), were tested to compare the different etching properties by varying the number of chlorine atoms in the organic molecule from one to four. From the safety point of view, it can be pointed out that all the chlorinated compounds we used are non-corrosive liquids under normal conditions. This implies that they can be contained in conventional stainless steel bubblers and consequently be introduced into the MOVPE reaction chamber through a standard metalorganic line. The bubbler temperature was varied from 0°C to 18°C depending on the vapour pressure of the organic compound, in order to obtain the same chlorine flow into the reactor.

As a phosphorus overpressure is necessary to prevent the thermal decomposition of the InP-based alloys, either phosphine ($PH_3$) or tertiarybutylphosphine (TBP) was used as the phosphorus source. In the case of TBP, the bubbler was maintained at 18°C varying the carrier gas flow (always argon) in the range 10 cc/min to 50 cc/min.

All the etching experiments were done simultaneously on four samples: two (100) InP substrates and two laser structures (previously grown by MOVPE on InP), patterned with SiN stripes along the [011] and [0$\bar{1}$1] directions. The laser structure we considered for both experimental runs and device realisation is a nine well partially strain compensated InGaAsP/InGaAsP SCH-MQW (separate confinement heterostructure-MQW) having $\lambda_{em}$ = 1.55 μm. Before loading into the MOVPE reactor, the samples were dipped in concentrated sulphuric acid to remove the surface oxides, then rinsed in deionised water and dried.

## C ETCHING RESULTS

The 2-chloropropane is the most volatile compound we used (p ≅ 430 torr at 20°C), compared to dichloromethane (p ≅ 350 torr at 20°C). Etching experiments using $C_3H_7Cl$ or $CH_2Cl_2$ were carried out by flowing the Cl-compound in the range 5 × $10^{-5}$ mol/min - 6 × $10^{-4}$ mol/min at a temperature ranging from 510°C to 590°C. As dichloromethane and 2-chloropropane give the same etching results on both InP and MQW laser structures, the following considerations have to be seen as valid for both chlorinated compounds. The standard etching procedure uses argon as carrier gas and 100 cc/min of pure $PH_3$. Best results have been obtained at a temperature of 555°C with 2-chloropropane flow of 3 × $10^{-4}$ mol/min or with dichloromethane flow of 2 × $10^{-4}$ mol/min. There is a large difference in the InP and laser structure etching rate. FIGURE 1 and FIGURE 2 show the results obtained after 30 min etching when stripes are oriented along the [011] direction. It should be noted that it is quite simple to calculate the etch rate (ER) of InP (in this case ER ≈ 1.5 μm/hr); on the contrary, the evaluation of the laser structure ER is very difficult due to the different compositions of the MQW layers, each of them having its own etch rate. The side-wall orientation results in a well defined {111} plane (55°) for both samples.

As expected, the resulting side-wall orientations of the stripes are strongly dependent on the direction of the dielectric mask. FIGURE 3 and FIGURE 4 show the etching behaviour in the case of stripes oriented along the [0$\bar{1}$1] direction for the InP (after 30 min etch) and the laser structure (after 15 min etch) respectively. We can observe the development of nearly vertical (88°) facets for both samples. It should be noticed that the side-wall slope evolution during etching in the case of the InP with stripes oriented along the [0$\bar{1}$1] direction is not well stabilised. In our experiments we have observed that at the beginning of etching (e.g. 15 min) the side-wall orientation is close to the {111} direction and sometimes the opposite side-walls are not perfectly specular. For longer etching time (e.g. 30 min) the side-wall orientation became stable and perpendicular to the (100) surface. Regarding the MQW structure (FIGURE 4), we can observe well defined vertical planes, developed at a relatively high etch rate, mainly concentrated near the leading edge of the dielectric mask. This is probably due

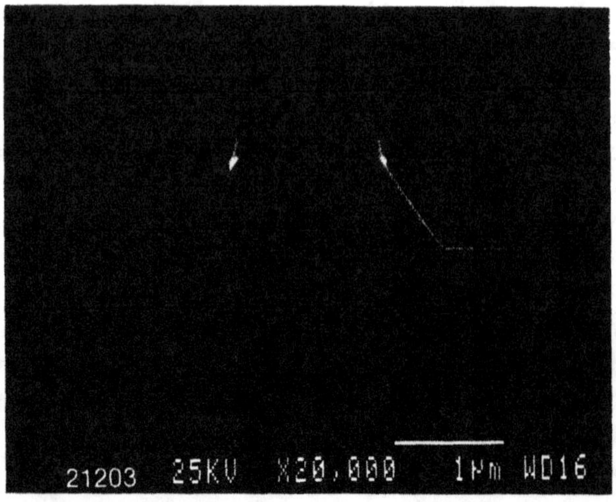


FIGURE 1 Etching profile of InP with SiN stripes parallel to the [011] direction.

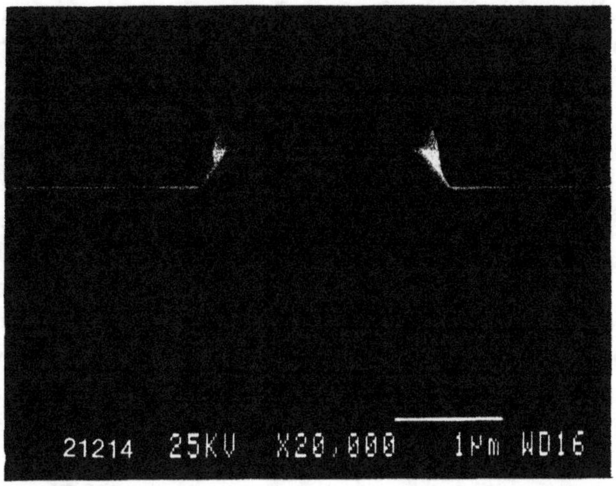


FIGURE 2 Etching profile of SCH-MQW laser structure ($\lambda_{em}$ = 1.55 μm) with SiN stripes parallel to the [011] direction (under $C_3H_7Cl$ or $CH_2Cl_2$ flow).

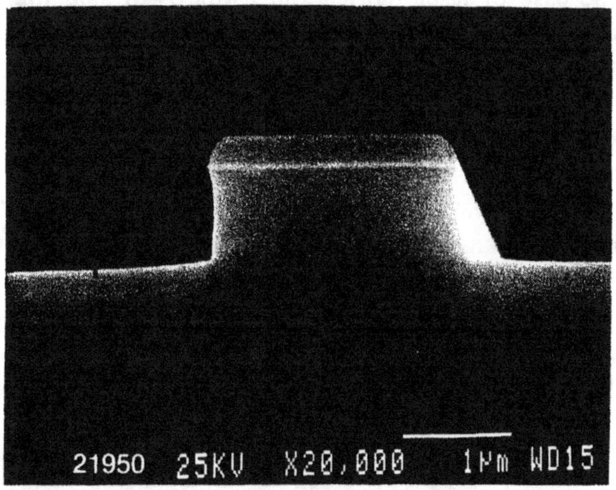


FIGURE 3 Etching profile of InP with SiN stripes parallel to the [0$\bar{1}$1] direction (under $C_3H_7Cl$ or $CH_2Cl_2$ flow).

FIGURE 4 Etching profile of SCH-MQW laser structure ($\lambda_{em}$ = 1.55 μm) with SiN stripes parallel to the [0$\bar{1}$1] direction (under $C_3H_7Cl$ or $CH_2Cl_2$ flow).

to the high Cl-compound concentration coming from the dielectric mask. The favourable etching conditions for this crystallographic orientation are also confirmed by the strong under-etching effect (direction parallel to the (100) surface), leading to the destruction of the stripe if the etching time is too long.

We did not observe any bumps or steps corresponding to the MQW stack for either the oblique or the vertical side-walls of the laser structure. In order to explore a wide range of operating conditions, TBP replaced $PH_3$, keeping the other etching conditions constant. What we observed was an increasing of the etching rate for all the samples. The morphology of the etched surface remained flat without any deposits. In order to verify how hydrogen may influence etching characteristics, a mixture of argon + hydrogen (50/50) instead of pure argon has been used as carrier gas. No differences have been observed in the etching behaviour leading to the conclusion that the process would be directly applicable to commercially available MOVPE systems that usually flow hydrogen as carrier gas.

SI-BH Fabry-Perot lasers have been fabricated by using both $CH_2Cl_2$ and $C_3H_7Cl$ to etch active stripes; device characterisation and process yield are reported in [5] and [6] respectively. Vertical walls as shown in FIGURE 4 are suitable for laser and passive waveguide butt-coupling in a single step by making, as in the case of laser fabrication, a regrowth step immediately after the in-situ etching. A further possibility is to perform, in a quite simple way and so for low cost applications, on-wafer etching of Fabry-Perot laser facets.

In the case of chloroform, the increasing number of chlorine atoms not only decreases the vapour pressure ($p \equiv 155$ torr at 20°C), but also has a very strong etching effect on InP and a deleterious effect on the InGaAsP-containing laser structure. By using chloroform we have carried out many experiments in the temperature range 500°C - 625°C and flowing the Cl-compound in the range $1 \times 10^{-5}$ mol/min - $6 \times 10^{-5}$ mol/min in the attempt to find usable etching conditions. To obtain an acceptable morphology, it was necessary to reduce the chloroform flow rate down to $3 \times 10^{-5}$ mol/min and simultaneously to increase the etching temperature up to 590°C. Under these conditions, the stripes must be etched along both [011] and $[0\bar{1}1]$ in InP (see FIGURE 5).

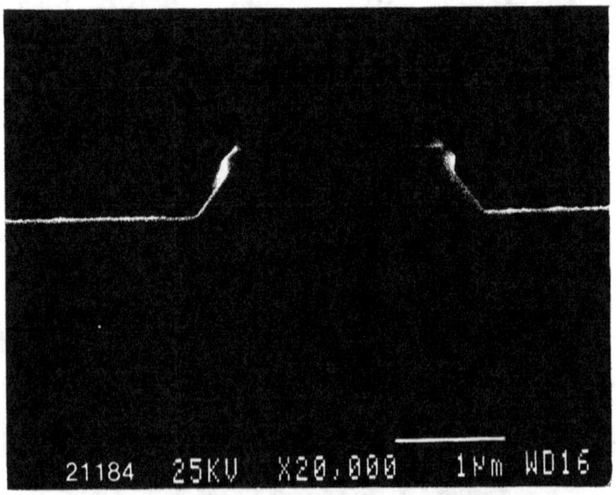


FIGURE 5 Etching profile of InP with SiN stripes parallel to the [011] direction (under $CHCl_3$ flow).

With these operating conditions, the etching produced a deposition of a thin white/grey film on InP while a thicker brown/reddish film, whose composition was found to be GaP by X-ray microanalysis (XMA), was on the surface of the laser structure containing the InGaAsP alloy. Chloroform was found to be unsuitable for etching laser stripes because the control of the stripe under-etching is very difficult and the undesired GaP layer is very likely to form, as well illustrated in FIGURE 6.

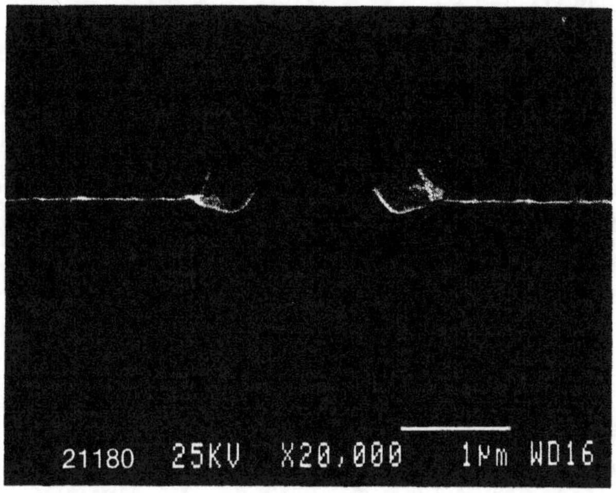


FIGURE 6 Etching profile of SCH-MQW laser structure ($\lambda_{em}$ = 1.55 μm) with SiN stripes parallel to the [011] direction (under $CHCl_3$ flow).

The substitution of $PH_3$ by TBP lowers the etching temperature but does not solve the problem of the GaP formation and neither does the introduction of hydrogen in the carrier gas.

It seems that the chloroform etching rate is higher than the desorption rate of the products from the surface of the InP-based compounds. This is particularly evident when a quite stable compound (i.e. GaP) may be formed, such as during etching of an InGaAsP alloy. If so, reaction products cannot desorb from the surface quickly, blocking further etching.

Carbon-tetrachloride, which is widely used as a liquid source for p-doping GaAs-based materials, is not suitable for etching stripes to be regrown in InP-based materials. In the range of working conditions considered, similar to those adopted for experiments using chloroform, we always obtained a grey/black carbon deposit on the surface and an unduly strong, uncontrollable, etching effect.

We conclude that $CCl_4$ is unsuitable for etching laser stripes, while an unusual behaviour has been observed when the unmasked surface of the sample is reduced. FIGURE 7 shows the etching result obtained on InP where the unmasked surface is a double, narrow (1 μm) stripe with dielectric mask all

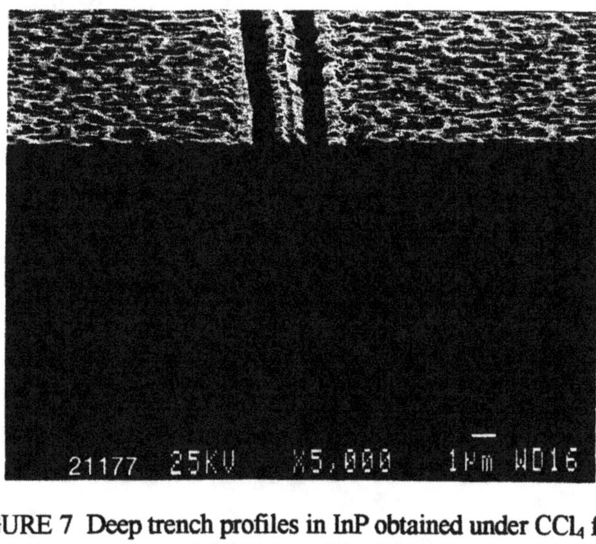


FIGURE 7 Deep trench profiles in InP obtained under $CCl_4$ flow (etch time = 15 min).

around. In this case the reactants have a large diffusion surface leading to a high concentration of Cl-compounds over the unmasked windows. The addition of the strong etching effect and the high concentration of carbon-tetrachloride produces trenches such as illustrated in FIGURE 7 in both InP and laser structures with an etch rate of nearly 40 μm/hr.

When the etch rate becomes so high, the development of crystallographic planes is inhibited and nearly vertical facets are produced, without undercutting effects, independently of the crystallographic direction of the stripes. This etching behaviour would be useful to obtain the separation of electrodes in the realisation of multi-section lasers. Under a narrow range of etching conditions, chloroform produces similar effects.

## D CONCLUSION

From the above considerations, we would like to make a simple model of the process chemistry. We believe that we have firstly to consider the reaction:

$$H_2 + Cl_2 \leftrightarrow 2\,HCl \tag{1}$$

and then the reaction considered as responsible for the etching of InP:

$$HCl + InP \rightarrow InCl + 1/2\,H_2 + 1/2\,P_2 \tag{2}$$

Reaction (1) indicates that the composition of the etching gas is a mixture of $Cl_2$ + HCl, with a variable percentage ratio depending not only on the temperature, but also strongly on the concentration of active hydrogen and active chlorine, probably in radical forms. The presence of free chlorine in the etching mixture may also explain the relative ease of etching the InGaAsP alloys.

Much work is still to be done for a better understanding of the chemistry and for a complete profitable use of the process, but the conclusion is that in-situ etching techniques represent a powerful means of mass production, improved yield and reliability of integrated photonic devices.

## ACKNOWLEDGEMENTS

The author wishes to thank R. Campi, R.Y. Fang, F. Gorgellino, G. Magnetti, M. Meliga, G. Morello, R. Paoletti and G.M. Rossi for various contributions to the results obtained in this work.

## REFERENCES

[1] Y. Inoue, K. Shimoyama, K. Fujii, N. Hosoi, H. Gotoh [ *J. Cryst. Growth (Netherlands)* vol.145 (1994) p.881-5 ]

[2] S. Ikawa, M. Ogura [ *IEEE Photonics Technol. Lett. (USA)* vol.9 (1997) p.719-21 ]

[3] C. Caneau, R. Bhat, M. Koza, J.R. Hayes, R. Esagui [ *J. Cryst. Growth (Netherlands)* vol.107 (1991) p.203-8 ]

[4] D.G. Knight, G. Kelly, R.A. Bruce [ *Proc. 7th EW-MOVPE* Berlin, Germany, June 1997 (TU Berlin, Germany, 1997) ]

[5] D. Bertone, R. Campi, G. Morello [ to be published in *J. Cryst. Growth (Netherlands)* ]

[6] D. Bertone et al [ *Proc. 24th ECOC* Madrid, Spain, September 1998 ]

# 7.6 Oxidation of InP

O.R. Monteiro

This Datareview appeared in *Properties of Indium Phosphide* (INSPEC, IEE, London, UK, 1991) and is reproduced here for the reader's convenience.

## A INTRODUCTION

Most work on the oxidation of InP has aimed at the use of this material for metal-insulator-semiconductor field effect transistors (MISFET). The evolution of the InP-based MISFET technology requires a thorough understanding of the InP/native oxide interface, as well as the dielectric layers produced on InP. The oxidation processes most frequently discussed in the literature can be divided into three groups: thermal, anodic and plasma assisted oxidation. This Datareview will treat each one of these groups separately. Prior to doing so, the work on native oxides is presented, because of its technological importance.

In treating oxidation of InP, the pure thermochemical approach [26] to predict oxide compositions has important limitations, since it assumes that thermodynamic rather than kinetic factors dominate either the growth or the reaction processes. Another shortcoming is the neglect of losses due to high volatility phases and thermal decomposition [5,6,17]. Hence the importance of experimental work for detailed characterisation of the oxides.

## B NATIVE OXIDES

Although the term native oxide has been sometimes used to describe any product of oxidation, here I shall limit this term only to those oxides that have been produced by the exposure of fresh surfaces to ambient conditions. Native oxides on (110) [13,14] and on (100) surfaces [33] have been characterised by photoelectron spectroscopy using ultra-violet light (UPS) and X-rays (XPS). Low exposures of cleaved (110) InP surfaces (i.e. less than $10^8$ L $O_2$) leads to oxygen chemisorption, whereas higher exposures promote the formation of an amorphous phosphate phase [13,14]. Addition of monolayers of alkaline metals to (110) InP greatly enhances the oxidation rate of such surfaces, leading to the formation of phosphates [2,29,30]. Monch [21] has suggested that the oxidation of (110) InP surfaces proceeds by activated adsorption of oxygen followed by field-aided transport across the initial chemisorbed layer. The composition of native oxides on (100) surfaces is still debatable. Wager et al [34] have suggested the surfaces consist predominantly of In-O compounds, tentatively identified as $In_2O_3$, with small amounts of P-O compounds, in the form of phosphates such as $InPO_4$, $In(PO_3)_3$ or $P_2O_5$. XPS has also been used to study the Fermi level movement on n-InP (110) surfaces subjected to several exposures of $O_2$. Pinning of the Fermi level at (110) InP surfaces has been correlated to the presence of a thin native oxide [4]. Clean (100) InP surfaces, on the other hand, have intrinsic states in the gap, and pin the Fermi level in the absence of any oxide.

## C THERMAL OXIDATION

Despite several works on the thermal oxidation of InP, the process is not fully understood yet. The application of the classical model of alloy oxidation to the oxidation of InP has been critically reviewed by Wilmsen [36]. Composition profiles presented in that work indicate that the oxide, formed at 450°C in dry air is basically $In_2O_3$, while P segregates at the interface. This result does not agree with the thermochemical approach of Schwartz [26], which predicted $InPO_4$ as the only reaction product. Strong experimental evidence based on X-ray diffraction [32], Raman spectroscopy [28] and transmission electron microscopy [23] points to $InPO_4$ as the main oxidation product. Structure and composition of the oxide layers are strongly dependent on the temperature for oxidation performed in pure oxygen or air at atmospheric pressure [23]. Below 350°C the oxide is predominantly amorphous [22], consisting of In, P and O [28], with some red P at the interface. At temperatures between 400°C and 600°C, the oxide film consists of mixtures of $In_2O_3$ and $InPO_4$ [23,35,38], and perhaps some $P_2O_5$ [35]. $In_2O_3$ and $P_2O_5$ would result from simultaneous incongruent vaporisation [17] and oxidation. Due to the high vapour pressures of phosphorus and its oxides, these compounds are lost to the gas phase resulting in pores and blisters in the oxide films [18]. Transmission electron microscopy has demonstrated that at the early stages, $InPO_4$ grows epitaxially on (100)-InP according to four twin-related crystallographic orientations [22,23].

Empirical expressions have been determined to describe the oxidation kinetics of InP [35,38]. At temperatures between 350°C and 450°C the oxide thickness has a complicated dependence on time [35] whereas between 450°C and 620°C a parabolic dependence was found [38]. Activation energies for the parabolic rate constants for the latter regime are 2.06 eV for both (100) and (111) surfaces [38], which was considered to be that for oxygen diffusion through the InP oxide. At higher temperatures, the volatilisation losses became too intense to allow any mathematical description of the oxide growth. Enhancement of oxidation rates is accomplished by using incident photons [7,9] or high pressures [10,11,27,37]. High pressures (up to 500 atm steam) result in a uniform $InPO_4$ layer [10], although its effect on phosphorus segregation at the interface is still debatable [27,37]. Beserman et al addressed the effect of impurities on the thermal oxidation of InP carried out between 400°C and 600°C [3]. Raman and Auger electron spectroscopy showed that doping enhances the formation of crystalline $InPO_4$, and favours the growth of crystalline phosphorus at the oxide/InP interface, although the oxidation rate remains unaltered. Unfortunately no attempt has been made to explain this phenomenon.

## D ANODIC OXIDATION

A variety of electrolytes has been used for anodic oxidation of InP: $H_3PO_4$-buffered KOH solutions, as well as citric acid and arsenic acid solutions with propylene glycol base [12]. Electron spectroscopy analysis has shown that the interfacial composition is the same for all electrolytes, namely $In_2O_3$ and $P_2O_5$, with no evidence of elemental P. The outermost layers of the oxide, however, depend upon the type of electrolyte, probably due to solubility differences. Franz [8] has proposed an electrolyte consisting of citric acid buffered to nearly neutral pH and highly diluted in glycerine, which produces a densely packed and very smooth oxide film. Anodic oxidation has also been attempted in pure water [24]. XPS revealed a double layer structure, with an indium-rich oxide at the surface and a phosphorus-rich $In(PO_3)_3$ glass-like oxide at the interface [24]. Photons have also been shown to affect the anodic oxidation of InP [16]. The relative stability of different InP surfaces with respect to anodic oxidation is (111) > (100) > (111).

Stability of anodic oxide films when exposed to air has been a usual problem for the utilisation of such films in practical applications. This instability has sometimes been attributed to the hygroscopic character of indium-rich layers, which have been frequently found at the surface of these oxides [31]. A comparison among aqueous electrolytes containing citric acid or tartaric acid or oxalic acid has shown that only the latter is promising for producing uniform, and therefore stable, films.

## E PLASMA OXIDATION

Standard conditions for plasma oxidation of InP are RF excitation (13.56 MHz) of low pressure oxygen-containing gas mixtures. This method has been used to produce MIS Schottky diodes [15] at 0.5 torr oxygen partial pressure and 100 W RF power. Auger electron spectroscopy has shown a depletion of P in the oxide layer, but no enrichment at the interface [15]. This result has been disputed by others [34], who have shown that the oxide layer consists primarily of $InPO_4$ with some $In_2O_3$ in the thicker layers. When plasma oxidation is carried out at temperatures higher than 350°C the thermal component of the process dominates. Attempts to correlate the surface roughness to the process parameters have not been successful yet [19,20]. Biasing of the InP surface during plasma oxidation has suggested that the oxidation is due to the interaction of negatively charged particles (negative ions or electrons) with the surface [25]. A new approach has been proposed by Al-Refaie [1], who used reactive sputtering of an InP target in a magnetically confined DC excited plasma of pure oxygen. This process produces films with uniform composition and equal amounts of In and P.

## F CONCLUSION

Oxidation of InP has been reviewed. Thermal oxidation carried out at low temperatures results in amorphous films containing In, P and O. At higher temperatures the product is a combination of $InPO_4$ and $In_2O_3$ although only the former is thermodynamically stable. Despite some contrary results, it seems that phosphorus always segregates at the InP/oxide interface. Despite all the work that has been done, a complete description of the oxidation process is not available yet. In order to make useful thermal oxides a greater uniformity is required, and that can be achieved only upon a better understanding of the processes that take place during the oxidation. Anodic and plasma oxide films have achieved some degree of compositional uniformity, despite the unsolved mechanisms, but neither have got to the point of practical application.

## REFERENCES


[1] S.N. Al-Refaie [ *Jpn. J. Appl. Phys. (Japan)* vol.27 no.2 (1988) p.273-9 ]

[2] M.H. Bakshi, P. Soukiassian, T.M. Gentle, Z. Hurych [ *J. Vac. Sci. Technol. A (USA)* vol.5 no.4 (1987) p.1425-7 ]

[3] R. Beserman et al [ *Appl. Phys. Lett. (USA)* vol.56 no.10 (1990) p.919-21 ]

[4] K.A. Bertness, T. Kendelewicz, R.S. List, M.D. Williams, L. Lindau, W.E. Spicer [ *J. Vac. Sci. Technol. A (USA)* vol.4 no.3 (1986) p.1424-6 ]

[5] A.R. Clawson, W.Y. Lum, G.E. McWilliams [ *J. Cryst. Growth (Netherlands)* vol.46 no.2 (1979) p.300-3 ]

[6] R.F.C. Farrow [ *J. Phys. D (UK)* vol.7 no.17 (1974) p.2436-48 ]

[7] M. Fathipour, P.K. Boyer, G.J. Collins, C.W. Wilmsen [ *J. Appl. Phys. (USA)* vol.57 no.2 (1985) p.637-42 ]

[8] G. Franz [ *J. Appl. Phys. (USA)* vol.63 no.2 (1988) p.500-5 ]

[9] M. Fukuda, K. Takahei [ *J. Appl. Phys. (USA)* vol.57 no.1 (1985) p.129-34 ]

[10] R.G. Gann, K.M. Geib, C.W. Wilmsen, J. Costello, G. Hrychowain, R.J. Zeto [ *J. Appl. Phys. (USA)* vol.63 no.2 (1988) p.506-9 ]

[11] K.M. Geib, R.R. Chang, D.L. Lile, C.W. Wilmsen, G. Hrychowain, R.J. Zeto [ *J. Vac. Sci. Technol. B (USA)* vol.7 no.5 (1989) p.1126-9 ]

[12] K.M. Geib, C.W. Wilmsen [ *J. Vac. Sci. Technol. (USA)* vol.17 no.5 (1980) p.952-7 ]

[13] G. Hollinger, G. Hughes, F.J. Himpsel, J.L. Jordan, J.F. Morar, F. Houzay [ *Surf. Sci. (Netherlands)* vol.168 (1986) p.617-25 ]

[14] G. Hughes, R. Ludeke [ *J. Vac. Sci. Technol. B (USA)* vol.4 no.4 (1986) p.1109-14 ]

[15] Y. Imai, T. Ishibashi, M. Ida [ *J. Electrochem. Soc. (USA)* vol.129 no.1 (1982) p.221-4 ]

[16] P.A. Kohl, C. Wolowodiuk, F.W. Ostermeyer [ *J. Electrochem. Soc. (USA)* vol.130 no.11 (1983) p.2288-93 ]

[17] W.Y. Lum, A.R. Clawson [ *J. Appl. Phys. (USA)* vol.50 no.8 (1979) p.5296-301 ]

[18] J.J. McLaren III, A. Nelson, K. Geib, R. Gann, C.W. Wilmsen [ *J. Vac. Sci. Technol. A (USA)* vol.1 no.3 (1983) p.1486-90 ]

[19] C. Michel, J.M. Guillot, B. Lepley, N. Dupont-Pavlovsky, K. Karnicka-Moscicka [ *J. Phys. D (UK)* vol.16 no.11 (1983) p.2229-37 ]

[20] C. Michel, B. Lepley, B. Bouchikhi, S. Ravelet [ *Rev. Phys. Appl. (France)* vol.18 no.12 (1983) p.745-9 ]

[21] W. Monch [ *Surf. Sci. (Netherlands)* vol.168 (1986) p.577-93 ]

[22] O.R. Monteiro [ *Thermal Oxidation of III-V Compound Semiconductors* PhD Thesis, LBL report no.26337 (1988) p.75-92 ]

[23] O.R. Monteiro, J.W. Evans [ *J. Electrochem. Soc. (USA)* vol.135 no.9 (1988) p.2366-9 ]

[24] Y. Robach, J. Joseph, E. Bergignat, G. Hollinger [ *J. Electrochem. Soc. (USA)* vol.136 no.10 (1989) p.2957-9 ]

[25] A.A. Saada, C. Michel, M. Remy, J.R. Cussenot [ *J. Phys. D (UK)* vol.21 no.10 (1988) p.1524-30 ]

[26] G.P. Schwartz, W.A. Sunder, J.E. Griffiths [ *J. Electrochem. Soc. (USA)* vol.129 no.6 (1982) p.1361-4 ]

[27] G.P. Schwartz, U.K. Chakrabarti, R.A. Laudise, A.J. Caporaso [ *J. Appl. Phys. (USA)* vol.65 no.5 (1989) p.2154-6 ]

[28] G.P. Schwartz, W.A. Sunder, J.E. Griffiths [ *Appl. Phys. Lett. (USA)* vol.37 no.10 (1980) p.925-7 ]

[29] P. Soukiassian, M.H. Bakshi, Z. Hurych [ *J. Appl. Phys. (USA)* vol.61 no.7 (1987) p.2677-81 ]

[30] P. Soukiassian, M.H. Bakshi, H.I. Starnberg, A.S. Bommannavar, Z. Hurych [ *Phys. Rev. B (USA)* vol.37 no.11 (1988) p.6496-9 ]

[31] J. van de Ven, J.J. Binsma, N.M.A. de Wild [ *J. Appl. Phys. (USA)* vol.67 no.12 (1990) p.7568-71 ]

[32] L.L. Vereikna [ *Khim. Svoistva Metody Anal. Tugoplavkikh Soedin (USSR)* Ed. G.V. Sasonov (Naukova Dumka, Kiev, 1969) p.37-41 ]

[33] J.F. Wager, D.L. Elsworth, S.M. Goodnick, C.W. Wilmsen [ *J. Vac. Sci. Technol. (USA)* vol.19 no.3 (1981) p.513-8 ]

[34] J.F. Wager, W.H. Makky, C.W. Wilmsen, L.G. Meiners [ *Thin Solid Films (Switzerland)* vol.95 no.4 (1982) p.342-50 ]

[35] J.F. Wager, C.W. Wilmsen [ *J. Appl. Phys. (USA)* vol.51 no.1 (1980) p.12-4 ]

[36] C.W. Wilmsen [ *Thin Solid Films (Switzerland)* vol.39 (1976) p.105-17 ]

[37] C.W. Wilmsen, K.M. Geib, R. Gann, J. Costello, G. Hrychowain, R.J. Zeto [ *J. Vac. Sci. Technol. B (USA)* vol.3 no.4 (1985) p.1103-6 ]

[38] M. Yamaguchi, K. Ando [ *J. Appl. Phys. (USA)* vol.51 no.9 (1980) p.5007-12 ]

# Subject Index

www.ingramcontent.com/pod-product-compliance
Lightning Source LLC
LaVergne TN
LVHW080033210926
848481LV00011B/3177